MICROBIOLOGICAL METHODS FOR ENVIRONMENTAL BIOTECHNOLOGY

A complete list of titles in the Society for Applied Bacteriology
Technical Series appears at the end of this volume.

THE SOCIETY FOR APPLIED BACTERIOLOGY
TECHNICAL SERIES NO. 19

MICROBIOLOGICAL METHODS FOR ENVIRONMENTAL BIOTECHNOLOGY

Edited by

J. M. GRAINGER

*Department of Microbiology
University of Reading
Berkshire, UK*

AND

J. M. LYNCH

*Glasshouse Crops Research Institute
West Sussex, UK*

1984

ACADEMIC PRESS, INC.
(Harcourt Brace Jovanovich, Publishers)
LONDON • ORLANDO • SAN DIEGO • NEW YORK
TORONTO • MONTREAL • SYDNEY • TOKYO

COPYRIGHT © 1984 BY THE SOCIETY FOR APPLIED BACTERIOLOGY
ALL RIGHTS RESERVED.
NO PART OF THIS PUBLICATION MAY BE REPRODUCED OR
TRANSMITTED IN ANY FORM OR BY ANY MEANS, ELECTRONIC
OR MECHANICAL, INCLUDING PHOTOCOPY, RECORDING, OR ANY
INFORMATION STORAGE AND RETRIEVAL SYSTEM, WITHOUT
PERMISSION IN WRITING FROM THE PUBLISHER.

ACADEMIC PRESS INC. (LONDON) LTD.
24-28 Oval Road,
London NW1 7DX

United States Edition published by
ACADEMIC PRESS, INC.
Orlando, Florida 32887

British Library Cataloguing in Publication Data

Microbiological methods for environmental
    biotechnology.----(The Society for Applied
    Bacteriology technical series; 19)
    1. Micro-organisms   2. Microbial ecology
    I. Grainger, J. M.   II. Lynch, J. M.   III. Series
    574.5'223   QR100

Library of Congress Cataloging in Publication Data
Main entry under title:

Microbiological methods for environmental biotechnology.

    (Technical series / Society for Applied Bacteriology ;
no. 19)
    Includes index.
    1. Industrial microbiology--Methodology. 2. Bio-
technology--Methodology. I. Grainger, J. M. II. Lynch
J. M. (James Michael) III. Series: Technical series
(Society for Applied Bacteriology) ; no. 19.
QR53.M47 1984   628.4'01'576   84-12357
ISBN 0-12-295040-2 (alk. paper)

PRINTED IN THE UNITED STATES OF AMERICA

84 85 86 87   9 8 7 6 5 4 3 2 1

Contributors

Numbers in parentheses indicate the pages on which the authors' contributions begin.

D. B. ARCHER (139), *Agricultural Research Council, Food Research Institute, Colney Lane, Norwich, Norfolk NR4 7UA, UK*

M. T. M. BALBA (275), *Biomass International, Milnthorpe, Cumbria LA7 7HB, UK*

H. H. M. BALYUZI (313), *Department of Physics, Queen Elizabeth College, University of London, London W8 7AH, UK*

M. J. BAZIN (19, 313), *Department of Microbiology, Queen Elizabeth College, University of London, London W8 7AH, UK*

J. E. BERINGER[1] (79), *Soil Microbiology Department, Rothamsted Experimental Station, Harpenden, Hertfordshire AL5 2JQ, UK*

P. BRODA (33), *Department of Biochemistry and Applied Molecular Biology, University of Manchester Institute of Science and Technology, Manchester M60 1QD, UK*

MELANIE J. BROWN[2] (197), *Public Health Engineering Laboratory, Imperial College of Science and Technology, London SW7 2BU, UK*

C. BUCKE (219), *Tate & Lyle Group Research & Development, Philip Lyle Memorial Research Laboratory, Whiteknights, Reading, Berkshire RG6 2BX, UK*

S. J. CHAPMAN (69), *Agricultural & Food Research Council Letcombe Laboratory, Wantage, Oxfordshire OX12 9JT, UK*

P. S. J. CHEETHAM (219), *Tate & Lyle Group Research & Development, Philip Lyle Memorial Research Laboratory, Whiteknights, Reading, Berkshire RG6 2BX, UK*

R. G. CHISHOLM (365), *Thames Water Authority, New River Head Laboratories, London EC1R 4TP, UK*

N. E. CROOK (323), *Glasshouse Crops Research Institute, Littlehampton, West Sussex BN17 6LP, UK*

P. F. ENTWISTLE (323), *Natural Environment Research Council, Institute of Virology, Oxford OX1 3SR, UK*

[1]Present address: Unit of Molecular Genetics, The Medical School, University of Bristol, University Walk, Bristol, UK.
[2]Present address: QMC Industrial Research Ltd., 229 Mile End Road, London E1, UK.

X. A. FERNANDES (159), *Water Research Centre, Stevenage Laboratory, Elder Way, Stevenage, Hertfordshire SG1 1TH, UK*

R. FORSTER[3] (375), *Water Research Centre, Medmenham Laboratory, Medmenham, Marlow, Buckinghamshire LS7 2HD, UK*

P. D. GILBERT (235), *Shell Research Limited, Sittingbourne Research Centre, Sittingbourne, Kent ME9 8AG, UK*

J. M. GRAINGER (1, 259), *Department of Microbiology, University of Reading, Reading, Berkshire RG1 5AQ, UK*

S. H. T. HARPER (69), *Agricultural & Food Research Council Letcombe Laboratory, Wantage, Oxfordshire OX12 9JT, UK*

C. HARRIES (119), *Microbial Biochemistry Department, Rowett Research Institute, Bucksburn, Aberdeen AB2 9SB, UK*

N. R. HARRIS (365), *Thames Water Authority, New River Head Laboratories, London EC1R 4TP, UK*

D. S. HAYMAN (95), *Soil Microbiology Department, Rothamsted Experimental Station, Harpenden, Hertfordshire AL5 2JQ, UK*

B. N. HERBERT (235), *Shell Research Limited, Sittingbourne Research Centre, Sittingbourne, Kent ME9 8AG, UK*

STEPHEN HILL (349), *Virology Branch, Agricultural Development and Advisory Service, Ministry of Agriculture, Fisheries and Food, Cambridge CB2 2DR, UK*

M. G. HILTON (139), *Agricultural & Food Research Council, Food Research Institute, Colney Lane, Norwich, Norfolk NR4 7UA, UK*

P. N. HOBSON (119), *Microbial Biochemistry Department, Rowett Research Institute, Bucksburn, Aberdeen AB2 9SB, UK*

P. M. HOTTEN (259), *Department of Microbiology, University of Reading, Reading, Berkshire RG1 5AQ, UK*

FRANCES R. HUNTER (323), *Department of Microbiology, University of Reading, Reading, Berkshire RG1 5AQ, UK*

I. R. JOHNSON[4] (183), *Welsh Water Authority, Tremains House, Bridgend, Breconshire, UK*

A. W. B. JOHNSTON (79), *John Innes Institute, Colney Lane, Norwich, Norfolk NR4 7UH, UK*

G. L. JONES (169), *Water Research Centre, Stevenage Laboratory, Elder Way, Stevenage, Hertfordshire SG1 1TH, UK*

K. L. JONES (259), *Department of Microbiology, University of Reading, Reading, Berkshire RG1 5AQ, UK*

---

[3]Present address: LSR Rome Toxicology Centre, Via Tito Speri 14, 00040 Pomezia, Rome, Italy.

[4]Present address: Cardiff Laboratories for Energy and Resources Ltd., Lewis Road, East Moors, Cardiff, Glamorganshire CF1 5EG, UK.

B. R. JORDAN (5), *Glasshouse Crops Research Institute, Littlehampton, West Sussex BN17 6LP, UK*

B. H. KIRSOP (139), *Agricultural & Food Research Council, Food Research Institute, Colney Lane, Norwich, Norfolk NR4 7UA, UK*

Y.-K. LEE[5] (313), *Department of Microbiology, Queen Elizabeth College, University of London, London W8 7AH, UK*

J. N. LESTER (197), *Public Health Engineering Laboratory, Imperial College of Science and Technology, London SW7 2BU, UK*

R. W. LOVITT (295), *Department of Microbiology, University College, Cardiff CF2 1TA, UK*

J. M. LYNCH (1, 69), *Glasshouse Crops Research Institute, Littlehampton, West Sussex BN17 6LP, UK*

S. E. MATCHAM (5), *Glasshouse Crops Research Institute, Littlehampton, West Sussex BN17 6LP, UK*

A. J. MCCARTHY (33), *Department of Biochemistry and Applied Molecular Biology, University of Manchester Institute of Science and Technology, Manchester M60 1QD, UK*

F. E. MOSEY (159), *Water Research Centre, Stevenage Laboratory, Elder Way, Stevenage, Hertfordshire SG1 1TH, UK*

A. R. PASKINS (169), *Water Research Centre, Stevenage Laboratory, Elder Way, Stevenage, Hertfordshire SG1 1TH, UK*

A. PATERSON (33), *Department of Biochemistry and Applied Molecular Biology, University of Manchester Institute of Science and Technology, Manchester M60 1QD, UK*

S. J. PIRT (313), *Department of Microbiology, Queen Elizabeth College, University of London, London W8 7AH, UK*

G. E. POWELL (139), *Agricultural & Food Research Council, Food Research Institute, Colney Lane, Norwich, Norfolk NR4 7UA, UK*

J. F. REES[6] (259), *Environmental Safety Group, Harwell Laboratory, Oxfordshire OX11 0RA, UK*

THOMASINE RUDD (197), *Public Health Engineering Laboratory, Imperial College of Science and Technology, London SW7 2BU, UK*

J. E. RUIZ SAINZ (79), *Departamento de Microbiología, Facultad de Biología, Universidad de Seville, Seville, Spain*

E. SENIOR (275), *Department of Bioscience and Biotechnology Applied Microbiology Division, University of Strathclyde, Glasgow G1 1XW, UK*

J. S. SLADE (365), *Thames Water Authority, New River Head Laboratories, London EC1R 4TP, UK*

[5]Present address: Department of Microbiology, National University of Singapore, Kent Ridge, Singapore 0511.
[6]Present address: BioTechnica Limited, 5 Chiltern Close, Cardiff CF4 5DL, UK.

D. A. STAFFORD (183), *Department of Microbiology, University College, Cardiff CF2 1TA, UK*

R. M. STERRITT (197), *Public Health Engineering Laboratory, Imperial College of Science and Technology, London SW7 2BU, UK*

R. SUMMERS (119), *Microbial Biochemistry Department, Rowett Research Institute, Bucksburn, Aberdeen AB2 9SB, UK*

M. K. THEODOROU (19), *Animal Nutrition and Production Division, The Grassland Research Institute, Hurley, Maidenhead, Berkshire SL6 5LR, UK*

A. P. J. TRINCI (19), *Department of Botany, University of Manchester, Manchester M13 9PL, UK*

D. R. TROLLOPE (393), *Department of Botany and Microbiology, University College of Swansea, Singleton Park, Swansea SA2 8PP, UK*

D. A. VEAL (69), *Agricultural & Food Research Council Letcombe Laboratory, Wantage, Oxfordshire OX12 9JT, UK*

M. R. WALACH (313), *Department of Microbiology, Queen Elizabeth College, University of London, London W8 7AH, UK*

J. W. T. WIMPENNY (295), *Department of Microbiology, University College, Cardiff CF2 1TA, UK*

D. A. WOOD (5), *Glasshouse Crops Research Institute, Littlehampton, West Sussex BN17 6LP, UK*

Preface

The chapters in this volume, which is number 19 in the Society for Applied Bacteriology Technical Series, are a record of the contributions to the Society's Demonstration Meeting held at Brunel University on 22 September 1982.

Environmental biotechnology was suggested as the meeting topic by Dr. R. S. Holdom who represents the Society on the European Federation of Biotechnology (EFB) and who is Chairman of the EFB Working Party on Environmental Biotechnology. The programme was compiled by Dr. J. M. Grainger assisted by Dr. J. C. Fry, Mr. N. W. LeRoux, Dr. J. M. Lynch, and Dr. E. B. Pike. The local arrangements at Brunel University were made by Dr. J. Douglas and other staff of the School of Biological Sciences, and Dr. Janet E. L. Corry planned the allocation of space and services for the demonstrations.

We are most grateful to all of these colleagues, without whose efficiency and co-operation the meeting could not have been arranged. To the contributors, whose demonstrations and written contributions made the meeting and the book possible, we express our sincere appreciation and our hope that this publication does justice to the effort, time, and co-operation that they willingly gave. Finally, we wish to thank Dr. R. W. A. Park, the Society's Hon Meetings Secretary, for his guidance and support through the planning of the meeting programme to its execution.

October 1984

J. M. GRAINGER
J. M. LYNCH

Contents

LIST OF CONTRIBUTORS	v
PREFACE	ix

Introduction: Towards an Understanding of
   Environmental Biotechnology 1
  J. M. GRAINGER AND J. M. LYNCH
    Text 1
    References 4

Methods for Assessment of Fungal Growth
   on Solid Substrates 5
  S. E. MATCHAM, B. R. JORDAN AND D. A. WOOD
    Growth of Fungi in Axenic Conditions on
      Solid Substrates 6
    Growth of Fungi in Non-Axenic Conditions on
      Solid Substrates 9
    Conclusion 14
    References 15

The Dynamics of Cellulose Decomposition 19
  M. K. THEODOROU, M. J. BAZIN AND A. P. J. TRINCI
    The Experimental System 20
    Mathematical Model 22
    Experimental Evaluation of the Model 24
    Discussion 29
    References 30

The Application of Molecular Biology to
   Lignin Degradation 33
  A. PATERSON, A. J. MCCARTHY AND P. BRODA
    Naturally Occurring Lignins 34
    Preparation of Synthetic and Model Lignin Compounds . . 39
    Isotope-Labelled Lignocellulose 41
    Structural Analysis of Lignin Polymers 42

Biodegradation	47
The Molecular Biological Approach	54
Discussion	62
Note Added in Proof	63
References	63

Approaches to the Controlled Production of Novel Agricultural Composts 69
J. M. LYNCH, S. H. T. HARPER, S. J. CHAPMAN AND D. A. VEAL

Isolation of Members of N_2-Fixing Cellulolytic Communities	70
Selection and Testing of Potential Compost Inocula	70
Growth of Mixed Populations on Solid Substrates in a Conventional Chemostat	71
Linked Culture Vessels	72
The Oxygen Diffusion Column	73
A Plant Bioassay for Effects of Microbial Degradation Products	76
Effects of Degradation Products on Soil Aggregate Stability	77
Discussion	77
References	78

Methods for the Genetic Manipulation of *Rhizobium* . . . 79
J. E. BERINGER, J. E. RUIZ SAINZ AND A. W. B. JOHNSTON

Growth Conditions	80
Gene Transfer between *Rhizobium* Strains	86
Physical Analysis of the Genes that Determine Symbiotic Properties in *Rhizobium*	89
Conclusion	92
References	92

Methods for Evaluating and Manipulating Vesicular–Arbuscular Mycorrhiza 95
D. S. HAYMAN

Occurrence	96
Identification	96
Quantification	101
Isolation	106
Cultures	107
Genetics	109
Inoculum Production	110
Field Inoculation Methods	111

Endophyte Screening	113
Conclusions	114
References	115

Single- and Multi-Stage Fermenters for Treatment of Agricultural Wastes 119
P. N. HOBSON, R. SUMMERS AND C. HARRIES

Anaerobic Digestion	120
Types of Digester	121
Theoretical Aspects of Stirred-Tank Digestion	125
Some Theoretical and Practical Results from Two-Stage Digesters	131
Conclusion	137
References	138

Methanogenesis in the Anaerobic Treatment of Food-Processing Wastes 139
B. H. KIRSOP, M. G. HILTON, G. E. POWELL AND D. B. ARCHER

Choice of Experimental Fermenter System	141
Procedures for Experimental Fermenter Systems	142
Methanogenesis in Fed Batch Fermentations	145
Isolation of Methanogenic Bacteria	149
Storage of Pure Cultures	152
Characterisation of Methanogenic Bacteria	153
Conclusion	154
Appendix	155
References	157

Mathematical Modelling of Methanogenesis in Sewage Sludge Digestion 159
F. E. MOSEY AND X. A. FERNANDES

The Anaerobic Digestion Process	159
Mathematical Modelling of the System	162
Computer Simulations	163
Field Measurements of H_2 in Digester Gases	166
Conclusions	167
References	168

Computer-Controlled Fermenters for Simulation of the Activated-Sludge Process 169
A. R. PASKINS AND G. L. JONES

Experimental Procedure	170

Performance of the System 176
Discussion 179
References 180

Control of the Activated-Sludge Process Using Adenosine Triphosphate Measurements 183
I. R. JOHNSON AND D. A. STAFFORD
Factors Affecting Performance of the
  Activated-Sludge Process 183
Measurement of Microbial Activity in the
  Activated-Sludge Process 184
Possible Uses for ATP Measurements in the
  Activated-Sludge Process 185
Procedure for ATP Assays 186
Application of ATP Measurements to an Operational
  Activated-Sludge Plant 188
Conclusions 194
References 195

Assessment of the Role of Bacterial Extracellular Polymers in Controlling Metal Removal in Biological Waste Water Treatment 197
J. N. LESTER, R. M. STERRITT, THOMASINE RUDD, AND MELANIE J. BROWN
Cultivation of Polymer-Forming Organisms 198
Extraction and Assay of Polymer 199
Metal Complexation by Soluble and Matrix
  Extracellular Polymers 207
Discussion 214
References 215

Immobilisation of Microbial Cells and Their Use in Waste Water Treatment 219
P. S. J. CHEETHAM AND C. BUCKE
Immobilisation of Microbial Cells in Calcium Alginate . . 221
Performance of Calcium Alginate–Cell Complexes . . . 224
Comparison with Alternative Methods of
  Cell Immobilisation 225
The Use of Immobilized Cells or Enzymes as
  Detoxifying Agents 227
Discussion 230
References 231

Isolation and Growth of Sulphate-Reducing Bacteria 235
B. N. HERBERT AND P. D. GILBERT
    Requirements for Growth 236
    Isolation Procedures 241
    Practical Aspects 243
    Culture Media 246
    Conclusions 253
    Appendix 253
    References 255

Estimation and Control of Microbial Activity in Landfill . . . 259
J. M. GRAINGER, K. L. JONES, P. M. HOTTEN AND J. F. REES
    Enzyme Activities 260
    Chemical Analysis 266
    Laboratory Experiments 267
    Field Experiments 268
    Conclusions 270
    Appendix 271
    References 272

The Use of Single-Stage and Multi-Stage Fermenters
to Study the Metabolism of Xenobiotic and
Naturally Occurring Molecules by
Interacting Microbial Associations 275
E. SENIOR AND M. T. M. BALBA
    Pertinent Research Areas 276
    Growth Media and Analytical Methods 278
    Enrichment for and Isolation of Interacting Microbial
        Associations and Component Monocultures 281
    Closed-Culture Fermenters 281
    Open-Culture Fermenters 283
    Conclusion 291
    References 291

The Investigation and Analysis of Heterogeneous
Environments Using the Gradostat 295
J. W. T. WIMPENNY AND R. W. LOVITT
    Description and Operation of the Gradostat 296
    Transfer of Materials in the Gradostat 299
    Growth of Organisms in the Gradostat 303
    Discussion 308
    References 311

Computer Control of a Photobioreactor to Maintain
  Constant Biomass during Diurnal Variation
  in Light Intensity 313
M. R. WALACH, H. H. M. BALYUZI, M. J. BAZIN,
Y.-K. LEE AND S. J. PIRT
  Computation of Specific Growth Rates 314
  Practical Considerations 316
  Instrumentation 317
  Simulating the Day/Night Cycle 317
  Experiments to Test the System 318
  Discussion 320
  References 321

Viruses as Pathogens for the Control of Insects 323
FRANCES R. HUNTER, N. E. CROOK AND P. F. ENTWISTLE
  Characteristics of the Baculoviridae 326
  Laboratory Practice 330
  Rearing Insects in the Laboratory 331
  Propagation and Purification of Virus 332
  Identification and Characterisation of Baculoviruses . . . 334
  Counting Baculoviruses 337
  Infectivity Assays 339
  Safety Testing and Registration of Baculoviruses . . . 340
  Utilisation of Baculoviruses for Pest Control 341
  Conclusion 344
  Appendix 344
  References 345

The ELISA (Enzyme-Linked Immunosorbent Assay)
  Technique for the Detection of Plant Viruses 349
STEPHEN HILL
  The Double Antibody Sandwich (DAS) ELISA . . . 350
  Practical Applications of ELISA 359
  Conclusion 361
  Appendix 1 361
  Appendix 2 362
  References 362

Detection of Enteroviruses in Water by
  Suspended-Cell Cultures 365
J. S. SLADE, R. G. CHISHOLM AND N. R. HARRIS
  Procedures for Suspended-Cell Plaque Assays 366

Comparison of the Cell Suspension and
  Monolayer Assays 370
Discussion 373
References 374

Mutagenicity Testing of Drinking Water Using Freeze-Dried Extracts 375
R. FORSTER
Sample Collection and Analysis 375
Preparation of Samples for Mutagenicity Testing 376
Assay for Mutagenic Activity 378
Mutagenic Activity in Drinking Water 385
Discussion 388
References 390

Use of Molluscs to Monitor Bacteria in Water 393
D. R. TROLLOPE
Preparation of Mollusc Tissue Suspension 394
Isolation of Bacteria 397
Monitoring of Environmental Changes 399
Other Specific Environmental Applications 404
Appendix 405
References 406

INDEX 409

Introduction: Towards an Understanding of Environmental Biotechnology

J. M. GRAINGER

Department of Microbiology, The University, London Road, Reading, Berkshire, UK

J. M. LYNCH

Glasshouse Crops Research Institute, Worthing Road, Littlehampton, West Sussex, UK

One of the first tasks in considering a biotechnological topic is to look at the meaning of biotechnology in the particular context under discussion. Bull *et al.* (1982) have proposed a common definition of biotechnology for member countries of the Organization for Economic Cooperation and Development (OECD):

> The application of scientific and engineering principles to the processing of materials by biological agents to provide goods and services.

However, in 1981 when the organizers of the Society for Applied Bacteriology 1982 Demonstration Meeting, of which this book is a record, began their work, a widely accepted definition was not available. Two definitions widely referred to at the time were

> The application of biological organisms, systems or processes to manufacturing and service industries (Anonymous 1980, i.e. the "Spinks Report")

and

> The integrated use of biochemistry, microbiology and engineering sciences in order to achieve the technological (industrial) application of the capabilities of micro-organisms, cultured tissue cells, and parts thereof (European Federation of Biotechnology, EFB).

These and other recent definitions and their sources, eleven in all, are listed by Bull *et al.* (1982).

Thus there were broad and narrow definitions of biotechnology, and none seemed to accommodate the organizers' views and prejudices. Therefore, without the presumption of burdening the literature with yet another definition, they evolved by plagiarism a working definition to indicate more clearly that biotechnology has a role in agriculture, i.e.,

> The application of organisms, biological systems or biological processes to manufacturing industries and agriculture and their service industries.

This point has been made in the definitions embodied in reports from Australia, Holland and the International Union of Pure and Applied Chemistry (see Bull *et al.* 1982), and also by Lynch (1983) in a definition of *soil biotechnology*:

> The study and manipulation of soil micro-organisms and their metabolic processes to optimize crop productivity.

In the past, prospects for health care and genetic engineering have tended to receive the most publicity despite the existence of other less newsworthy areas of biotechnology that are of at least equal importance and of proven value; one such is waste treatment, the largest current application of biotechnology in terms of volume. Therefore, the decision of the Society for Applied Bacteriology to devote the 1982 Demonstration Meeting to microbiological methods for environmental biotechnology was timely and was widely welcomed in the United Kingdom and also on the continent of Europe. Indeed, this appears to have been the first meeting of its type to be devoted to environmental biotechnology.

The EFB Working Party on Environmental Biotechnology, one of its nine working parties, has defined *environmental biotechnology* as

> The specific application of biotechnology to the management of environmental problems, including waste treatment, pollution control and integration with non-biological technologies (Anonymous 1983).

The importance attached to environmental issues has been underlined also by the European Economic Community (EEC) organization for Forecasting and Assessment in Science and Technology (FAST) in supporting environmental biotechnology as one of the three main topics in its Bio-Society programme.

The pivotal positions that microbiology and techniques drawn from it have in environmental biotechnology are reflected in the chapters of this book, the authors of which are drawn from industry, universities, government research institutes and other research organizations. The purpose of the programme organizers was to consider environmental biotechnology in a broader context than hitherto. A wide range of processes in environmental protection and waste management, including relevant aspects of food production and exploitation of renewable sources of energy, was displayed, thereby showing existing roles of the activities of a wide range of microorganisms and microbiological techniques and pointing to possible future developments. It was also an opportunity to emphasize the importance of the integration of different disciplines and to demonstrate the extent of the collaboration that exists between different research groups in the UK.

The range of groups of microorganisms discussed encompasses algae, moulds, bacteria and viruses, studied in pure and mixed culture and in association with plants. The value of the use of microbial associations in fundamental research is increasingly appreciated and the perhaps more immediate benefits of improving some plant–microorganism associations, i.e. symbiotic nitrogen fixation and mycorrhiza, are demanding much attention. A wide selection of techniques is presented for studying this range of microorganisms under various circumstances. Each microbial group, and frequently the special physiological types within a group, may demand a different approach to cultivation and to the measurement and control of growth and activity; also a large-scale process may need a different procedure from that which is suitable in the laboratory.

Field-scale and laboratory fermenter systems are shown in continuous and batch operation, and as single-stage and multi-stage systems under aerobic and anaerobic conditions. Application to large-scale operation of knowledge gained from fundamental research is illustrated in the areas of mathematical modelling, simulations, plant design and computer control. The possibilities and potential value of commercial exploitation of methods in molecular biology, genetical manipulation, immobilized cell technology and immunology are considered and described.

Microbial activities and microbiological techniques are set in the context of the processes in which they are involved, thereby revealing interfaces between areas that otherwise may seem disparate. By drawing on processes in waste management and non-traditional agricultural practice, areas of common interest are shown in the treatment, safe disposal and re-use of a variety of "waste" materials (sewage, sewage sludge, municipal refuse, farm waste, and food-processing waste), the soil ecosystem, and the composting process through such topics as lignin and cellulose degradation, sulphate reduction and CH_4 generation. The detection in and removal from water of disease agents for animals (including man), mutagenic substances, heavy metals, and other chemical pollutants are considered because they are problems in the water industry in the solution of which microbiological methods and the activities of microorganisms have a central role. Detection and control of plant disease and pests, improvement in plant growth, and production of single-cell protein are included because these approaches to the improvement of the world supply and distribution of protein also include elements of environmental protection.

Clearly environmental biotechnology offers many prospects to society and it is hoped that the methods described in this volume will equip those who are engaged in either research or teaching with some means with which to study and advance such prospects.

References

ANONYMOUS 1980 *Biotechnology. Report of a Joint Working Party*. London: HMSO.
ANONYMOUS 1983 *Environmental Biotechnology: Future Prospects*. Vol. I Summary and Recommendations from the Versailles Workshop, 6–8 October 1982, organized by the EFB Environmental Biotechnology Working Party. Brussels: Forecasting & Assessment in Science & Technology.
BULL, A. T., HOLT, G. & LILLY, M. D. 1982 *Biotechnology. International Trends and Perspectives*. Paris: Organization for Economic Cooperation and Development.
LYNCH, J. M. 1983 *Soil Biotechnology: Microbiological Factors in Crop Productivity*. Oxford: Blackwell.

Methods for Assessment of Fungal Growth on Solid Substrates

S. E. MATCHAM, B. R. JORDAN AND D. A. WOOD

Glasshouse Crops Research Institute, Littlehampton, West Sussex, UK

Measurement of the growth of filamentous fungi in solid substrates provides an interesting challenge in several areas of microbiology and biotechnology. Examples of the type of materials utilised as solid substrates by fungi for growth include plant tissue in various forms such as living plant tissue, wood pulp and paper products, textiles, foodstuffs, leaf litter, plant residues from agriculture, soils and composted plant residues. Solid substrates also encompass animal tissues, living or dead, and mixed cultures of microorganisms. The latter includes sewage sludges and other industrial processes utilising mixed cultures. In some of these substances, fungi are present as a monoculture growing in axenic conditions. In other environments, one or more fungal species will be found associated with other classes of microorganisms, e.g. in composts or soils. There, assay methods have attempted to utilise fungal specific compounds as the basis of growth assessment. Increased interest in biotechnology and the use of bioconversion processes means that questions concerning growth rate measurement, total fungal biomass yield, the partitioning of fungal biomass between living, senescent, and dead hyphae and the quantity of extracellular products in solid substrates will need to be more accurately assessed.

The growth kinetics of filamentous fungi in liquid media have been thoroughly explored by several workers (Righelato 1975), and many methods are available for quantitatively determining fungal growth in stirred or static cultures or on the surface of agar media (Calam 1969). Such methods include wet or dry weight determinations, packed cell volume, absorbance measurements, total protein, total N, substrate consumption, CO_2 production and O_2 consumption (Calam 1969). Many of these methods will obviously be impracticable where fungi are growing in solid substrates with the fungal mycelium tightly intermingled with the substrate.

This chapter will attempt to review assay procedures for estimating fungal growth in solid substrates and to describe the applications and limitations of these methods.

Growth of Fungi in Axenic Conditions on Solid Substrates

Kjeldahl Nitrogen Determination

This method has been widely used for estimating protein yield and hence fungal growth in solid substrates. There are numerous variations on the types of digestion mixtures and the titration methods available (e.g. Mallette 1969; Anonymous 1975). A recent example of its use for growth estimation in solid substrate fermentations can be found in Ulmer *et al.* (1980). The method estimates total N in a given sample and will therefore measure both soluble and insoluble N. Soluble N can be removed by the use of acid precipitation (Ulmer *et al.* 1980). Fungal chitin contains N and therefore contributes to apparent protein N. This fact is not often taken into account when published data on the protein content of fungal solid substrate fermentations are presented. In addition, the method gives no index of the ratio of living to dead fungal material.

Nucleic Acids

Assays for either DNA or RNA can be made after using appropriate acid treatments (Mallette 1969; Hanson & Phillips 1981). The method will be applicable if the quantity of non-fungal nucleic acid is very low in the initial substrate, e.g. cellulosic substrates or other carbohydrate polymers. More chemically complex substrates may contain materials which will interfere with the colorimetric estimations for DNA or RNA as well as having high initial levels of nucleic acids.

Adenosine Triphosphate (ATP)

Assays for the quantity of extractable ATP have been extensively used for determining *total* biomass in various complex ecological environments such as sediments or soils (Holm-Hansen 1973; Jenkinson *et al.* 1979). ATP estimation can be used to determine fungal biomass in solid substrates in the absence of other living organisms. It gives an estimate of the quantity of viable hyphae but not of dead hyphae. ATP content of individual organisms may also vary markedly between species; therefore ATP content for such organisms must be initially measured in pure cultures (Fosberg & Lam 1977).

Respiration

Sugama & Okazaki (1979) estimated mycelial dry weight by following CO_2 evolution from colonised substrates. They found CO_2 evolution to be directly proportional to mycelial growth during the first 44 h of colonisation of rice grains by *Aspergillus oryzae* in the preparation of koji; CO_2 evolution was assayed continuously by titration after the gas had been trapped in NaOH solution. Mycelial dry weight was determined by digesting away the starch substrate with amylases and washing and weighing the mycelium. This type of assay would be of value for other axenic cultures growing in other chemically homogenous substrates. Harima & Humphrey (1980) measured Q_{O_2} (O_2 consumption) and Q_{CO_2}, (CO_2 evolution) for *Trichoderma* QM914 growing in glucose and were able to use these rates for calculating specific growth rate, cell concentration and other parameters. Unfortunately they did not extend this analysis to growth of *Trichoderma* on celluloses (see also Theodorou *et al.*, this volume).

Respiration rate (CO_2 evolution) has been used to measure *total* fungal biomass in soils (Anderson & Domsch 1975). In order to separate fungal and bacterial contributions to total respiration, inhibitors of prokaryote or eukaryote specificity were used. The results give an approximate estimate of the contribution of the two groups.

Oxygen consumption has also been utilised for biomass estimation in koji fermentation studies (Okazaki & Sugama 1979). Consumption of oxygen was measured by a manometric method coupled to a pressure detector. Respiratory quotient (CO_2 evolved:O_2 consumed) remained reasonably constant throughout growth.

Substrate Consumption/Weight Loss

This method has been extensively used in biodegradation studies of solid materials, e.g. wood or textiles (Bravery 1975). By making certain assumptions about fungal growth in the substrate it is possible to arrive at estimates of fungal biomass. The types of assumptions made would include an estimate of the yield coefficient, wt fungus produced/wt substrate consumed (Solomons 1975), that the bulk of volatilised substrate is CO_2, and that there is little diffusion of soluble materials from the mycelium into the substrate. This method will give only an approximate estimate of biomass and is also a relatively insensitive assay (Table 1).

TABLE 1

Sensitivity, duration and instrument requirements of assays for fungal growth in solid substrates

Assay method	Lower limit of mycelium detectable (µg/g dry wt substrate)	Duration of assay procedure (h)	Instrumentation
Kjeldahl nitrogen	100–200	10–36	—
Nucleic acids	100–200	2–4	Colorimeter or spectrophotometer
ATP	20–200	1	Fluorimeter or scintillation counter
Respiration	100–300	Continuous assay	Gas chromatograph or gas analyser
Substrate consumption	1000–5000	Minutes	—
$^{15}NH_4$ incorporation	—	—	Mass spectrometer
Agar film, hyphal length	25–200	1	Light microscope
Chitin	200–1000	8	Spectrophotometer or amino acid analyser
Ergosterol	25–200	2–4	UV spectrophotometer, GLC or HPLC
Fluorescent antibody	25–200	1	Fluorescence microscope
ELISA, RIA	1–10	2–4	Colorimeter or scintillation counter
Growth-linked enzymes	200–500	Up to 2	Spectrophotometer, O_2 electrode
Electrophoretically separable proteins	100–200	>24	High-speed centrifuge, electrophoresis apparatus

^{15}N Incorporation

A recent study has attempted to overcome the problem of significant levels of non-fungal protein in the initial substrate (Wissler *et al.* 1982). These authors measured the incorporation of [^{15}N]ammonium sulphate into a solid cellulosic substrate inoculated with *Chaetomium cellulolyticum*. Using mass spectrometry the abundance of ^{15}N in the substrate before and after fermentation was measured. The ratio of $^{15}N:^{14}N$ was used to calculate the quantity of fungal protein produced. The method has been correlated with conventional Kjeldahl assays and fungal chitin assays.

Growth of Fungi in Non-Axenic Conditions on Solid Substrates

Agar Film–Hyphal Length Measurement

Use of measurement of length of hyphae as a means of estimating fungal biomass was first proposed by Jones & Mollison (1948). The method consists of dispersing the solid substrate in some suitable diluent, with or without molten agar, and layering suitable volumes onto a haemocytometer slide chamber of fixed depth. Hyphal length measurements can then be made using eye piece graticules (Olson 1950). Biomass is calculated from total hyphal length and average diameter. Many samples must be analysed to obtain statistically significant data (Frankland *et al.* 1978). Various modifications have been proposed including the use of gridded membrane filters (Hanssen *et al.* 1974). In complex solid substrates some overlap of the size of fungal hyphae with those of other microorganisms will be inevitable. This will be particularly noticeable where large populations of actinomycetes are present (e.g. Sparling *et al.* 1982). The hyphal length assay will not distinguish between living or dead hyphae unless some staining procedure for viable cells is used (see below).

Chitin Estimation

Chitin is an insoluble linear polymer of α-1,4-linked N-acetylglucosamine units produced by most fungi and all insects. It is found in variable quantities in the cell walls of most fungi (Rosenberger 1976), but not in green plants, most animals or other microorganisms. These properties make it a very useful fungal-specific compound for biomass estimation in many solid substrates colonised by fungi, and where insect bodies or other sources of glucosamine such as bacteria are unlikely to be found.

The method was first used for estimating fungal biomass by Arima & Uozomi (1967) to measure the amount of *Aspergillus oryzae* in fermented rice (rice-koji). Glucosamine levels obtained after acid hydrolysis of the substrate were correlated with nucleic acid levels. After sulphuric acid hydrolysis and Dowex-50 ion-exchange chromatography, eluted glucosamine was measured by the Elson–Morgan reaction. Further improvement in the sensitivity and specificity of the assay came with the use of the reagent 3-methyl-2-benzothiazolinone hydrazone hydrochloride monohydrate (MBTH) by Tsuji *et al.* (1969). Later workers modified the assay by using an initial alkaline deacetylation step followed by glucosamine determination with MBTH. Ride & Drysdale (1972) and Aidoo *et al.* (1981), for example, give a useful correlation of the various hydrolysis and assay methods. The method has now been used to estimate fungal biomass in living plant tissue (Ride &

Drysdale 1972), decaying wood (Swift 1973), leaf litter (Frankland et al. 1978), food products (Jarvis 1977), wood (Braid & Line 1981) and cereal grains (Donald & Mirocha 1977; Nandi 1978).

One of the problems of the assay was observed by its first users, Arima & Uozomi (1967), i.e. chitin content of fungal cells does not remain in direct proportion to fungal growth. Other workers have observed similar effects (Sharma et al. 1977; Whipps & Lewis 1980). A further problem is that some complex organic substrates such as soils or composts contain high levels of glucosamine from insects or bacteria or both, which give very high background levels in the assay (Pusztai 1964). We have observed that mushroom compost (composted wheat straw) contains sufficient interfering substances before inoculation with the cultivated mushroom *Agaricus bisporus* to show subsequently, after inoculation and heavy colonisation by mushroom mycelium, only an apparent twofold increase in fungal biomass. Other indices of fungal biomass such as ergosterol assay, growth-linked enzyme assay or substrate consumption show a manyfold increase in *Agaricus* mycelium.

Chitin assay also gives an estimate of total fungal material but does not distinguish between fungal species in a substrate or between living or dead hyphae.

Ergosterol

Ergosterol (ergosta-5,7,22-trien-3β-ol) is the predominant sterol of most fungi (Weete 1974) and is probably a component of fungal cell membranes. It is not found to any significant extent in most green plants. Its use as a measure of fungal biomass was described by Seitz et al. (1977, 1979) to measure the extent of fungal colonisation of cereal grains. Various procedures can be used to remove ergosterol for quantitative assay, and a flow chart summarising some of them is shown in Fig. 1.

Ergosterol can be quantitatively measured by gas chromatography, HPLC, or UV spectrophotometry. The last is very convenient since plant sterols absorb poorly at the absorption maxima for ergosterol (Fig. 2).

Seitz et al. (1979) achieved good correlation of ergosterol production with fungal chitin levels during colonisation of cereal grains. As with other fungal products the specific ergosterol content (units ergosterol/unit dry wt fungus) may change with the physiological status of the fungus (Seitz et al. 1979). In addition, fungi may vary markedly between species in their ergosterol content, thus mixed fungal populations will show an "average" sterol content. A further complication of spectral assay might arise if the fungus produces other sterols which have similar spectral properties. Woods (1971) has observed that *Saccharomyces cerevisiae* produces a sterol, 24(28)-dehydroergosterol, which has spectral properties similar to ergosterol but its production levels

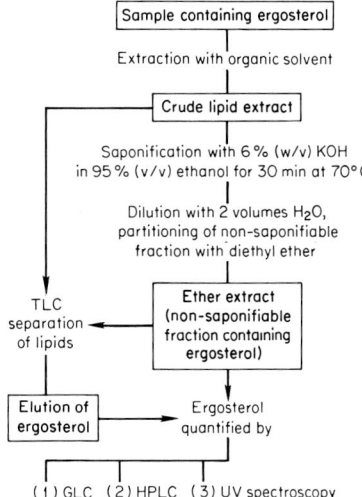

FIG. 1. Protocol for the extraction and assay of ergosterol.

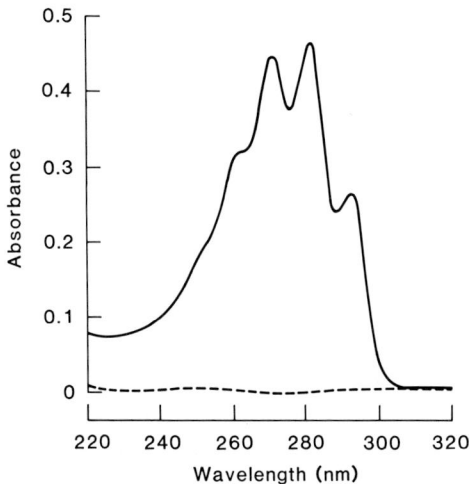

FIG. 2. Ultraviolet spectra of ergosterol (continuous line) and sitosterol (broken line) in ethanol.

are more sensitive to cultural manipulations. Since no data are available to show the levels of ergosterol in various ages of fungal tissue, it is not known if ergosterol assays measure content of live, senescent or dead hyphae. Despite these reservations ergosterol measurement seems a useful and rapid technique for estimating fungal biomass in many types of solid substrate.

Fluorescent Antibodies

The use of fluorescent antibodies as diagnostic tools for microbes and their antigens has a long history of use (Weir 1978; Bohlool & Schmidt 1980). Recent work has attempted to use fluorescent antibodies to measure the biomass of specific fungi in complex environments (Frankland *et al.* 1981). Antisera were raised to wall or cytoplasm antigens of the basidiomycete *Mycena galopus* and fluorescent antibodies were used to measure the biomass of hyphae of the latter in fungal-colonised leaf litter. Frankland *et al.* (1981) obtained good correlation with previous methods of estimating hyphae in situations where *Mycena* was the dominant fungus. One of the difficulties of the method is the specificity of the antiserum. Although precautions can be taken to avoid non-specific staining by the use of rhodamine–gelatin conjugates or pre-adsorption to remove non-specific antibody, some cross-reactions are observed with other fungi outside the genus *Mycena*. This may have been due to the fact that the dominant cell wall antigens are not highly species-specific, unlike their corresponding bacterial counterparts. This problem might be overcome by the use of monoclonal antibodies. The method also shows that fungal cells react with fluorescent antibody at certain preferred sites which probably reflects physiological and biochemical differences between parts of hyphae. It is not yet clear whether the method will distinguish living from dead hyphae.

Baker & Mills (1982) have recently developed a method for enumerating viable cells of the bacterium *Thiobacillus ferrooxidans* in natural environments which may provide a means of distinguishing between live and dead hyphae. The technique combined the use of fluorescent antibody with tetrazolium staining to estimate the proportion of live and dead cells.

Enzyme-Linked Immunosorbent Assay (ELISA) and Radioimmunoassay (RIA)

These methods have also been used extensively as diagnostic aids in microbial detection, particularly for viruses and bacteria. The procedures have been described in detail (Weir 1978; see Hill, this volume). The benefits of the assay methods are that by linking enzymes or radiolabels to specific antibodies, the "signal" produced by antibody–antigen interactions can be greatly amplified and thus produce a highly sensitive assay (see Table 1 for comparison of sensitivities of various assays). Their use to measure biomass of fungi in solid substrates has so far been limited. Casper & Mendgen (1979) used ELISA to measure fungal pathogens in infected plant tissue but gave no values for the possible fungal biomass. Another example of ELISA, given by Johnson *et al.* (1982) in a similar application, also made no calculation of fungal biomass. Some cross-reaction of the antisera with other fungal genera was observed at higher antibody levels. RIA has also been used in plant pathology to measure fungal infection (Savage & Sall, 1981). ELISA and RIA lend themselves to detection of very small quantities of fungal tissue (see Table 1) but whether better quantitative data emerge remains to be determined.

Growth-Linked Enzymes

In certain special cases fungal extracellular enzymes can be used to measure biomass. Wood (1979) found that the extracellular enyzme, laccase (a polyphenoloxidase), could be used for biomass estimation of the mycelium when the edible mushroom *Agaricus bisporus* colonised composted wheat straw. It was observed that production of laccase paralleled colonisation of this medium by fungal hyphae (Wood & Goodenough 1977). Good correlation was obtained between growth (dry matter) in complex liquid media and laccase production (Dijkstra *et al.* 1972; Wood 1979). Good correlation was also obtained between hyphal linear growth in composted straw and laccase production (Wood 1979; see also Fig. 3). In this environment laccases from other related species are unlikely to be found. Wood (1979) also showed that extracts produced from the colonised compost cross-reacted with antiserum

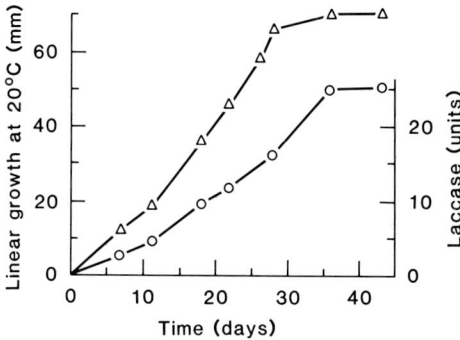

FIG. 3. Linear growth (△) and laccase production (○) of *Agaricus bisporus* on compost.

against purified *Agaricus bisporus* laccase and that only a few closely related species gave cross-reactions.

Correlation of fungal growth and enzyme production was also observed by Aidoo *et al.* (1981) for *Aspergillus oryzae* growing on rice. The two saccharifying enzymes amylase and amyloglucosidase were produced in parallel with increases in chitin during substrate colonisation.

Finding a growth-linked enzyme relies on finding a suitable candidate enzyme which increases in direct proportion to fungal growth, is stable and has a relatively simple assay. Cultural conditions are well known to affect enzyme production levels. Wood (1980) showed that laccase production by *Agaricus bisporus* differed by up to three- or fourfold between complex media (malt extract) and a defined medium. Such changes in enzyme production would produce similar proportional changes in calculated fungal biomass. Despite these limitations enzyme assays give a rapid means of monitoring growth rate and biomass.

Ribosomal Protein or Other Electrophoretically Separable Proteins

Marshall & Partridge (1981) observed that certain ribosomal proteins of *Fusarium moniliforme* could be electrophoretically separated from host (*Zea mays*) ribosomal proteins. From the detection sensitivities of Coomassie Blue-stained proteins on polyacrylamide gel, an estimate was made of fungal biomass present in the host tissue. As presently constituted this would be an extremely lengthy assay (Table 2) but, as pointed out by the authors, identification and isolation of species-specific proteins could lead to serological methods of assay. Antibodies to such proteins may overcome the problem of insufficient antibody specificity found with fluorescent antibody and ELISA or RIA techniques.

Conclusion

Several techniques are available for estimating various growth parameters of filamentous fungi growing in solid substrates, e.g. doubling times, biomass. All of these methods are of necessity indirect and must be correlated with fungal growth in liquid media. The assay methods of choice will depend on a number of factors which include the nature of the substrate.

The simplest class of solid substrates to assay are those containing monocultures of fungi in axenic environments on chemically simple media, e.g. cellulose fermentations. With increasing microbiological and chemical complexity the number of suitable assay techniques becomes more limited.

Some prejudgement of the sensitivity and accuracy required of the tech-

TABLE 2

Suitability of methods for use with axenic or non-axenic cultures

	Use with		
Assay method	Axenic cultures	Non-axenic cultures	General comments
Kjeldahl nitrogen	+	−	Interference from non-protein nitrogen (chitin)
Nucleic acids	+	−	Interference from non-fungal nucleic acids
ATP	+	−	Measures living tissue only; content varies with species and physiological state
Respiration	+	−	Variation with physiological state
Substrate consumption	+	−	Insensitive
$^{15}NH_4$ incorporation	+	−	—
Agar film, hyphal length	+	+	Hyphal size overlap with actinomycetes
Chitin	+	+	Cell wall composition varies with age and physiological state
Ergosterol	+	+	Varies with age and species
Fluorescent antibody	+	+	Cross-reactions occur with antigens of other genera
ELISA, RIA	+	+	Cross-reactions occur with antigens of other genera
Growth-linked enzymes	+	+	Species specific, production may be medium dependent
Electrophoretically separable proteins	+	+	Lengthy assay, used for developing serological methods

nique is called for, e.g. the current ELISA and RIA assays are very sensitive in detecting fungal mycelium but poor in giving quantitative data on fungal biomass. The problem may also call for either measurement of total fungal biomass or for that of an individual species in a mixed population with other fungi or other classes of microorganisms.

The current assay methods of choice for fungal biomass in complex environments appear to be those for ergosterol and chitin, although these have limitations. It is essential to correlate the data from two or more distinct types of growth assay on solid substrates to obtain reliable information on the likely magnitude of fungal growth.

References

AIDOO, K. E., HENDRY, R. & WOOD, B. J. B. 1981 Estimation of fungal growth in a solid substrate fermentation system. *European Journal of Applied Microbiology and Biotechnology* **12**, 6–9.

ANDERSON, J. P. E. & DOMSCH, K. H. 1975 Measurement of bacterial and fungal contributions to respiration of selected agricultural and forest soils. *Canadian Journal of Microbiology* **21**, 314–322.

ANONYMOUS 1975 In *Official Methods of Analysis* ed. Horwitz, W., 12th ed. Washington D.C: Association of Official Analytical Chemists.

ARIMA, K. & UOZOMI T. 1967 A new method for the estimation of mycelial weight in koji. *Agricultural and Biological Chemistry* **31**, 119–123.

BAKER, K. H. & MILLS, A. L. 1982 Determination of the number of respiring *Thiobacillus ferrooxidans* cells in water samples by using combined fluorescent antibody–2-(*p*-iodophenyl)-3-(*p*-nitrophenyl)-5-phenyltetrazolium staining. *Applied and Environmental Microbiology* **43**, 338–344.

BOHLOOL, B. B. & SCHMIDT, E. L. 1980 The immunofluorescent approach in microbial ecology. *Advances in Microbial Ecology* **4**, 203–241.

BRAID, G. H. & LINE, M. A. 1981 A sensitive chitin assay for the estimation of fungal biomass in hardwoods. *Holzforschung* **35**, 10–15.

BRAVERY, A. F. 1975 Microbiological assay of chemicals for the protection of wood. In *Some Methods for Microbiological Assay* eds. Board, R. G. & Lovelock, D. W., Society for Applied Bacteriology Technical Series No. 8, pp. 203–216. London & New York: Academic Press.

CALAM, C. T. 1969 The evaluation of mycelial growth. In *Methods in Microbiology* eds. Norris, J. R. & Ribbons, D. W., Vol. 1, pp. 567–591. London & New York: Academic Press.

CASPER, R. & MENDGEN, K. 1979 Quantitative serological estimation of a hyperparasite, detection of *Verticillium lecanii* in yellow rust infected wheat leaves by ELISA. *Phytopathology Zeitschrift* **94**, 89–91.

DIJKSTRA, F. I., SCHEFFERS, W. A. & WIKEN, T. O. 1972 Submerged growth of the cultivated mushroom *Agaricus bisporus*. *Antonie van Leeuwenhoek* **38**, 329–340.

DONALD, W. W. & MIROCHA, C. J. 1977 Chitin as a measure of fungal growth in stored corn and soybean seed. *Cereal Chemistry* **54**, 466–474.

FOSBERG, C. W. & LAM, K. 1977 Use of adenosine 5′-triphosphate as an indicator of the microbiota biomass in rumen contents. *Applied and Environmental Microbiology* **33**, 528–537.

FRANKLAND, J. C., LINDLEY, D. K. & SWIFT, M. J. 1978 A comparison of two methods for the estimation of mycelial biomass in leaf litter. *Soil Biology and Biochemistry* **10**, 323–333.

FRANKLAND, J. C., BAILEY, A. D., GRAY, T. R. G. & HOLLAND, A. A. 1981 Development of an immunological technique for estimating mycelial biomass of *Mycena galopus* in leaf litter. *Soil Biology and Biochemistry* **13**, 87–92.

HANSON, R. S. & PHILLIPS, J. A. 1981 In *Manual of Methods for General Bacteriology* eds. Gerhardt, P., Murray, R. G. E., Costilow, R. N., Nester, E. W., Wood, W. A., Krieg, N. R. & Phillips, G. B., pp. 328–364. Washington D.C: American Society for Microbiology.

HANSSEN, J. F., THINGSTAD, T. F. & GOKSOYR, J. 1974 Evaluation of hyphal lengths and fungal biomass in soil by a membrane filter technique. *Oikos* **25**, 102–107.

HARIMA, T. & HUMPHREY, A. E. 1980 Estimation of *Trichoderma* QM 914 biomass and growth rate by indirect determination. *Biotechnology and Bioengineering* **82**, 821.

HOLM-HANSEN, O. 1973 The use of ATP determinations in ecological studies. *Bulletin of the Ecological Research Committee* **17**, 215–222.

JARVIS, B. 1977 A chemical method for the estimation of mould in tomato products. *Journal of Food Technology* **12**, 581–591.

JENKINSON, D. S., DAVIDSON, S. A. & POWLSON, D. S. 1979 Adenosine triphosphate and microbial biomass in soil. *Soil Biology and Biochemistry* **11**, 521–527.

JOHNSON, M. C., PIRONE, T. P., SIEGEL, M. R. & VARNEY, D. R. 1982 Detection of *Epichloe typha* in tall fescue by means of enzyme-linked immunosorbent assay. *Phytopathology* **72**, 647–650.

JONES, P. C. T. & MOLLISON, J. E. 1948 A technique for the quantitative estimation of soil micro-organisms. *Journal of General Microbiology* **2**, 54–69.
MALLETTE, M. F. 1969 Evaluation of growth by physical and chemical means. In *Methods in Microbiology* eds. Norris, J. R. & Ribbons, D. W., Vol. 1, pp. 521–566. London & New York: Academic Press.
MARSHALL, M. R. & PARTRIDGE, J. R. 1981 *Fusarium moniliforme* detection in corn (*Zea mays* L.) stalk tissue: identification of specific ribosomal proteins by polyacrylamide gel electrophoresis. *Physiological Plant Pathology* **18**, 133–141.
NANDI, T. 1978 Glucosamine analysis of fungus-infected wheat as a method to determine the effect of anti-fungal compounds in grain preservation. *Cereal Chemistry* **55**, 121–126.
OKAZAKI, N. & SUGAMA, S. 1979 A new apparatus for automatic growth estimation of mold cultured on solid media. *Journal of Fermentation Technology* **57**, 413–417.
OLSON, F. C. W. 1950 Quantitative estimates of filamentous algae. *Transactions of the American Microscopical Society* **69**, 272–279.
PUSZTAI, A. 1964 Hexosamines in the seeds of higher plants. *Nature, London* **201**, 1328–1329.
RIDE, J. P. & DRYSDALE, R. B. 1972 A rapid method for the chemical estimation of filamentous fungi in plant tissue. *Physiological Plant Pathology* **2**, 7–15.
RIGHELATO, R. C. 1975 Growth kinetics of mycelial fungi. In *The Filamentous Fungi* eds. Smith, J. E. & Berry, D. R. Vol. 1, pp. 79–103. London: Arnold.
ROSENBERGER, R. F. 1976 The cell wall. In *The Filamentous Fungi* eds. Smith, J. E. & Berry, D. R., Vol. 2, pp. 328–344. London: Arnold.
SAVAGE, S. D. & SALL, M. A. 1981 Radioimmunosorbent assay for *Botrytis cinerea*. *Phytopathology* **71**, 411–415.
SEITZ, L. M., MOHR, H. E., BURROUGHS, R. T. & SAUER, D. B. 1977 Ergosterol as an indicator of fungal invasion in grains. *Cereal Chemistry* **54**, 1207–1217.
SEITZ, L. M., SAUER, D. B., BURROUGHS, R., MOHR, H. E. & HUBBARD, J. D. 1979 Ergosterol as a measure of fungal growth. *Phytopathology* **69**, 1202–1203.
SHARMA, P. D., FISHER, P. J. & WEBSTER, J. 1977 Critique of the chitin assay technique for estimation of fungal biomass. *Transactions of the British Mycological Society* **69**, 479–483.
SOLOMONS, G. L. 1975 Submerged culture production of mycelial biomass. In *The Filamentous Fungi* eds. Smith, J. E. & Berry, D. R., Vol. 1, pp. 249–264. London: Arnold.
SPARLING, G. P., FERMOR, T. R. & WOOD, D. A. 1982 Evaluation of methods for measuring microbial biomass in composted wheat straw and the possible contribution of biomass to the nutrition of *Agaricus bisporus*. *Soil Biology and Biochemistry* **14**, 609–611.
SUGAMA, S. & OKAZAKI, N. 1979 Growth estimation of *Aspergillus oryzae* cultured on solid media. *Journal of Fermentation Technology* **57**, 408–412.
SWIFT, M. J. 1973 The estimation of mycelial biomass by determination of the hexosamine content of wood tissue decayed by fungi. *Soil Biology and Biochemistry* **5**, 321–332.
TSUJI, A., KINOSHITA, T. & HOSHINO, M. 1969 Analytical chemical studies on amino sugars. Determination of hexosamines using 3-methyl-2-benzothiazobone hydrazone hydrochloride. *Chemical and Pharmaceutical Bulletin* **17**, 1505–1510.
ULMER, D. C., TENGERDY, R. P., MURPHY, V. G. & LINDEN, J. C. 1980 Solid-substrate fermentation of manure fibres for SCP production. *Developments in Industrial Microbiology* **21**, 425–434.
WEETE, J. D. 1974 *Fungal Lipid Biochemistry: Distribution and Metabolism*. New York: Plenum Press.
WEIR, D. M. 1978 *Handbook of Experimental Immunology*. Oxford: Blackwell.
WHIPPS, J. M. & LEWIS, D. H. 1980 Methodology of a chitin assay. *Transactions of the British Mycological Society* **74**, 416–418.
WISSLER, M. D., TENGERDY, R. P. & MURPHY, V. G. 1982 Biomass measurement in solid

substrate fermentations using ^{15}N mass spectrometry. *Society for Industrial Microbiology News* **32**, 41.

WOOD, D. A. 1979 A method for estimating biomass of *Agaricus bisporus* in a solid substrate, composted wheat straw. *Biotechnology Letters* **1**, 255–260.

WOOD, D. A. 1980 Production, purification and properties of extracellular laccase of *Agaricus bisporus*. *Journal of General Microbiology* **117**, 327–338.

WOOD, D. A. & GOODENOUGH, P. W. 1977 Fruiting of *Agaricus bisporus:* changes in extracellular enzyme activities during growth and fruiting. *Archives of Microbiology* **114**, 161–165.

WOODS, R. A. 1971 Nystatin resistant mutants of yeast: alterations in sterol content. *Journal of Bacteriology* **108**, 69–73.

The Dynamics of Cellulose Decomposition

M. K. Theodorou

Animal Nutrition and Production Division, The Grassland Research Institute, Hurley, Maidenhead, Berkshire, UK

M. J. Bazin

Microbiology Department, Queen Elizabeth College, University of London, London, UK

A. P. J. Trinci

Department of Botany, University of Manchester, Manchester, UK

In the natural environment many situations exist where it is not possible, or at least very difficult, to estimate microbial activity either because appropriate techniques do not exist or because sampling destroys the system under study. In particular, it is often difficult to estimate the way in which microbial biomass and products change with respect to time. One approach to studying the dynamics of microbial ecosystems has been to build model systems and determine the behaviour of the models under defined conditions. The models may be either physical analogues of the natural phenomenon or sets of mathematical equations. In the latter case, mathematical models can be employed usefully to specify precisely assumptions and hypotheses, and subsequent mathematical analysis can generate experimentally testable predictions.

Lignocellulose is the most abundant renewable carbon resource available and therefore is a potentially important substrate for biotechnological conversion to useful products such as ethanol and single-cell protein. In the form of wheat-straw stubble, lignocellulose is an agricultural waste product which when removed by burning causes an environmental hazard. Therefore, lignocellulose and cellulose degradation is of considerable practical importance.

One aspect of particular interest is the dynamics of cellulose decomposition. In the work reported here, both experimental and mathematical models were used to study the growth of and production of cellulolytic enzymes by *Trichoderma reesei* in an environment such as the soil.

Microbial activity in the soil takes place in an open ecosystem through which the nutrient supply is usually discontinuous. The distinguishing characteristics of such an ecosystem are the unidirectional flow of nutrients and metabolic products and the adhesion of the microbial flora to, or entrapment between, the solid particulate material. These two properties give rise to concentration gradients which do not occur in well-mixed ecosystems. For an experimental model to represent the dynamic aspects of a soil ecosystem, it is necessary to incorporate these two fundamental properties. For a theoretical model to reflect the same properties, changes with respect to two independent variables, time and depth, must be considered simultaneously. This is in contrast to the dynamics of a well-mixed system such as a chemostat where the only independent variable is time.

In this chapter, first we describe the experimental system and then derive kinetic equations, based on a set of clearly defined assumptions which form the basis of a mathematical model describing the dynamics of the system. The results are given of experiments which tested the efficiency of the mathematical model close to steady state and after an environmental perturbation which shifted it far from steady state.

The Experimental System

A continuous flow column similar to those used by Bazin & Saunders (1973) to study nitrification formed the basis of our experimental system. The glass column had an internal diameter of 1.5 cm and was filled to a height of 12 cm with a mixture of glass beads (0.4-cm diameter) and 0.5 or 1.25 g ashless cellulose floc (Whatman Biochemicals Ltd., Maidstone, UK); the void volume of the column was ca. 11 ml. The column was surrounded by a water jacket and the temperature was maintained at 25 ± 1°C. Inflowing and outflowing solutions passed through stainless steel tubes set in silicon bungs at the top and the bottom of the column. Sterile filtered air and sterile medium were passed into the top of the column at rates of 100 ml/h and 2.0 ml/h, respectively. Flow rates were regulated using Watson–Marlow (MHRE2) peristaltic pumps. A diagram of the experimental column is given in Fig. 1. The composition of the inflowing medium was based on that used by Mandels & Reese (1957) for the growth of *Trichoderma viride*. The original Mandels and Reese medium, which was formulated empirically and has been used extensively to study *Trichoderma* spp., was modified (Theodorou

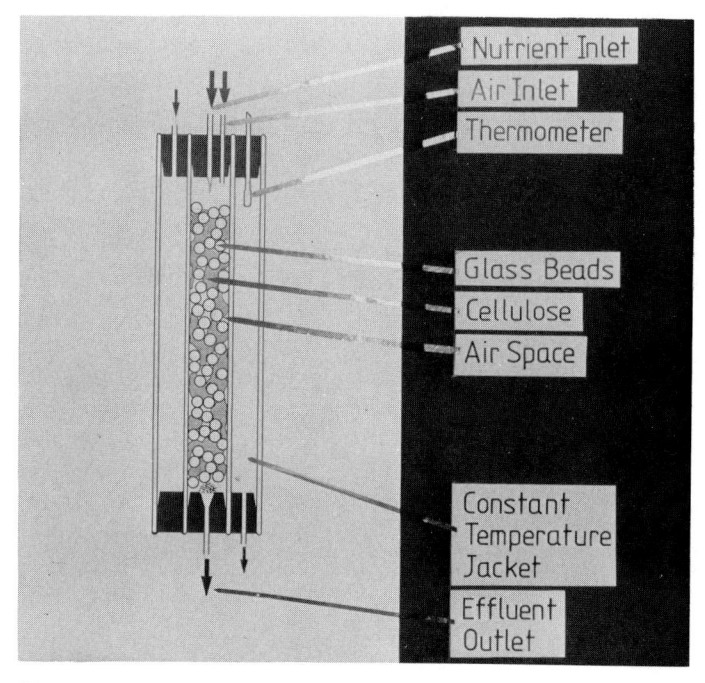

FIG. 1. Diagram of the continuous-flow experimental column for studying cellulose degradation.

et al. 1981) by incorporating a chelating agent, to prevent the formation of precipitates, and by omitting peptone and urea, thus making it a fully defined mineral salts medium.

Effluent solution from the column was collected in a refrigerated (4°C) fraction collector at 4-h intervals. The pH of each fraction was recorded, the volume measured and the sample was stored at −20°C for subsequent analysis. The samples were thawed and centrifuged, and the supernatant liquid was assayed for carboxymethylcellulase (CMC-ase) activity and β-glucosidase activity (Theodorou *et al.* 1980). Reducing sugar (Nelson 1944) and soluble protein (Lowry 1951) content were also analysed. The CO_2 content of the effluent gas was measured by gas chromatography. Effluent gas was collected in 10-ml volumes by connecting a flow-through syringe to the effluent outlet of the column. At least 30 ml gas were allowed to flow through the syringe before a sample was taken.

Columns were inoculated with a volume of spore suspension approximately equal to the void volume of the column and the nutrient flow was started 12 h later.

Two experiments were performed. In the first, carbon-free medium was allowed to pass down the column and microbial activity was estimated in the effluent from the bottom of the column. In the second experiment, perturbation of nutrient supply was achieved by amending the input medium with a soluble carbon source for one 300-h period by switching between reservoirs containing mineral salts medium and mineral salts medium + glucose, 1 mg/ml.

Mathematical Model

Because the dependent variables (biomass, cellulose, enzymes, substrates, etc.) change with respect to both time and distance down the column, the system is properly described by a set of partial differential equations (De Wiest 1969). Some microbiological aspects of the dynamics of soil columns have been studied using such equations, e.g. nitrification in the absence of nitrifier growth (Cho 1971) and simplified models of nitrification (Saunders & Bazin 1973). In general, however, analysing sets of non-linear partial differential equations is difficult; mathematical mistakes are easy to make (McLaren 1973) and the application of numerical techniques for solving partial differential equations is considerably harder than for ordinary differential equations. Therefore, we have used a conceptually simple method which was used by Prosser & Gray (1977) to describe nitrification in a soil column and which describes the system in terms of ordinary differential equations. In this method we suppose that the column is divided into a number of compartments, each of which can be considered to be a well-mixed chemostat system. Flow is simulated by letting the output of one compartment be the input to the one below it. The larger the number of compartments considered, the closer the system approximates a smooth change with respect to distance down the column. The advantage of this approach is that numerical techniques for solving the ordinary differential equations describing the activity in each compartment are relatively easy to apply using a digital computer.

The Kinetics of Cellulose Digestion by Trichoderma reesei

The enzymatic decomposition of cellulose by microorganisms is still incompletely understood. We based our model on the following assumptions:

(1) The "cellulase" complex secreted by *T. reesei* is made up of three enzymes which react as follows:

$$\text{Crystalline cellulose} \xrightarrow{\text{endocellulase}} \text{Amorphous cellulose}$$

$$\text{Amorphous cellulose} \xrightarrow{\text{exocellulase}} \text{Cellobiose}$$

$$\text{Cellobiose} \xrightarrow{\text{cellobiase}} \text{Glucose}$$

(2) The rate of production of each enzyme is first order with respect to fungal biomass.
(3) The mass of enzymes produced is negligible compared to the total biomass present.
(4) Each enzymatic reaction has the potential of being product inhibited by competitive inhibition.
(5) Growth of *T. reesei* is limited by the concentration of glucose present and can be described by the Monod (1942) saturation function.
(6) The yield of fungal biomass produced per unit of glucose consumed is constant.
(7) A constant fraction of the biomass in each compartment is washed out.

Following from these assumptions, the kinetic equations describing the system can be specified. From assumption (2) the change in the concentration of each enzyme (E_i) is described by

$$dE_i/dt = kX \tag{1}$$

where k is a first-order rate constant and X is the biomass concentration of *T. reesei*. From assumption (4) the change in the concentration of each substrate (C_i) is

$$\frac{dC_i}{dt} = \frac{V_i C_i E_i}{K_i(1 + P_i/I_i) + C_i} \tag{2}$$

where V_i, K_i and I_i are constants representing the maximum reaction velocity, the saturation constant and the inhibition constant of the reaction, respectively, and P_i is the concentration of the product.

Assumption (5) implies

$$dX/dt = \mu_m XG/(K_s + G) \tag{3}$$

where μ_m is the maximum specific growth rate, K_s is the saturation constant and G is the concentration of glucose present. Assumption (3) implies that a term for the loss of biomass to enzymes need not be included. From assumption (6),

$$dG/dt = \mu_m XG/Y(K_s + G) \tag{4}$$

In order to describe the dynamics of the system, all that is required in addition to these kinetic equations are the flow terms for the passage of soluble materials and biomass in and out of the compartments. For each compartment this results in eight ordinary non-linear differential equations which are given in full by Theodorou *et al.* (1981). For a 10-compartment model this means that a total of 80 differential equations must be solved. Fortunately, with the exception of the equations for the first compartment, which has no input from a compartment above it, all the remaining 72 equations can be solved numerically by an iterative procedure so that only 16 equations have to be specified in a computer program.

Experimental Evaluation of the Model

In the analysis of well-mixed microbial ecosystems such as chemostats, one of the first steps is to determine the steady-state behaviour, i.e. the properties of the system when it stops changing with respect to time. In our experimental system, steady state is achieved only after all the carbon source, i.e. cellulose, is exhausted, that is to say when no biological activity at all occurs. The system does not come to a non-trivial steady state, but is always changing with respect to time, albeit slowly. The steady-state behaviour of homogeneous single-species ecosystems is quite well described by kinetic models of the Monod (1942) type; on the other hand, if such systems are perturbed far away from equilibrium, e.g. by rapid enrichment of the nutrient supply, their behaviour is only poorly described by such models (Bazin *et al.* 1976). The goal of the experiments to be described was to determine how well a model based on Monod-type kinetics would describe a heterogeneous, open microbial ecosystem represented by the experimental system under constant environmental conditions and after perturbation of the nutrient supply. Switching from a mineral salts medium to one containing glucose and back produced a square-wave pulse of glucose in the nutrient feed.

Figure 2 is a simulation of the behaviour of the lowest compartment of a 10-compartment model after homogeneous inoculation of a column with actively growing *T. reesei*. The characteristic features predicted by the model are the relatively rapid decrease in both crystalline and amorphous cellulose (*a* and *b*), the relatively rapid appearance and disappearance of detectable amounts of glucose and cellobiose (*c* and *d*), the rapid increase in biomass followed by a slower increase and then a relatively slow decline (*e*) and the similarity in the way all three enzyme concentrations change (*f*, *g* and *h*). This predicted behaviour was compared with that of the effluent from an experimental column. It was expected a priori that because an inoculum of spores was used, the experimental system would exhibit a marked lag period which would not be observed in simulations.

Not all of the variables of the mathematical model could be measured

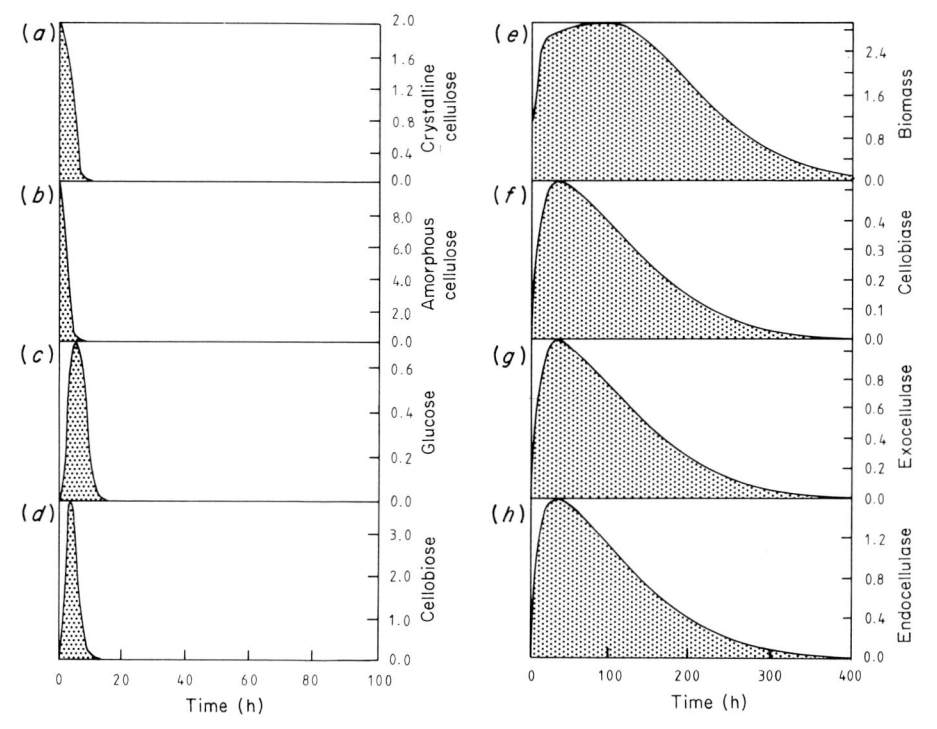

FIG. 2. Simulation of the degradation of cellulose in the lowest compartment of a 10-compartment model. All units are expressed as mg/ml; equations used for coefficients are given in full by Theodorou et al. (1981).

experimentally. Therefore, the predicted results were compared with the behaviour of variables considered to approximate to them. Biomass accumulation was estimated in terms of percentage CO_2 in effluent gas, glucose + cellobiose concentration was measured in terms of reducing sugar equivalents, cellobiase activity was determined as β-glucosidase and total enzyme (endocellulase, exocellulase and cellobiase) was taken as the total protein concentration. The simulated behaviour of the lowest compartment of the column was compared with the effluent solution from the experimental column.

Figure 3 shows the properties of the effluent from a column which initially contained 0.5 g cellulose and was supplied with glucose-free medium for 500 h. Apart from the time lags in the experimental variables which were expected, the overall qualitative behaviour was in good agreement with that predicted by the model. In particular, the similarity between the reducing sugar activity in the effluent (Fig. 3e) and the predicted behaviour of glucose

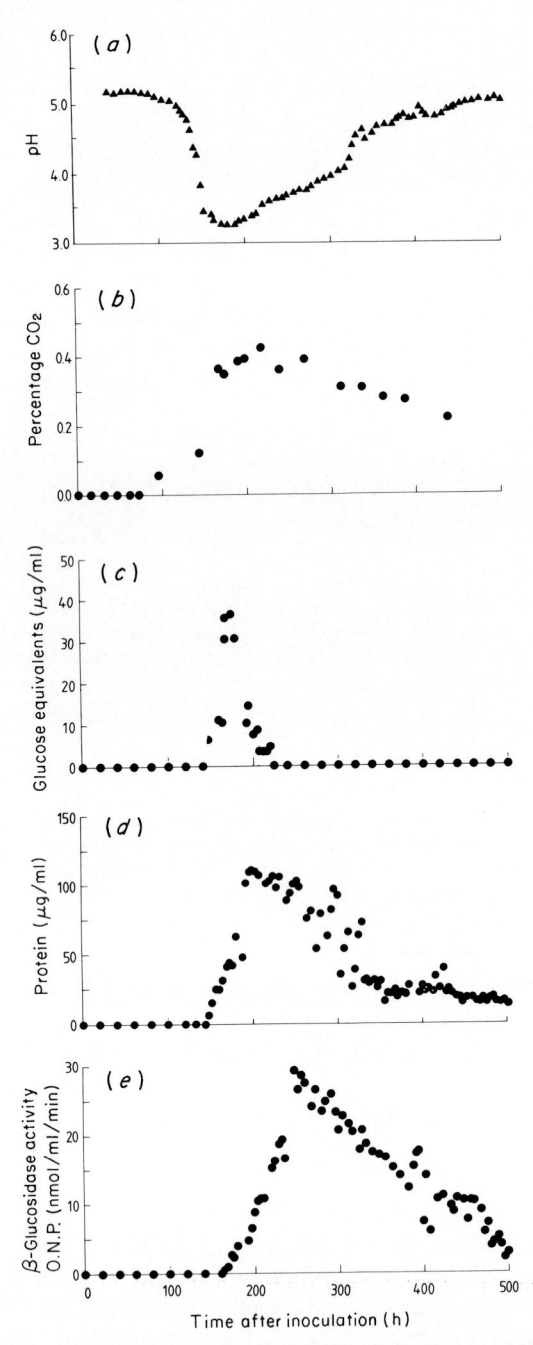

FIG. 3. The characteristic properties of the effluent from an experimental column. The column initially contained 0.5 g cellulose and was supplied with a glucose-free medium for the first 500 h after inoculation.

and cellobiose, both of which were predicted to appear as pulses (Fig. 2c and d), was especially striking. No attempt was made to fit the model quantitatively to the experimental data as the kinetic equations only approximate to the behaviour of the system which is still not known in detail. In this light it is remarkable that the data matched the predicted results so well.

Figure 4 shows a simulated experiment in which glucose was included in the input medium as a 100-h pulse. Figure 5 shows the data from the equivalent experiment with a 300-h pulse of glucose. In this case marked differences between predicted and experimental results occurred. The simu-

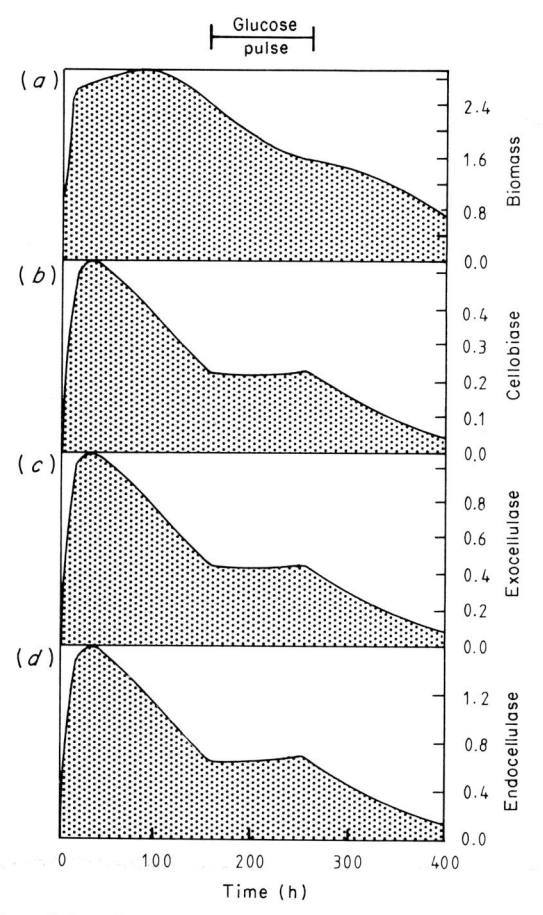

FIG. 4. Simulation of the effect in the lowest compartment of a 10-compartment model of adding a square-wave pulse of glucose in the input medium. Glucose was added to the first compartment 150–250 h after the onset of the simulation. All units are expressed as mg/ml.

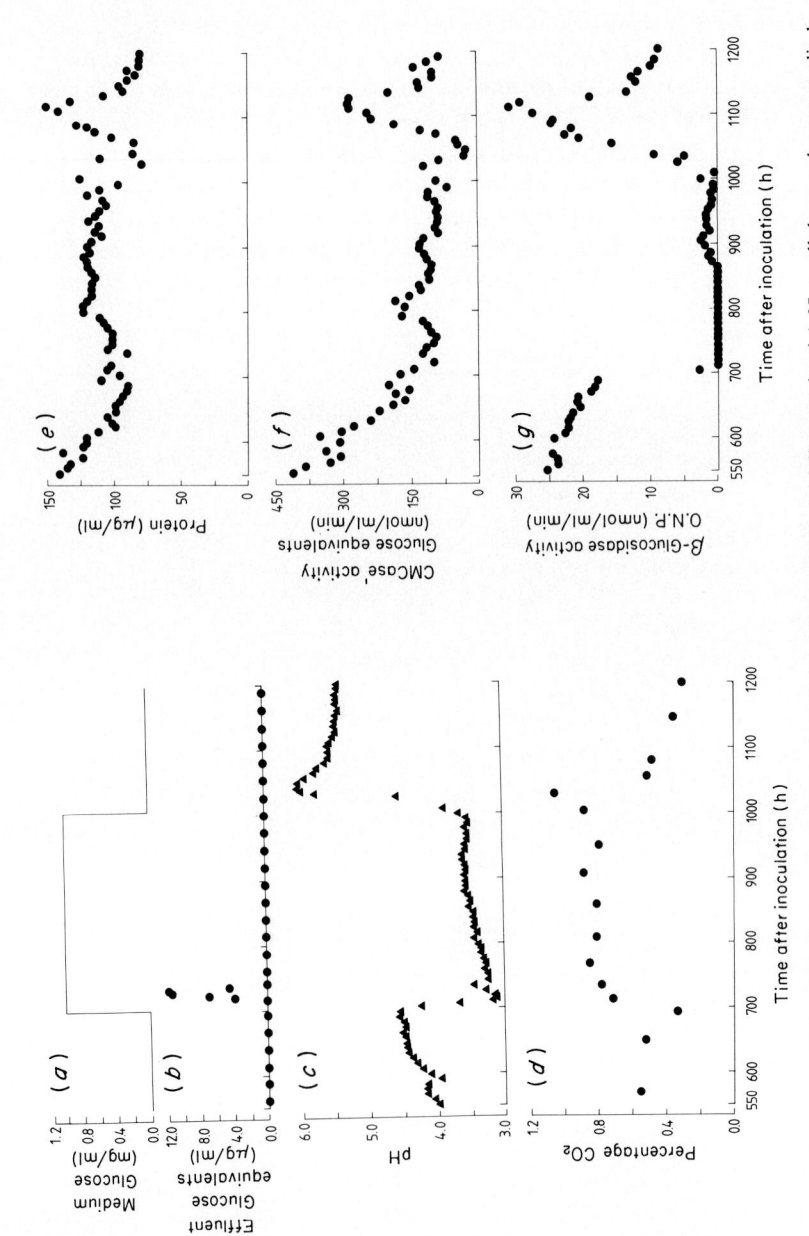

FIG. 5. Properties of the effluent from an experimental column. The column initially contained 1.25 g cellulose and was supplied with glucose-free medium except for a 300-h pulse of glucose 700–1000 h after inoculation. Data are presented for the last 650 h of a 1200-h experiment.

lation predicted that enzyme concentration would stabilise during the addition of glucose (Fig. 4*b*–*d*), but β-glucosidase (equivalent to cellobiase) activity declined in the experiment (Fig. 5*g*). Also, the simulation predicted a decline in biomass (Fig. 4*a*), but CO_2 production increased in the experiment (Fig. 5*b*); in this case the difference between the results expected and those obtained is presumably due to an increase in CO_2 production caused by respiration stimulation during the glucose pulse, indicating that CO_2 production is only a poor estimate of biomass under these experimental conditions.

Discussion

When medium containing no glucose was fed to a column containing cellulose and *T. reesei*, the behaviour of the effluent corresponded quite well with that predicted in the lowest compartment of a 10-compartment model. On the other hand, when the input medium was amended with a square-wave pulse of glucose, the predicted and experimental behaviour diverged significantly. We conclude, therefore, that while the system is changing slowly it can be described in terms of kinetics of the Monod (1942) type. Sudden changes, however, result in behaviour which does not correspond to such kinetics.

These results agree with those of other workers using both theoretical and experimental approaches (Bazin *et al.* 1976; Cooney *et al.* 1981; Barford *et al.* 1982; Condrey 1982; Cunningham & Maas 1982; Pickett 1982) in that Monod-type kinetics are inadequate in describing microbial growth under transient conditions. This inadequacy might be explained by the form of the Monod function itself which is often equated to the Michaelis–Menten function for enzyme kinetics to which it is mathematically equivalent. In deriving the Michaelis–Menten expression (e.g. Bull 1964) we usually consider the following reactions to take place:

$$E + S \rightarrow ES \rightarrow E + P$$

where E is an enzyme with substrate S, P is the product of the reaction, and ES is a transient enzyme–substrate product. If an analogy is made to the conversion of a substrate into microbial biomass, we might suppose the following equation to hold:

$$A + S \rightarrow AS \rightarrow A + X$$

where S is substrate, A represents active sites (say, on the surface of the microorganisms), X is the biomass produced by the reaction, and AS an active site–substrate complex. In the derivation of the Michaelis–Menten equation the enzyme–substrate concentration is assumed to be constant. Under such an assumption the familiar enzyme kinetic function

$$V = V_{max} S/(K_m + S) \tag{5}$$

can be derived. In order to derive the Monod function in Eq. (3) using the same reasoning, it is necessary to suppose that the concentration of active site–substrate complex, analogous to the enzyme–substrate complex, also remains constant. If the concentration of active sites increases as biomass increases, which it presumably does, then as the microbial population increases in size so will the number of active sites and the concentration of active site–substrate complexes. Therefore, for the production of biomass, kinetics of the Michaelis–Menten type are clearly not suitable unless it can be assumed that the active site–substrate complex concentration remains constant. This will occur only under steady-state conditions. Thus it is not surprising that at steady state, functions of the Monod type often adequately describe microbial growth, but not under transient conditions of growth.

The Monod function may give a reasonable description of microbial growth in systems which appear to be near to steady state. An example of such a system is fed-batch cultures which obey "quasi-steady-state" kinetics of the Monod type, as they have been called by Pirt (1975). The activity of *T. reesei* in our first experiment can be described in a similar manner. On the other hand, adding or withdrawing glucose from the feed medium resulted in a situation which moved the system far from steady state. Then Monod kinetics would not be expected to describe the results and, indeed, they did not.

It would appear, therefore, that for soil ecosystems changing relatively slowly and not subjected to significant environmental perturbations, it might be possible to analyse and even predict behaviour in terms of Monod kinetics. For situations where the environment changes in such a way as to move the system away from steady state, a more complex structured model is likely to be needed.

References

BARFORD, J. P., PAMMENT, N. B. & HALL, R. J. 1982 Lag phases and transients. In *Microbial Population Dynamics* ed. Bazin, M. J., pp. 55–89. Boca Raton: CRC Press.

BAZIN, M. J. & SAUNDERS, P. T. 1973 Dynamics of nitrification in a continuous flow system. *Soil Biology and Biochemistry* **5**, 531–543.

BAZIN, M. J., SAUNDERS, P. T. & PROSSER, J. I. 1976 Models of microbial interactions in the soil. *CRC Critical Reviews in Microbiology* **4**, 463–498.

BULL, H. B. 1964 *An Introduction to Physical Biochemistry*. Philadelphia: Davis Company.

CHO, C. M. 1971 Convective transport of ammonium with nitrification in soil. *Canadian Journal of Cell Science* **51**, 339–350.

CONDREY, R. E. 1982 The chemostat and Blackman kinetics. *Biotechnology and Bioengineering* **24**, 1705–1709.

COONEY, C. L., KOPLOV, H. M. & HÄGGSTRÖM, M. 1981 Transient phenomena in continuous culture. In *Continuous Culture of Cells* ed. Calcott, P. H., Vol. 1, pp. 143–168. Boca Raton: CRC Press.

CUNNINGHAM, A. & MAAS, P. 1982 The growth dynamics of unicellular algae. In *Microbial Population Dynamics* ed. Bazin, M. J., pp. 167–188. Boca Raton: CRC Press.

DE WIEST, J. M., ed. 1969 *Flow Through Porous Media*. London: Academic Press.

LOWRY, O. H. 1951 Protein measurements with Folin phenol reagent. *Journal of Biological Chemistry* **193**, 265–275.

MCLAREN, A. D. 1973 Nitrification in soil; systems approaching steady state: a correction. *Soil Society of America Proceedings* **37**, 336–337.

MANDELS, M. & REESE, E. T. 1957 Induction of cellulases in *Trichoderma viride* as influenced by carbon sources and metals. *Journal of Bacteriology* **73**, 269–278.

MONOD, J. 1942 *Recherches sur la Croissance des Cultures Bacteriennes*. Paris: Hermann.

NELSON, N. 1944 A photometric adaptation of the Somoggi method for the determination of glucose. *Journal of Biological Chemistry* **153**, 375–380.

PEITERSON, N. & ROSS, E. W. 1979 Mathematical model for enzymatic hydrolysis and fermentation of cellulose by Trichoderma. *Biotechnology Bioengineering* **21**, 997–1017.

PICKETT, A. M. 1982 Growth in a changing environment. In *Microbial Population Dynamics* ed. Bazin, M. J., pp. 91–124. Boca Raton: CRC Press.

PIRT, S. J. 1975 *Principles of Microbe and Cell Cultivation*. Oxford: Blackwell.

PROSSER, J. I. & GRAY, T. R. G. 1977 Use of finite difference method to study a model system of nitrification at low substrate concentrations. *Journal of General Microbiology* **102**, 119–128.

SAUNDERS, P. T. & BAZIN, M. J. 1973 Non-steady-state studies of nitrification in soil: theoretical considerations. *Soil Biology and Biochemistry* **5**, 545–557.

THEODOROU, M. K., BAZIN, M. J. & TRINCI, A. P. J. 1980 Cellulose degradation in a structured ecosystem which is analogous to soil. *Transactions of the British Mycological Society* **75**, 451–454.

THEODOROU, M. K., BAZIN, M. J. & TRINCI, A. P. J. 1981 Growth of *Trichoderma reesei* and the production of cellulolytic enzymes in a continuous flow column. *Archives of Microbiology* **130**, 372–380.

The Application of Molecular Biology to Lignin Degradation

A. Paterson, A. J. McCarthy and P. Broda

Department of Biochemistry and Applied Molecular Biology, University of Manchester Institute of Science and Technology, Manchester, UK

Lignin is an abundant natural polymer and it is a possible valuable raw material resource for the future. Currently lignocellulosic materials are pulped to remove lignin from the commercially important cellulose; this is an expensive process and causes pollution. However, biological selective removal of lignin from plant material could radically alter the economics of pulping and the production of fuel alcohol from crops and wastes. At present the value of lignin is as a fuel since, when burnt, its high calorific value makes a major contribution to the large energy requirements of pulp mills. Lignocellulosic materials also have a relatively high cost for any potential industrial process since the large volumes required are expensive to collect. Lignin is also intrinsically interesting because it is a non-ordered polymer of aromatic carbon residues showing no stereospecificity, unlike the other major polymers such as cellulose, proteins and nucleic acids, a feature that has contributed to the slow progress in understanding lignin biodegradation.

Most lignin research has been funded by paper and pulping concerns, for which reason there is a quite detailed understanding of the reactions which take place during the pulping and bleaching operations in cellulose manufacture (Gierer 1982). This understanding required the elucidation of the structure of lignins and therefore the best-defined lignins are those of the woody plants, especially spruce, birch and beech, which are of commercial importance. A more limited amount of information is available on the lignins of grasses and bamboos, where the polymer, although basically similar, has distinctive characteristics. Study of the organisms involved in biodegradation has been predominantly confined to the white-rot fungi. However, these organisms are unrelated to any of the fungi much studied by physiologists and geneticists, and therefore little is yet known about their biology.

The purpose of this chapter is to describe our approach to the elucidation of the mechanism of lignin degradation and to define the different steps of the process and their control by use of genetics. We are primarily interested in the structure and degradation of straw and grasses and their component lignocelluloses and lignins, since this is of biotechnical interest in Britain and has been relatively neglected.

Naturally Occurring Lignins

Composition of the Plant Cell Wall

In the plant cell wall, lignin (see next paragraph) and hemicellulose components form essentially two-dimensional block co-polymers, referred to as "glycolignin" (Sarkanen 1980). Hemicelluloses are polymers with both pentose and hexose monomers. Xylans are predominantly polymers of the pentose D-xylose with single-unit side chains of arabinose; they form a major hemicellulose component of plants. The other important components are the mannans, of which galactoglucomannan is the principal type, and arabinogalactans which are also present in significant quantities in certain plant species (Whistler & Richards 1970). Cellulose, the most important structural molecule of the wall, is a polymer of β-1,4-linked glucose units forming elongated chains organised in parallel into elementary microfibrils that contain ca. 40 glucan chains and are of 3.5-nm diameter. These elementary microfibrils are themselves organised in a regular manner, giving cellulose a crystalline structure which lignin lacks. Pectins (methyl esters of polygalacturonic acid) have an important role in the intercellular matrix but are present only in small quantities in lignified tissue.

Lignin Biosynthesis

Lignin exists *in situ* as molecules of large but indefinite size, linked to hemicelluloses probably by ester or ether bonds. It is formed by enzyme-catalysed oxidative addition of three monomers—*p*-coumaryl alcohol, coniferyl alcohol and sinapyl alcohol—to chains of the growing polymer. In precursor synthesis, cinnamic acids, which are synthesised from aromatic amino acids, are hydroxylated at ring 4, 3 and 5 positions. The intermediates with 3 and 5 hydroxylations are methylated at these positions by methyl transferases and then carboxyl groups are reduced to give the corresponding alcohols (Fig. 1). For different species the ratio of monomers available for

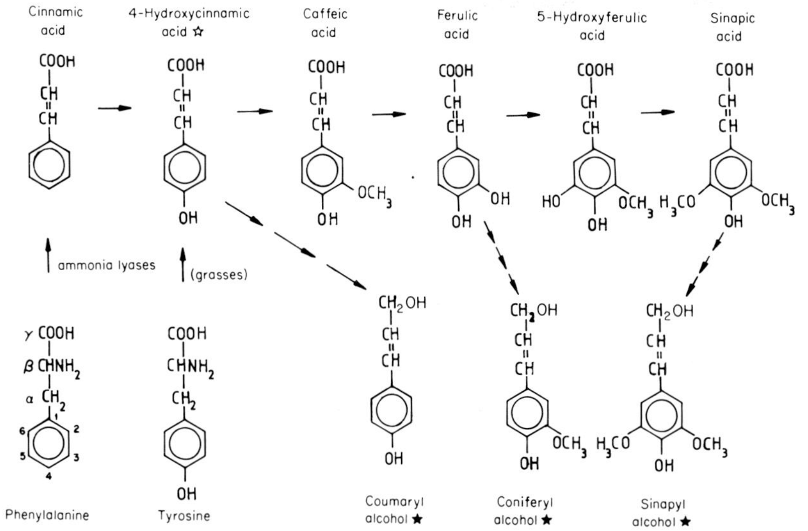

FIG. 1. Biosynthesis of lignin monomers (adapted from Higuchi 1980). (★), A monomer in all lignins; (☆), esterified to alcohols in grass lignins.

the polymerisation reaction is determined by the ratio of methyl transferase activities. The lignins formed are of two major types, guaiacyl (softwood) and guaiacyl-syringyl (hardwood). Grass lignins are thought to be of guaiacyl-syringyl type with esterified cinnamic acids, a similar feature to poplar lignins which have esterified *p*-hydroxybenzoic acid groups (Table 1). Lignification has recently been discussed in detail by Higuchi (1980).

Lignins are synthesised *in situ* after the cellulose microfibrils have been assembled and serve to separate them. They are distributed unevenly between primary and secondary cell walls and the middle lamellae, and in these regions may differ in their content of intermonomer C—C bonds (degree of condensation).

Polymerisation is initiated by the removal of electrons from phenoxyl anions by phenol oxidases, generating phenoxyl radicals. The electron system of the phenoxyl radical allows delocalisation of the electron and formation of a hybrid of a number of mesomeric structures (Fig. 2a). The subsequent attack by the phenoxyl radical on the terminal phenolic hydroxyl of the polymer yields an ether bond between the β carbon of the radical and the 4' phenolic hydroxyl. Repeated endwise addition in this manner generates the growing polymer so that the β aryl ether (β-*O*-4) bond is the major type in lignins (Fig. 2b,c). The possibility of forming other bonds also exists; when unsaturated C—C bonds have been removed the formation of β-*O*-4 bonds can no longer occur and C—C bonds are formed instead (Fig. 2d). The most

TABLE 1

Percentage of monomer components in different lignins

Plant	Monomer		
	Coniferyl	Sinapyl	Coumaryl
Softwood (gymnosperm) e.g. spruce, pine	80	6	14
Hardwood (angiosperm) e.g. birch, eucalyptus	46	46	8
Grasses (angiosperm) e.g. maize, bamboo	40	40	20

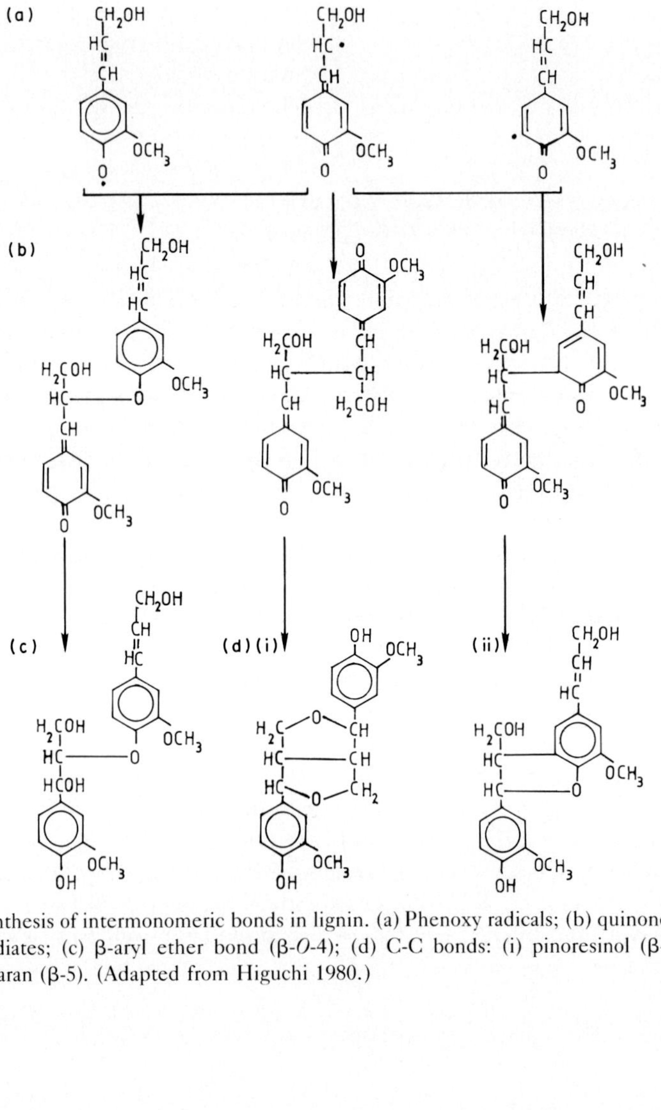

FIG. 2. Synthesis of intermonomeric bonds in lignin. (a) Phenoxy radicals; (b) quinone methide intermediates; (c) β-aryl ether bond (β-*O*-4); (d) C-C bonds: (i) pinoresinol (β-β), (ii) phenylcoumaran (β-5). (Adapted from Higuchi 1980.)

important of these are the pinoresinol, benzofuran, β-5 and biphenyl types. Thus lignins differ in the proportions of monomers in the polymer and in the ratios of bond types; the latter property is, in part, a reflection of the former.

Extraction of Lignins

Different procedures have been developed that yield extracted lignins differing in molecular weight and degrees of chemical modification (Kolodzlejski *et al.* 1982). Oligomers which contain bonds important in the lignin polymer, commonly referred to as model compounds, have also been used extensively to study the separate reactions occurring in lignin breakdown. However, even the best purified preparations are only an approximation to the heterogeneity of natural lignins in native lignocarbohydrate complexes. Purified and model compounds have to be chosen carefully for experiments, with adequate consideration of their relevance to the native polymer.

Milled-Wood Lignin (Björkman's Lignin)

Milled-wood lignin (MWL) is obtained by extraction of ball-milled plant material with neutral dioxane (Björkman 1956). It is generally accepted as the best approximation to native lignin but differs in that it has a lower molecular weight, typically 10,000–15,000 (equivalent to 50–80 monomers), owing to cleavage of both C—C and C—O bonds during milling.

Straw, chopped to a manageable size, is passed through a Wiley or a Kek mill to give a powder of coarse flour consistency. Two passages through a Kek mill is adequate. The flour is extracted in a Soxhlet apparatus with acetone and subsequently acetone:water (9:1 by volume) until the extractant is clear. Extraction removes waxes and compounds which may interact with lignins during milling. The residue is dried over P_2O_5 and then transferred to a vibratory ball mill, two types of which are in common use. We use a Siebteknik mill (Tema Machinery Ltd., Banbury, Oxfordshire) which consists of two stainless steel pots filled with steel ball bearings and mounted on a sprung carriage that is vibrated by an eccentric cam driven by an electric motor. The NBS mill (now obsolete) is similar in principle but the pot is coupled to a motor drive by a flexible coupling which is supported by a wishbone spring such that drive rotation generates a whipping motion of the pot on the spring. In each case the ball bearings impact violently against each other and the walls of the pot, rapidly reducing particle size and disrupting the lignocarbohydrate matrix. After milling for 72 h in a 0.6-litre pot, the average particle is ca. 1 μm diameter and the domains of lignin and carbohydrate have been broken up; this allows subsequent chemical extraction of lignin.

Lignin is extracted from the fine powder with an aqueous solution of dioxane (95% v/v); the complete process takes 4–5 days although some material is extracted after 24 h. Some workers add a larger proportion of water to the extraction liquid to promote dissolution of hydrophilic bonds. Reagents used in this procedure should be of highest purity since contamination may react with the lignin. Convenient purifications for organic solvents are described by Perrin *et al.* (1980).

Lignin is extracted from the lignocarbohydrate and low-molecular-weight contaminants by the liquid–liquid extraction procedure described by Lundquist *et al.* (1977; and Fig. 3). The resulting lignin preparation is essentially free from non-covalently bound carbohydrate.

Kraft Lignins

Kraft lignins are precipitated by acidification of the "black liquor" obtained from the cooking process in alkaline kraft pulping. The precipitated material is dried and marketed as "Indulin AT" (Westvaco Corp., North Charleston, South Carolina, USA). It is contaminated with carbohydrates, low-molecular-weight phenolic compounds and inorganic material to various

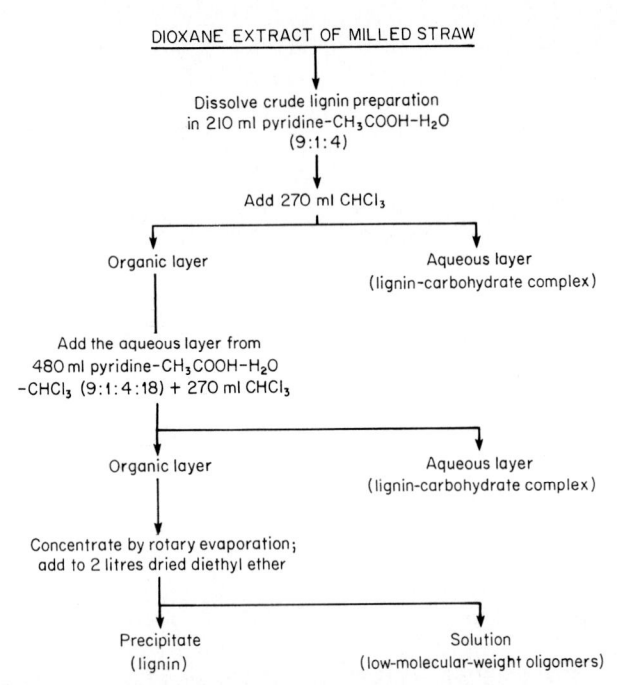

```
DIOXANE EXTRACT OF MILLED STRAW
          ↓
Dissolve crude lignin preparation
in 210 ml pyridine-CH₃COOH-H₂O
           (9:1:4)
          ↓
   Add 270 ml CHCl₃
      ↓         ↓
Organic layer   Aqueous layer
              (lignin-carbohydrate complex)
   ↓
Add the aqueous layer from
480 ml pyridine-CH₃COOH-H₂O
-CHCl₃ (9:1:4:18) + 270 ml CHCl₃
      ↓         ↓
Organic layer   Aqueous layer
              (lignin-carbohydrate complex)
   ↓
Concentrate by rotary evaporation;
add to 2 litres dried diethyl ether
      ↓         ↓
Precipitate    Solution
  (lignin)    (low-molecular-weight oligomers)
```

FIG. 3. Liquid–liquid extraction to purify lignin from dioxane-soluble portion of milled straw. (Adapted from Lundquist *et al.* 1977.)

extents, but purification by liquid–liquid extraction (Fig. 3) yields a lignin of molecular weight > 3000, equivalent to 14 monomers (Lundquist & Kirk 1980). Kraft lignins have a very small content of β-O-4 bonds and methoxyl groups, and much modified sidechains. Since most C—O bonds of the native polymer have been destroyed, the number of C—C intermonomer bonds has been increased, and many phenolic groups have been introduced to the ring, Kraft lignin is unrepresentative of native lignins. It is, however, convenient to prepare in large quantities.

Preparation of Synthetic and Model Lignin Compounds

Dehydrogenation Polymers

Dehydrogenation polymers (DHP) are synthetic lignins (Freudenberg 1956) that represent the basic polymer without bound carbohydrates or esterified groups; therefore they are good substrates for experiments if there is a ready source of the constituent monomer(s). They are generally prepared using solely coniferyl alcohol, although mixed monomer DHPs have been described (Schweers & Faix 1973).

There are three types of DHP which differ in molecular weight range and proportion of C—C bonds. All types are produced by a similar reaction mechanism to that involved in the natural synthesis, i.e. the generation and polymerisation of phenoxyl radicals. It is important that endwise, not bulk, polymerisation (addition of dimers or trimers) predominates otherwise the ether:C—C bond ratio decreases, with consequent loss of solubility and divergence from the natural structure (Brunow & Wallin 1981). The following methods for producing DHPs are those described by Tanahashi & Higuchi (1981).

(1) *Zutropfverfahren* (large ether bond content, molecular weight > 1000). The following solutions are prepared:
   Solution A.   300 mg coniferyl alcohol are dissolved in 10 ml acetone and 100 ml 0.1 M phosphate buffer (pH 6.1) are added.
   Solution B.   0.5 ml H_2O_2 (30% v/v) are added to 100 ml phosphate buffer.
   Solution C.   5 mg peroxidase (ca. 125–200 purpurogallin units/mg) are dissolved in phosphate buffer.

Solutions A and B are added separately to solution C at ambient temperature, over a period of 20 h with stirring, in the dark under N_2. After completion of the addition, stirring is continued for a further 16 h. The precipitate is then collected, washed and lyophilised.

The modification of Brunow & Wallin (1981), in which the coniferyl alcohol concentration in the mixture is maintained $< 10^{-4}\,M$, gives a yield of >80% and a structure, as assessed from NMR spectra, similar to that of milled wood lignin.

(2) *Zulaufverfahren* (large C—C bond content, molecular weight < 1000).

Solutions A, B and C are prepared as for *Zutropfverfahren* but are mixed together at once, in the dark under N_2 at ambient temperature. Stirring is continued for 36 h and the precipitate collected, washed and lyophilised. The rapid reaction due to the large concentration of coniferyl alcohol favours bulk polymerisation; thus many coniferyl alcohol dimers and oligomers are formed.

(3) *Higuchi dialysis sac method* (large C—C bond content, high molecular weight).

Peroxidase (5 mg) is dissolved in 15 ml phosphate buffer (pH 6.1) and the solution placed in a dialysis sac, which is sealed and immersed in a beaker containing 300 mg coniferyl alcohol in 1 litre phosphate buffer. Hydrogen peroxide (0.4 ml, 30% v/v) is then added under N_2 and the solution stirred in the dark at ambient temperature for 48 h. Because DHP forms on the inner wall of the dialysis sac and blocks the pores, a second sac identical to the first is introduced into the solution and the reaction allowed to continue for a further 48 h. The insoluble white precipitate, scraped from the inside of each dialysis sac, is washed and lyophilised. This type of DHP, although convenient to prepare, does not seem to reflect the structure of milled wood lignins and is of disputed value.

Model Lignin Compounds

Model compounds (dimers and trimers) have been produced by purely chemical syntheses (for a review, see Higuchi & Nakatsubo 1980). Such compounds have provided information on the nature of the chemical degradation of lignins which takes place during pulping and have also been used extensively to examine the modifications made by ligninolytic microorganisms (e.g. Enoki *et al.* 1981).

Guaiacyl-glycerol-β-coniferyl Ether and Dehydrodiconiferyl Alcohol

The advantage of the following biochemical method is that simultaneously a β-*O*-4 (β aryl ether) and a C—C (phenylcoumaran) model compound are formed by radical generation followed by dimerisation. The method was originally described by Katayama & Fukuzumi (1977).

A solution of 30 g coniferyl alcohol in the minimum of acetone and 4 mg horseradish peroxidase (Sigma) are added to 2 litres distilled water. With vigorous stirring, 1 litre 0.6% (v/v) H_2O_2 is added dropwise over a period of 1 h. The mixture is stirred until no coniferyl alcohol can be detected by thin-layer chromatography (TLC) on silica gel using 5% (v/v) methanol in methylene chloride as mobile phase, or by high-pressure liquid chromatography (HPLC) with a C-18 column using 70% (v/v) methanol as eluant. The reaction mixture is then acidified to pH 2 with concentrated HCl and extracted with three 2-litre portions of ethyl acetate. The combined extracts are washed with a saturated solution of NaCl, dried over anhydrous Na_2SO_4 and the solvent removed by rotary evaporation. The residue is applied to a column of silica gel (5 × 80 cm, 850 g) and is eluted using a gradient of benzene–acetone, 10:1 to 1:1. The guaiacyl-β-coniferyl ether elutes at a solvent ratio of 3:1 and can be further purified by preparative TLC or HPLC as described above. The yield is ca. 15% (Iwahara *et al.* 1980). Dehydrodiconiferyl alcohol elutes at a benzene–acetone ratio of 6:1 and can be further purified by recrystallisation from acetone by addition of *n*-hexane, to give a 34% yield (Iwahara *et al.* 1980).

Coniferyl Alcohols

Coniferyl alcohol (*p*-hydroxycinnamyl alcohols) is the starting material for the preparation of synthetic lignins (DHP) and certain model compounds. The development of syntheses for specifically labelling designated carbon atoms in this compound with ^{14}C and ^{13}C has been important in that it has simplified the analysis of lignin biodegradation. The experience in this laboratory is that the synthesis of these compounds is not straightforward. The synthetic sequences of Nakamura *et al.* (1974) and Nakamura & Higuchi (1976) are convenient starting points for syntheses.

Specifically labelled DHPs have been produced and used in experiments to study biodegradation, e.g. Kirk *et al.* (1975), ^{14}C; Haider & Trojanowski (1975), ^{14}C; Ellwardt *et al.* (1981), ^{13}C. Labelling with ^{14}C produces substrates which can be used to assay the total catabolism of lignins, whereas ^{13}C labelling allows the detection of the mechanism of attack on the polymer using ^{13}C NMR. Model compounds labelled with ^{14}C may also facilitate the identification and study of enzymes involved in the degradative pathway (Weinstein *et al.* 1980).

Isotope-Labelled Lignocellulose

Methods have been devised for specific labelling of carbon atoms in lignin *in vivo* with ^{14}C and ^{13}C. The methods developed for labelling lignins in woody plants have been discussed by Crawford *et al.* (1980) and Crawford (1981). In the UK [^{14}C]phenylalanine, of high specific activity, is available from Amersham International (Amersham, Buckinghamshire) and ^{13}C-labelled compounds are available from CEA (Fluorochem Ltd., Glossop, Derbyshire) and MSD Isotopes (Cambrian Chemicals, Croydon, Surrey). The commercial availability of intermediates of lignin biosynthesis with isotope-labelled structures has facilitated the preparation of labelled lignins by synthesis *in vivo*. A method has also been developed to label precursors on methoxyl positions, which are not available commercially, using isolated methylases and S-[^{14}C]adenosylmethionine (Marigo et al. 1981).

In this laboratory, isotope-labelled grass lignocellulose is produced by the following procedure. Wheat seedlings are grown from seed in compost, on a windowsill for ca. 14 days. Five microlitres (0.5 μCi) uniformly labelled [^{14}C]phenylalanine (450 mCi/mmol) are injected ca. 0.5 cm above soil level with a glass needle. Once the solution has been taken up, the plants are grown for a further 3–4 weeks before harvesting. Growth through day–night cycles is important for incorporation of the precursor into the polymeric lignin in the form required, because growth in continuous light favours unwanted incorporation of precursors as esterified groups.

The harvested stems and leaves are frozen in liquid N_2, ground using a pestle and mortar and are lyophilised. The powder is then washed with distilled water at 80°C, dried, and extracted with benzene–ethanol (1:1) and then ethanol until the extractants are clear and colourless. The wash with water at 80°C is repeated. The extractive-free lignocellulose is treated three times with protease (*Streptomyces griseus* P5130 Type 6, Sigma) after which no further significant release of ^{14}C can be detected. This gives an extractive-free lignocellulose, free from contaminating protein.

Finally, weighed portions are combusted in an ignition tube and the $^{14}CO_2$ produced is trapped in an ethanolamine-containing scintillation cocktail. Values in the order of 3×10^4 dpm/mg lignocellulose are usually obtained. The extraction procedures, protease treatment and method of labelling ensure that the ^{14}C is present solely in the lignin component of the lignocellulose.

$^{14}CO_2$ Trapping-Scintillation Cocktail

1,4-Di-2-(5-phenyloxazolyl)benzene (POPOP) (0.32 g) is dissolved in 100 ml toluene. 2,5-Diphenyloxazole (PPO) (0.5 g) is added with 30 ml ethanolamine. The result is a two-phase solution and addition of ca. 30 ml methanol is required to produce a single phase. During use the phases may separate and further addition of methanol may be necessary.

Structural Analysis of Lignin Polymers

Structural analysis of lignins which have been attacked by microorganisms does not in general give specific information on the mechanism of enzymic action, since lignin polymers are not ordered and different modifications are simultaneously introduced. However, spectroscopic and degradative methods give valuable and complementary information.

Nuclear Magnetic Resonance (NMR) Spectroscopy

Nuclear magnetic resonance spectroscopy gives information on the chemical environments of particular nuclei in monomer units incorporated in the lignin polymer. The two species of nucleus which can be most usefully examined are ^{1}H and ^{13}C. Equipped with suitable computers, NMR spectrometers can produce difference spectra (e.g. Brunow and Lundquist 1980) where only data of significance to the experimenter is displayed, allowing accurate assessment of changes in polymer structure.

Proton magnetic resonance (^{1}H NMR) spectroscopy is useful because ^{1}H nuclei are present as 98% of the total available in nature. Information obtained as chemical shifts of ^{1}H species is produced by detecting the degree to which bonding electrons interact with externally applied magnetic fields to shield the various protons in the monomer. Samples held in a magnetic field are irradiated with pulses of energy in the radio-frequency range; a relatively wide band of frequencies is used and nuclei absorb at their resonance frequencies and move to higher energy levels. Once the pulse of energy has ceased, the nuclei revert to their original state emitting the absorbed energy. The energy emitted is a composite of resonance frequencies of all the nuclei of the sample and by detecting the frequencies and the intensities of the emissions at these frequencies, as the signal decays, the computer of the spectrometer can calculate the proportions of ^{1}H nuclei in different chemical environments. Differences in the frequency of energy emission, as compared with the frequency emitted by reference compounds, give characteristic positions of signal on spectra (chemical shifts) and the intensity of signals gives information on the number of nuclei in that chemical environment in the polymer. ^{1}H nuclei on adjacent carbon atoms have a magnetic effect upon each other, referred to as "spin–spin coupling". This provides data on neighbouring hydrogen species. The very large number of ^{1}H environments present in the lignin polymer, in combination with overlapping signals and extensive coupling, results in very broad peaks on spectra (Fig. 4). However, pulse Fourier transform NMR using high-resolution spectrometers (typically 270 mHz and above) allows estimation of the proportions of bond types in different lignin samples (Lundquist 1980) and therein lies its value. Detailed and relatively accurate information on the modifications introduced into lignin model compounds can also be obtained.

Samples to be examined by ^{1}H NMR are normally acetylated by treatment with acetic anhydride in pyridine for 8–24 h at ambient temperature. Acetylation by this method leads to removal of contaminants and increases the

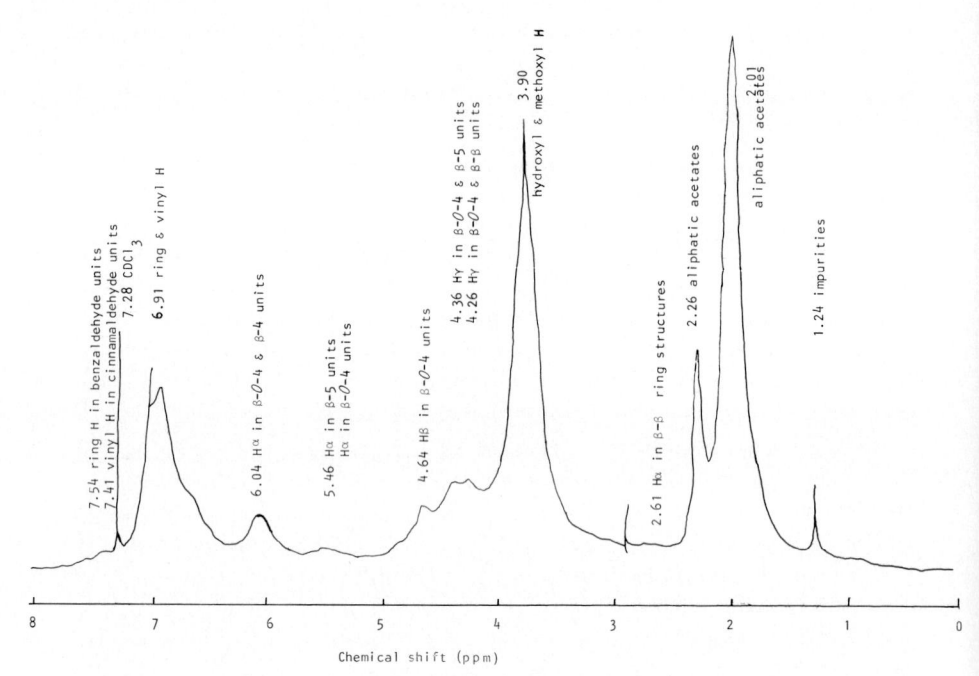

FIG. 4. ¹H pulse Fourier transform NMR spectrum of acetylated milled wood lignin from spruce, using 300 mHz high-resolution spectrometer, 10 ppm chemical shift range, 100 scans. (Sample provided by K. Lundquist; interpretation by L. Ramsey.)

solubility of lignins, improving the signal:noise ratio in NMR spectra. After the reaction, lignin is washed extensively in ether to remove by-products. Samples of >50 mg/ml [²H]chloroform (CDCl$_3$) are obtainable using acetylated material. Non-acetylated samples are dissolved in [²H]dimethylsulphoxide (DMSO)–D$_2$O (9:1) or dioxane-d_8–D$_2$O (5:1). Residual H$_2$O is often a problem with non-acetylated samples since the signal from the water protons interferes with those of the lignin.

¹³C carbon nuclei are ca. 1% of those in nature. The development of ¹³C pulse Fourier transform NMR techniques has allowed the achievement of very good signal:noise ratios and the peaks which can be obtained on spectra may be very sharp and well resolved in comparison with those obtained on ¹H spectra (Fig. 5). The area under peaks is not proportional to the number of ¹³C nuclei in that environment, in contrast to ¹H spectra. Because ¹³C nuclei are present in low abundance, the possibility of spin–spin coupling between ¹³C nuclei is very limited, and since ¹H species can be decoupled from ¹³C (heteronuclear decoupling) experimentally, reasonably accurate assignments of modifications to lignin samples can be made. However, the low abundance of ¹³C necessitates the analysis of large samples of lignin for

FIG. 5. ^{13}C pulse Fourier transform NMR spectrum of milled wood lignin from barley straw using 100 mHz high resolution spectrometer, 220 ppm chemical shift range, 419,300 scans. (Interpretation by L. Ramsey.)

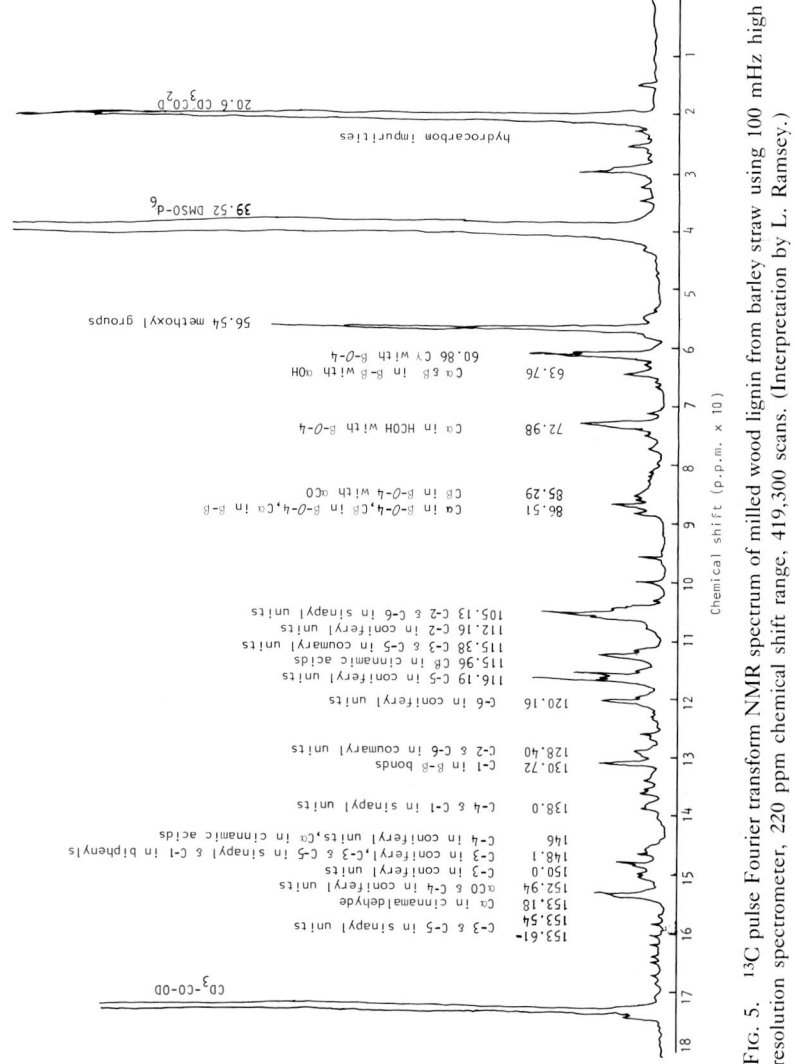

prolonged times. Use of lignins artificially enriched with ^{13}C permits smaller samples and shorter analysis times, and the technique is likely to become more important as preparation of ^{13}C lignins becomes cheaper. Odier & Monties (1980) have used ^{13}C NMR to analyse modifications introduced into poplar lignin during attack by gram-negative bacteria. ^{13}C-enriched DHP has been used by Ellwardt et al. (1981) to study modifications introduced into lignins by different fungi and bacteria.

Acid Hydrolysis

Refluxing with 0.2-M HCl in dioxane–water (9:1) hydrolyses specific types of bonds in lignin, leading to production of low-molecular-weight compounds which can be separated, identified and used to give information on the structure of the polymer (Lundquist 1976). The low-molecular-weight products can be identified by mass spectrometers linked to gas chromatographs or following isolation using preparative HPLC. Acid hydrolysis has been used to confirm and identify modifications of lignin by *Streptomyces viridosporus* (Crawford et al. 1982).

Other Degradative Methods

Oxidative degradation of lignin by alkaline nitrobenzene or permanganate and analysis of products are important techniques for characterising native and degraded lignins. The techniques are extensively discussed by Chang & Allan (1971).

Elemental analyses are also used in the study of degraded lignins. The technique allows estimation of the proportion of C, H and O atoms in lignins, giving information as to whether attack is oxidative or reductive (Crawford et al. 1982).

Ultraviolet and Infrared Spectroscopy

The UV spectra of partially degraded lignins give information on changes in aromatic and unsaturated bond content. Ionisation difference spectra are produced by comparing the UV spectra of native and borohydride-reduced samples (Kirk & Chang 1975), and these in particular can give information on the number of carbonyl groups in partially degraded lignin.

Infrared spectroscopy allows the detection of the content of hydroxyl, methoxyl, methyl, carbonyl and unsaturated C—C bond types in lignin samples. Intensity changes in absorption at particular wavelengths provide supportive evidence for observations based upon other data, e.g. attack on lignin by a streptomycete increases the carbonyl bond intensity at 1650–1700 cm (Crawford et al. 1982). The information is rather general in nature since absorption bands are rather wide, but information can be useful in the assessment of changes (e.g. Chang *et al.* 1980) and in the detection of carbohydrate contamination of lignin samples.

Gel-permeation Chromatography

Gel-permeation chromatography (GPC), or size-exclusion chromatography, is the method of choice for examining the molecular weight range of lignin preparations. Sephadex gels (Connors 1978), controlled pore glass (Hütterman 1978) and, increasingly, HPLC–GPC columns (Faix *et al.* 1981) are used to examine differences and changes in samples which have been subjected to different treatments. Changes in the molecular weight of lignosulphonates during attack by *Fomes annosus* were studied by Haars & Hütterman (1980) using GPC on Sepharose CL 6B.

In this laboratory, we use both conventional and HPLC GPC. Zorbax PSM 60S with a silica microsphere bead [Du Pont (UK) Stevenage, Hertfordshire] is an HPLC column suitable for analytical work with tetrahydrofuran (THF) and dimethylformamide (DMF) as solvents. For preparative work, Sephadex G50 with 0.1-M NaOH as eluant is convenient. The presence of salts, carbohydrate and sample aggregation greatly influence the results of GPC analyses, therefore acetylation of samples is advisable.

Biodegradation

Whereas previously only certain higher fungi were thought to degrade lignin, development of ^{14}C-labelling techniques has resulted in the identification of a much wider range of organisms capable of modifying the polymer. A promising new technique for attaching fluorescent dyes to lignins may allow ready identification of ligninolytic organisms and provide an additional impetus to the search for them (Haars *et al.* 1982).

Biodegradation by Fungi

A number of white-rot fungi have been studied quite extensively to elucidate the mechanism of lignin degradation and its regulation. White-rot fungi have been distinguished from other fungi which attack wood by their ability to give a positive Bavendamm reaction (Bavendamm 1928), which is produced by excretion of phenol oxidase (Higuchi 1953). *Phanerochaete chrysosporium*, *Sporotrichum pulverulentum* and *Polystictus (Coriolus) versicolor* have been most studied. In addition, *Fusarium solani* has been studied by Higuchi and co-workers (reviewed by Higuchi 1982). Much of the work published before 1978 has been discussed in Kirk *et al.* (1981). From investigation of ligninolytic fungi, a working hypothesis for the mechanism of lignin degradation can be constructed.

In fungi, lignins do not appear to act as sole carbon sources or primary growth substrates and it has been postulated that they may obtain little net energy from lignin (Kirk 1981). Lignin catabolism is most rapid when the fungi are grown in thin-layer cultures in an atmosphere with elevated O_2 content. This has suggested to a number of workers that molecular oxygen plays an important role in degradation of the polymer. Lignin degradation is certainly an oxidative process, since degraded lignins have a greater oxygen content than undegraded lignins. The onset of ligninolytic activity in *Phanerochaete chrysosporium* and *Sporotrichum pulverulentum* is not the result of the presence of the substrate as is cellulolytic activity. It is a response to limitation of one of three nutrients essential for growth, i.e. utilisable carbon (carbohydrate), nitrogen, or sulphur. During ligninolytic activity, a secondary metabolite, 3,4-dimethoxycinnamyl alcohol, is produced biosynthetically from phenylalanine (Shimada *et al.* 1981) and is converted by α-β C—C cleavage to veratryl alcohol, which accumulates in the medium (Lundquist & Kirk 1978). This simultaneous production of a secondary metabolite and catabolic activity in response to nutrient limitation suggests that lignin metabolism is itself a secondary metabolic activity in these fungi. This view is supported by the isolation of pleiotropic mutants of *Phanerochaete chrysosporium* defective in phenol oxidase activity, ligninolytic activity, veratryl alcohol synthesis, and fruiting body formation after UV mutagenesis of conidia (Gold *et al.* 1982). All the phenotypes can be regained with reversion of the mutation by the same mutagenic treatment. These mutants appear to be completely defective in secondary metabolic activities as opposed to other types of pleiotropic mutants in fungi which are the proposed result of inactivation of common regulator genes (Eriksson & Goodell 1974). The onset of secondary metabolism can be delayed or even reversed by addition of glutamate to the culture (Fenn & Kirk, 1981). This suggests that the enzymes of nitrogen metabolism have a key role in the regulation of the transition from primary to secondary metabolic activity. The evolution of this physiological regulation may be the result of the low nitrogen content of wood.

Study of the modification of lignin substructure models has given much information on fungal attack on the lignin polymer. The initial attack appears to be side-chain modification, involving hydroxylation or formation of a carbonyl group at the α-C atoms. These reactions may be mediated by such enzymes as alcohol oxidases (Iwahara *et al.* 1980) and laccases (Shimada 1980). Once either of the configurations has been reached, an activated oxygen species can attack the monomer unit, producing a variety of cleavage reactions. The formation of radicals during lignin degradation was initially observed by Caldwell & Steelink (1969), as was the occurrence of benzoquinones in the culture medium of ligninolytic fungi. Originally these authors postulated that this was the result of laccase action. However, although laccases are known to generate radicals, two lines of evidence suggest that they are not the primary depolymerising enzymes: first, etherified phenols cannot act as substrates for laccases (Ishihara 1980), and second, when the chelating agent thioglycolic acid is added to ligninolytic cultures, there is an accumulation of low-molecular-weight material but no inhibition of polymer degradation (Haars & Hütterman 1980). It is probable that thioglycolic acid interacts with copper atoms in laccase and thus inhibits its activity. Laccases may play an important role in metabolism of dimers, however, where the free phenolic hydroxyls necessary for their activity are found, the study of phenol oxidaseless mutants has shown that they are defective in ligninolytic activity except when laccases are added to the medium (Ander & Eriksson 1976). The "activated oxygen species" hypothesis suggested by Hall (1980) is attractive because during delignification in bleaching processes, electrophilic attack by radicals gives similar products to biological delignification (Gierer 1982). It is also attractive because it explains why ligninolytic cultures can degrade many different lignin-like preparations (e.g. Kraft lignins, DHP, polyguaiacol, grass and wood lignins) with little regard to the monomer constitution of the polymer and the stereospecificity of monomers (Kirk 1981). The key reactions of biological delignification have been postulated to be as follows (Higuchi 1982): (a) demethylation of methoxyl groups; (b) hydroxylation at aromatic ring C-2; (c) aromatic ring cleavage; (d) $C\alpha$-$C\beta$ cleavage.

Attack by activated oxygen species could result in all of these reactions taking place. The $C\alpha$-$C\beta$ cleavage reaction is quantitatively the most important in lignin degradation, resulting in formation of vanillic acid, dehydrodivanillic acid and similar compounds which accumulate in the medium (Chen *et al.* 1982). In fact, these low-molecular-weight aromatic acids are partly, at least, the product of quinone transformation by the enzyme cellobiose quinone oxidoreductase. The accumulated evidence strongly suggests that the major depolymerisation reaction in lignin degradation is produced by activated oxygen species generated by fungi, and this is our working hypothesis.

The activated oxygen species hypothesis is also supported by some experimental evidence. In the presence of an active oxygen-trapping molecule, anthracene-9,10-bisethane sulphonic acid, lignin degradation is repressed and the trapping agent consumed (Nakatsubo *et al.* 1981). It was proposed that $^1O_2$ (singlet oxygen) is the attacking species generated by fungi. However, experiments using ^{18}O as a tracer do not support this hypothesis (T. K. Kirk, personal communication). A possible alternative is that the hydroxyl radical (·OH) depolymerises lignin. In biological systems where superoxide

(O_2^-) is used as a biocide, as in phagocytosing leucocytes, the actual killing agent is postulated to be the hydroxyl radical (Halliwell 1982; Krinsky 1979). The electron potential of ·OH is greater than that of O_2^- and therefore it is a more reactive species. Hydroxyl radical can be generated by electron transfer from transition metals (as found in metalloenzymes) to H_2O_2 or by the interaction of O_2^- with H_2O_2 (Halliwell 1982). The presence of H_2O_2 in the culture medium of ligninolytic organisms may be the result of alcohol oxidase activity or active excretion of H_2O_2 by fungi. Generation of radical species may be complicated since, once generated, the radical may interact with other molecules to produce radical species, e.g. the biradical oxygen ·O_2. Such species of free radical are very reactive, attacking any molecule in proximity.

Eriksson (1981) has proposed that the free radical species involved in depolymerisation is generated by a cellobiose oxidase excreted by *Sporotrichum pulverulentum* (Ayers *et al*. 1978), but the evidence has not been published in detail.

It seems certain, however, that two oxidative enzymes of carbohydrate metabolism are involved in lignin degradation by fungi, i.e. cellobiose quinone oxidoreductase (Westermark & Eriksson 1974a) and glucose oxidase (Green 1977). These enzymes oxidise their respective substrates to the corresponding lactones, with concomitant reduction of quinones to phenols. The reaction removes quinones, produced by laccase activity or activated oxygen species, which are inhibitors of hydrolytic cellulose-degrading enzymes. The oxidation products, the lactones, are also inhibitors but are removed by lactonase activity in culture supernatants. These enzymes can act at reduced O_2 concentrations such as occur under pellicles of mycelium.

A schematic presentation of a possible interaction between these enzymes and oxygen species is shown in Fig. 6. Once monomeric compounds have been formed, they are utilised either by fungi or by bacteria associated with the fungi. A number of intracellular pathways for vanillic acid metabolism have been described in *Sporotrichum pulverulentum* (Buswell *et al*. 1982).

Biodegradation by Bacteria

More information is available on lignin degradation by actinomycetes than by eubacteria. This is primarily due to the work of Haider & Trojanowski (1980) and the Crawfords (Crawford & Crawford 1980) studying $^{14}CO_2$ release from labelled lignin preparations. The realisation that actinomycetes are also active cellulose degraders and that they have a major role in primary degradation of plant material in natural environments is stimulating interest in their ligninolytic potential. In this laboratory we are studying the correlation between the ability to clear native lignocellulose overlays on agar plates

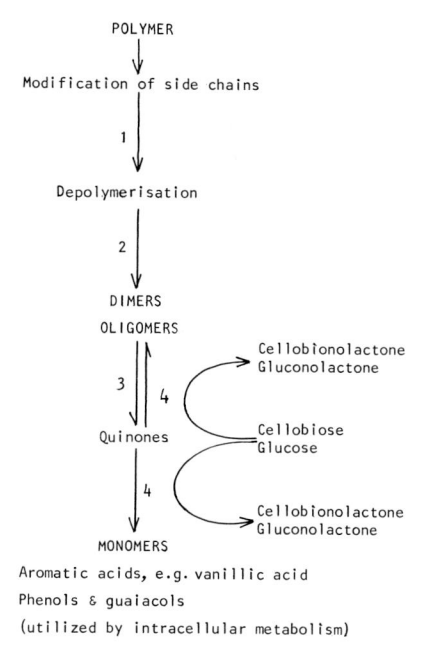

```
                    POLYMER
                       ↓
          Modification of side chains
                       
                    1  ↓
                       
                 Depolymerisation
                       
                    2  ↓
                       
                    DIMERS
                   OLIGOMERS
                    ↑ ↑        Cellobionolactone
                  3 ↓ ↑ 4  ⟶   Gluconolactone
                    ↓ ↑
                   Quinones    Cellobiose
                               Glucose
                    ↓  4
                    ↓       ⟶  Cellobionolactone
                               Gluconolactone
                   MONOMERS
          Aromatic acids, e.g. vanillic acid
          Phenols & guaiacols
          (utilized by intracellular metabolism)
```

FIG. 6. Schematic working hypothesis for the pathway of lignin degradation by fungi. Key to enzymes and radical species implicated: (1) Aromatic alcohol oxidases:laccases; (2) ·OH; $^1O_2$; laccases; cellobiose oxidase; (3) ·OH; $^1O_2$; aromatic alcohol oxidases; laccases (also uncharacterised enzymes); (4) cellobiose–quinone oxidoreductase:glucose–quinone oxidoreductase.

with release of $^{14}CO_2$ from (^{14}C-labelled lignin) wheat lignocellulose. Thermophilic actinomycetes, potentially useful for cellulose bioconversion processes, are particularly interesting.

Actinomycetes, through their filamentous growth form, are the most likely bacteria to be able to attack and degrade plant material, since the formation of a stable branching hyphal structure allows penetration of lignified cell walls. Lignin degradation by actinomycetes has been reviewed by Crawford & Crawford (1980) and Crawford et al. (1982). Actinomycetes appear to be able to solubilise lignins without extensively breaking down the polymer (Crawford et al. 1983). This, amongst other data, suggests that the mechanism of actinomycete degradation may differ from that used by fungi.

The eubacteria *Pseudomonas* and *Xanthomonas* have also been identified as able to attack lignin (Odier & Monties 1980). These bacteria produce marked changes in ^{13}C NMR spectra and micrographs of sections of poplar wood attacked by bacteria show reduction in lignin density (Odier et al. 1981). Strains of *Pseudomonas putida* are also able to attack and degrade lignin model compounds containing the β aryl ether bond by a mechanism that is similar to that of *Fusarium solani*, despite the absence of phenol oxidase (laccase) activity in bacteria (Fukuzumi & Katayama 1977).

Numerous reports of lignin degradation by eubacteria have not been substantiated by rigorous methods such as $^{14}CO_2$ release or ^{13}C NMR and therefore must be interpreted with caution. It seems likely that a wide range of bacteria are involved in lignin breakdown but most only in mixed culture with other bacteria and fungi. Such mutualistic relationships are of obvious importance in decay and in natural environments but will be difficult to analyse in the laboratory.

Enzymes Implicated in Lignin Degradation

Remarkably few enzymes have been implicated in lignin degradation although extracellular enzymes have been shown to solubilise ^{14}C-labelled lignins during aerobic incubation with *Coriolus versicolor* cell-free culture supernatants (Hall *et al.* 1980). In the extracellular proteins of this organism grown on cellulose, there are 30 different protein species, present in only very small concentrations. There are a number of possible reasons for the low protein content: adsorption to substrates; inclusion of extracellular proteins in the glucan layer surrounding the fungal hyphae; the mode of excretion of enzymes used by fungi, i.e. export in apical vesicles.

Lignin preparations very readily adsorb to proteins or microorganisms, resulting in apparent loss of lignin when bacterial cultures are given the finely divided powder. In fungi, extracellular enzymes are found both in the culture fluids and in the glucan layer external to the cell wall (Dickerson & Baker 1979); treatment of the glucan of *Sphacelia sorghi* with exo-(1,3)-β-D-glucanase results in release of fructofuranosidase, β-glucosidase, acid phosphatase and β-1,3-glucanase, suggesting that many enzymes are bound to this matrix. Ligninolytic cultures of *Phanerochaete chrysosporium* and *Sporotrichum pulverulentum* produce considerable amounts of extracellular carbohydrate.

In the fungus *Saprolegnia monoica*, the enzymes α-1,3-glucanase and cellulase are mainly associated with the edges of colonies on agar media. In *Sporotrichum pulverulentum* ligninolytic activity is also associated with the growth centre, producing rings of enzymic activity during colony growth (Westermark & Eriksson 1974b). Ligninolytic activity is also observed only around the hyphal tip when electron micrographs of partially degraded wood are examined (Eriksson *et al.* 1980). Thus, it seems plausible that ligninolytic enzymes are secreted in apical vesicles in a manner similar to that of *Saprolegnia monoica* (Févre 1977). If activated oxygen species are involved in lignin degradation, then the half life of secreted enzymes involved in their generation may be very short. Enzymes may also be secreted as precursors

(Eriksson 1981) which are then modified by extracellular proteases yielding the active enzyme, a phenomenon which may make interpretation of analyses difficult.

The following enzymes may be involved in ligninolytic activity.

Aromatic alcohol oxidases or dehydrogenases which have activities towards the sidechains of lignin-related compounds have been described in *Fusarium solani* (Iwahara *et al.* 1980), *Polystictus versicolor* (Farmer *et al.* 1959) and *Rhodococcus erythropolis* DSM 1069 (Eggeling & Sahm 1980; Jaeger *et al.* 1981). The best described enzymes are those from the fungi, both of which oxidise aromatic alcohols to aldehydes with reduction of molecular oxygen to H_2O_2. That of *Polystictus* has a specificity for monomeric alcohols whilst the *Fusarium* enzyme oxidises the γ-HCOH group in dehydroconiferyl alcohol to the corresponding aldehyde and shows O_2 uptake when given milled-wood lignin as a substrate. Assays for these enzymes are described in the original reports.

Laccases are copper-containing phenol oxidases that have a wide specificity for the reducing substrate although this must have a free phenolic group (Ishihara 1980). Dioxygen is required as the oxidising substrate and is reduced to two molecules of water. The products of laccase action are often 3-methoxybenzoquinone or 2,6-dimethoxybenzoquinone, the products of alkyl aromatic cleavage. Laccases are often present in fungi in multiple forms, some produced constitutively and some inducible, differing both in amino acid composition and surface glycosylations. Laccases have been extensively reviewed by Reinhammar (1984). Various assays have been described (e.g. Fåhraeus & Reinhammar 1967; Froehner & Eriksson 1977) and a spot test has been described by Wood (1980). The action of laccase alone on lignin preparations results in the formation of both higher and lower molecular weight products since the generated radicals and quinones can undergo spontaneous polymerisation.

Quinone-reducing oxidoreductases have been reported in *Sporotrichum pulverulentum* and *Polyporus (Coriolus) versicolor;* in *Polyporus versicolor* both cellobiose quinone oxidoreductase (CBQ) (Westermark & Eriksson 1974 a,b) and glucose oxidase (Green 1977) have been described. Assays for these enzymes are described in the original reports. The enzymes accept as cosubstrates the quinones commonly observed in the culture supernatants of ligninolytic fungi. The cellobiose quinone oxidoreductase of *Sporotrichum pulverulentum* is a glycosylated flavoprotein (Westermark & Eriksson 1974a).

Cellobiose oxidase has also been isolated from culture filtrates of *Sporotrichum pulverulentum* (Ayers *et al.* 1978). This enzyme is also glycosylated but has both bound haem and flavin groups. Only 50% of the O_2 consumed by the enzyme during oxidation of cellobiose is incorporated into the product, cellobionic acid, but in contrast to other enzymes H_2O_2 is not produced by the reaction. Eriksson (1981) has postulated that O_2^- is generated by this enzyme during the oxidation but the evidence supporting this has not been published.

Free radical species are known to be produced by fungal enzymes but no enzymes excreted by ligninolytic fungi have been categorically shown to produce such attacking species. The suggestion of Eriksson (1981), mentioned above, that O_2^- was involved was based upon data obtained using the method of Beauchamp & Fridovich (1971). This assay, also suitable for detecting enzymic activity in electrophoretograms, follows the reduction of tetrazolium blue by the change in absorption at 560 nm. Hydroxyl radicals appear to be produced by fungi degrading lignin (Forney *et al.* 1982; Kelley & Reddy 1982; Kutsuki & Gold 1982). Hydroxyl radicals can be detected because they generate ethylene in the presence of 2-keto-4-thiomethyl butyric acid or methional. In *Phanerochaete chrysosporium* the radicals are only produced under the conditions in which cultures are ligninolytic, and mutants defective in the ability to degrade lignin do not produce the radicals. Ligninolytic activity is also inhibited by hydroxyl radical scavengers and by the assay substrate (Kutsuki & Gold 1982). The experimental data indicate that the radicals are produced in lignin degradation but do not conclusively show that they effect the process. Forney *et al.* (1982) have proposed that ˙OH is generated by reduction of H_2O_2 extracellularly, a plausible mechanism.

The Molecular Biological Approach

The vigour of molecular biology lies in its generality of application. Although each organism has its individual physiology, biochemistry and genetics, the application of a limited range of recombinant DNA techniques can frequently produce an understanding of a system where application of classical techniques over many years has failed. The requirement of molecular genetics is either the isolation of mutant phenotypes, or the ability to induce a change in phenotype in response to a particular stimulus. Given either of these, analyses using recombinant DNA technology, protoplast fusion, immunological techniques and physical chemistry can be made. In the study of lignin degradation, the basis for the physical chemical analyses of changes in lignin structure has already been established in the development of ^{1}H and ^{13}C NMR analysis.

Actinomycetes are prokaryotes and, as such, have a comparatively well-understood regulation of gene expression and transfer of information from gene to enzymic activity. There has been much progress in the development of recombinant DNA techniques, allowing actinomycete DNA to be cloned into standard strains, since this is of interest to the antibiotics industry. Although in actinomycetes very little is known about the enzymes involved in carbohydrate metabolism or especially lignin degradation, the prospect of using recombinant DNA techniques remains attractive. Since *Streptomyces* strains and certain nocardioform bacteria appear to be involved in the turnover of lignins and humic acids in the soil, information on these processes would be very welcome.

The basidiomycetes *Phanerochaete chrysosporium* and *Sporotrichum pulverulentum* are also attractive as experimental systems since lignin degradation by these organisms has been well studied. The reason that so few enzymes have been implicated may be due to the technical problems discussed in the previous section (*Enzymes implicated in lignin degradation*). The use of cloning techniques could allow the production of large quantities of enzymes that are otherwise present in fungal culture supernatants in very small amounts; this would facilitate study of their activity towards lignins and model compounds.

Eukaryotic Gene Expression and Construction of Gene Libraries

The structure of fungal genes appears to be similar to that of other eukaryotes and therefore different from that of bacteria. Most fungal genes analysed to date contain both protein-coding and non-coding sequences, referred to as "exons" and "introns", respectively. The eukaryotic RNA polymerase initiates transcription at sites not recognised by bacterial polymerase and produces an RNA species complementary to both exons and introns, i.e. nuclear RNA. Nuclear RNA is processed by excision of introns (splicing), addition of a polyadenylate sequence [poly(A) tail] to the 3' end and formation of a methylated cap at the 5' end, and then exported to the cytoplasm where it is referred to as messenger RNA (mRNA). This codes directly for the polypeptides of the enzyme as does a cDNA copy of the mRNA which can by synthesised *in vitro* using the enzyme RNA-dependent DNA polymerase (reverse transcriptase).

The DNA of the genome is the information repository of the organism and codes for all its enzymes. The DNA can be randomly fragmented and incorporated into vector molecules which replicate the individual fragments in *Escherichia coli* or *Saccharomyces cerevisiae*. This collection of DNA fragments is referred to as a gene library. To be certain of having a complete set of genes in the library, a relatively random method of DNA scission and a statistically representative number of clones, including these fragments, must be used.

Two vector systems have been constructed for preparing gene libraries, bacteriophage λ derivatives and cosmids. The temperate bacteriophage λ has been extensively studied and the viral DNA can be packaged *in vitro* in its protein coat to generate virions. Modified bacteriophage λ with extensive deletions of genome and modification to DNA sequence and regulation of development can be used to clone eukaryotic DNA. Cosmids (plasmids which have been constructed so that they can be packaged in bacteriophage λ protein coats) can be used to clone DNA and replicate it in either *Escherichia coli* or *Saccharomyces cerevisiae*. Once genes have been cloned into either of these vectors, individual genes can be located using specific "probes", i.e. radiolabelled RNA or cDNA species complementary to the gene.

Identification of Genes Involved in Ligninolytic Activity

The onset of secondary metabolism results in changes in the RNA and protein profiles of the organism. The pleiotropic mutants of *Phanerochaete chrysosporium* described by Gold *et al.* (1982) show no evidence of secondary metabolic activity and are therefore likely to lack the RNA species specific to the cell in that state. The change in cultures from primary to secondary metabolism can be followed by monitoring the concentration of veratryl alcohol in the culture supernatant. These two observations can be used as the basis of a technique referred to as "cascade hybridisation". In this technique [^{32}P]cDNA is made, in this case from the mRNA of cells in secondary metabolism, and the species of interest enriched by repeated hybridisation with mRNA from a different population of cells, i.e. either cells in the primary phase or mutants unable to produce secondary metabolic activities. The sequences of RNA common to primary and secondary phase cells are removed as DNA–RNA hybrids, leaving the species specific for secondary metabolism and hence ligninolytic activity; these can be used to probe the gene library and identify genes coding for secondary metabolic functions. Using this approach, Zimmerman *et al.* (1980) obtained 350 clones homologous to RNAs produced specifically by conidiating cultures of *Aspergillus nidulans*. The number of clones which hybridised with the probe cDNA corresponded to less than 3% of the gene library, and less than 0.05% of the total poly(A)$^+$ RNA of conidiating cultures was presented in the final cDNA probe.

FIG. 7. Two-dimensional (O'Farrell) gel electrophoretograms of total proteins from *Sporotrichum pulverulentum* showing the difference between cells in primary growth phase (a) and in secondary metabolic phase (b). (a) Culture grown for 6 days in medium containing 24 mM (NH$_4$)$_2$HPO$_4$. (b) Culture grown for 6 days in medium containing 2.4 mM (NH$_4$)$_2$HPO$_4$. The photographs are of autoradiograms detecting ^{35}S-labelled polypeptides. IEF, Isoelectric focussing; pI, isoelectric point; SDS–PAGE, sodium dodecyl sulphate polyacrylamide gel electrophoresis; MW, molecular weight. (Analyses provided by M. J. MacDonald.)

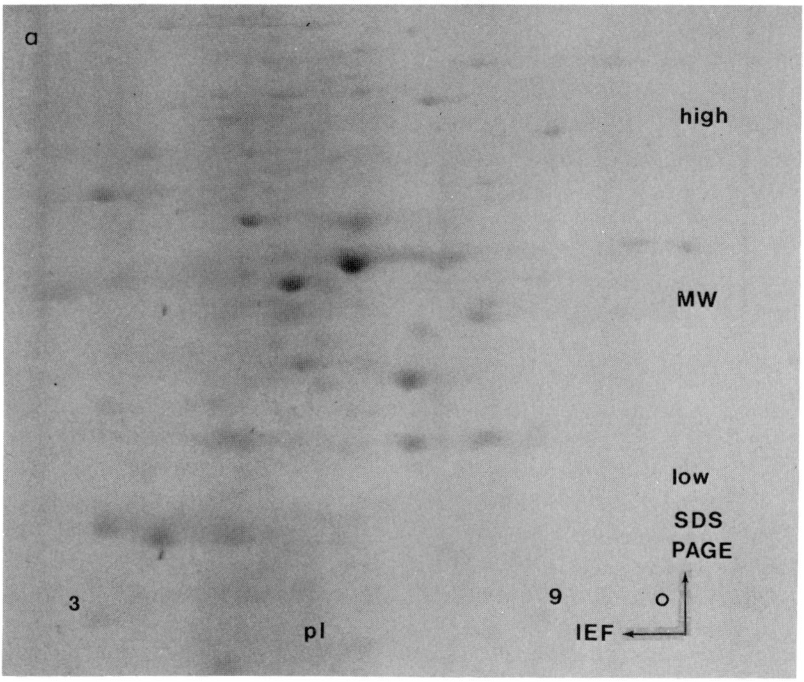
a
high
MW
low
SDS
PAGE
3
pI
9
o
IEF

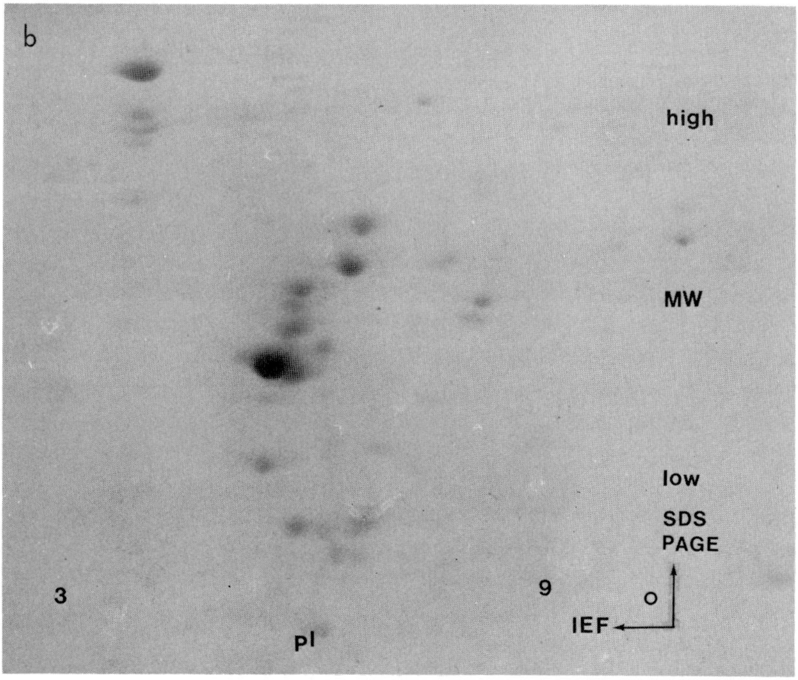
b
high
MW
low
SDS
PAGE
3
pI
9
o
IEF

Identification of the Proteins Coded for by Cloned Genes

Once clones corresponding to the genes induced during the onset of ligninolytic activity have been located, the corresponding proteins can be identified by *in vitro* translation of mRNA species transcribed from genes on the cloned fragment. Specific mRNAs can be purified by hybridising the cloned DNA to mRNA extracted from ligninolytic cultures and recovering the specifically bound mRNA from DNA–RNA hybrids.

The total protein profile of organisms can be mapped using a two-dimensional electrophoresis system (O'Farrell 1975). In O'Farrell gels, the first dimension is an isoelectric focussing gel and the second (at 90° to the first) is a denaturing gel containing sodium dodecyl sulphate (SDS). Thus each polypeptide has a unique map position, a function of surface charge of the enzyme and molecular weight of the subunit. The value of the technique is that a previously located ^{35}S-labelled polypeptide can be identified in picogram amounts, permitting the identification of the products of *in vitro* translation. Another use of O'Farrell gels is to study changes in the cell protein profile during the onset of secondary metabolism (Fig. 7).

A second method of polypeptide identification is by immunological means. Only small quantites of the purified proteins, with antigenic glycosylations removed, are required. Antibody to the proteins can be evoked in rabbits and the immunoglobulins purified from the serum. ^{125}I-labelled immunoglobulins can be used to detect very small amounts of the antigenic protein, as produced by genes on cloned DNA or by *in vitro*-translated mRNA (Clarke *et al.* 1979; Erlich *et al.* 1979). In these techniques, immunoglobulin-coated plastic (or immunoglobulins adsorbed to the inside of the well of a plastic microtitre tray) is exposed to the putative antigen and the positions of the antigen–antibody complexes are located by exposing these complexes to ^{125}I-labelled antibody, which forms a further complex with immobilised antigen. Such complexes, bound to the plastic, can be located by autoradiography.

Our approach is to use gene libraries of ligninolytic organisms to locate the genes coding for enzymes involved in lignin degradation, using the strategy described in Fig. 8. Once we have identified the genes, a number of approaches becomes possible, all leading to further understanding and exploitation of the ability of microorganisms to degrade plant materials.

Recombinant DNA Techniques

Criteria for the choice of vectors are discussed extensively by Williams & Blattner (1980) and cosmids and their use by Hohn & Hinnen (1980).

The vector currently in use in this laboratory is λgtWES.T5-622 (Fig. 9)

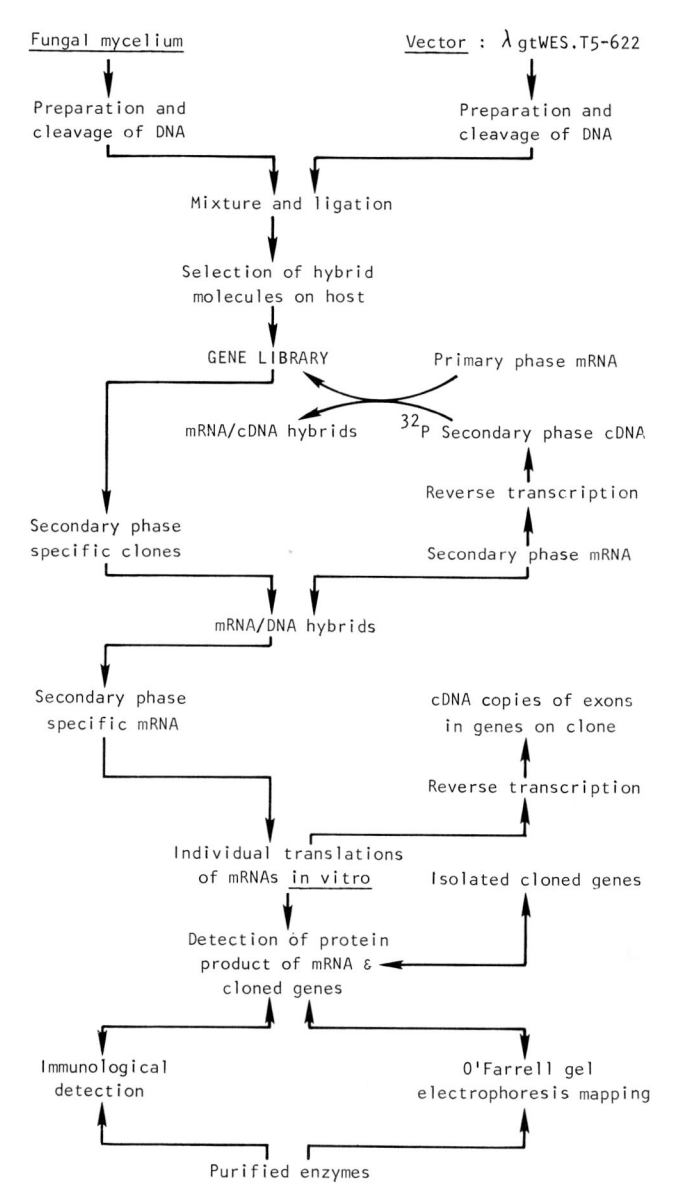

FIG. 8. A strategy for applying molecular genetics in the study of lignin degradation by white-rot fungi.

(a) λgtWES.T5-622

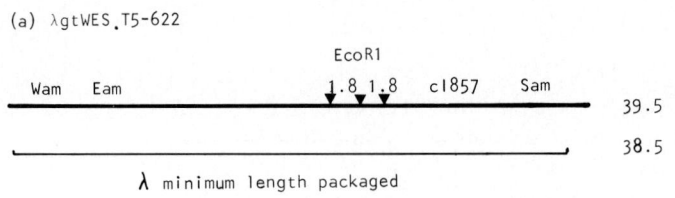

```
                        EcoR1
Wam    Eam            1.8 1.8    c1857      Sam
▬▬▬▬▬▬▬▬▬▬▬▬▬▬▬▼▬▬▼▬▼▬▬▬▬▬▬▬▬▬▬▬▬▬▬▬▬▬▬   39.5
▬▬▬▬▬▬▬▬▬▬▬▬▬▬▬▬▬▬▬▬▬▬▬▬▬▬▬▬▬▬▬▬▬▬▬       38.5
```

λ minimum length packaged

T5 inserts prevent replication in hosts containing the plasmid ColIb.

(b) pYc1

2μ yeast replicator

ampicillin resistance

amp

BamH1

histidine biosynthesis

pYc1
10

his3

ORI

Xho1

bacteriophage λ sequence

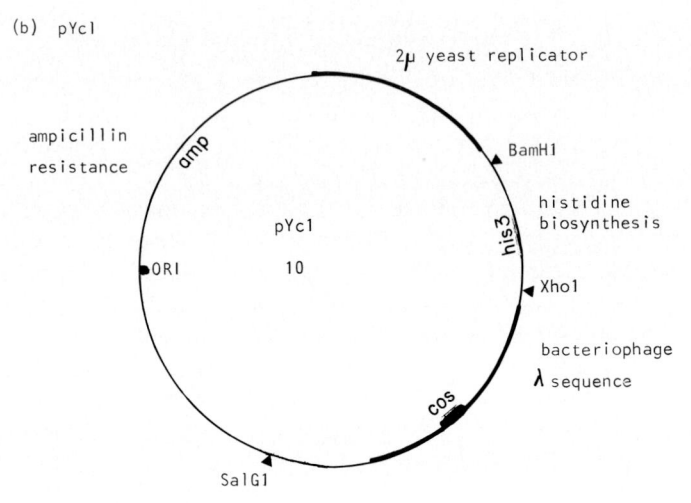

cos

SalG1

FIG. 9. Physical and genetic maps of cloning vectors used in construction of gene libraries. (a) The bacteriophage λ vector : λgtWES.T5-622. (b) the yeast cosmid vector: pYc1. Numbers refer to distances in kilobases. *Eco*R1, *Bam*H1, *Sal*G1, *Xho*1: nucleotide sequences cleaved by restriction endonucleases of that name.

(Davison *et al.* 1979). It allows positive selection for included DNA sequences replacing the T5 fragments. DNA fragments between approximately 2.1 and 16.7 kb will be accepted by the vector. A cosmid, pYC1, is also used in the construction of gene libraries. This cosmid (Fig. 9) accepts DNA sequences of 27 to 42 kb.

Practical details of the following are fully described by Maniatis *et al.* (1982): construction of hybrid DNA molecules; preparation of cDNA; location of clones using probes.

Preparation of Fungal Nucleic Acids

Preparation of DNA and RNA from fungi can be a problem since large amounts of polysaccharide tend to associate with the nucleic acids. Nuclease activity must also be minimised. The following techniques have proved successful with *Phanerochaete chrysosporium* and *Sporotrichum pulverulentum* RNA preparation modified from Poulson (1977).

Mycelium at the appropriate growth phase is harvested by filtration through muslin, washed with ice-cold sterile distilled water, blotted dry with filter paper and lyophilised. The dried mycelium is ground to a fine powder at 4°C using a pestle and mortar and extracted for 1 h at 4°C with the following buffer: sodium triisopropylnaphthalenesulphonate (TNS), 1% (w/v) (Eastman Kodak Corporation); *p*-aminosalicylic acid (PAS), 6% (w/v); *n*-butanol, 6% (v/v); 1,2-di(2-aminoethoxy)ethane N,N,N',N'-tetraacetic acid (EGTA), 0.25 M; Tris, 0.2 M (pH 8.5). Debris is removed by centrifugation and the pellet is washed with extraction buffer. The pooled supernatants are extracted with an equal volume of the following mixture: phenol, 50 g; water, 10 ml; 8-hydroxyquinoline, 0.05 g; *m*-cresol, 7 ml. After 1 min at room temperature a volume of chloroform is added and the phases are then separated by centrifugation. Approximately 50% of the aqueous phase is removed and added to an equal volume of chloroform. The phenol–chloroform phase is discarded and the aqueous and interphase are extracted twice with chloroform. All aqueous phases are pooled and extracted again with chloroform. Finally, the aqueous phase is centrifuged at 10,000 g for 45 min at 4°C to remove carbohydrate, and the purified RNA is precipitated with ethanol. Poly(A)-containing RNA can then be enriched by chromatography on poly(U) Sephadex (BRL Ltd Cambridge, UK) or oligo(dT) cellulose (Sigma).

DNA Preparation (Modified from Specht et al. 1982)

Cells, grown to linear phase (20 h at 30°C) from spores, are harvested by filtration through gauze, washed twice with distilled water, once in TSE buffer (NaCl, 0.15 M; Na$_2$EDTA, 0.1 M; Tris, 0.05 M; pH 8.0) and lyophilised. The dried cells are ground using a pestle and mortar at room temperature to give a fine powder which is then extracted with TSE buffer (25 ml/g dried cells) containing SDS, 2% (w/v) and toluene, 20% (v/v). The extraction mixture is shaken at approximately 30 rev/min at room temperature for 72 h and debris then removed by centrifugation at 1000 g for 15 min at room temperature. The supernatant is added to an equal volume of isoamyl alcohol–chloroform (1:24, v/v) and shaken for 10 min. The mixture is then centrifuged at 10,000 g for 1 h and nucleic acid precipitated from the aqueous phase by addition of 2 volumes of ethanol. The precipitate is collected by centrifugation at 10,000 g for 15 min at 4°C, the pellet resuspended in 6 ml TPE buffer (NaH$_2$PO$_4$, 0.036 M; Na$_2$EDTA, 0.002 M; Tris, 0.03 M; pH 7.8) and 0.3 ml RNase solution (4 mg/ml) added (pretreat RNase solution by heating at 80°C for 10 min to inactivate DNases). The preparation is incubated at 37°C for 2 h, 0.5 ml protease K solution (4 mg/ml) are added and incubation is continued overnight at 37°C. The extraction with isoamyl alcohol–chloroform is repeated and the ethanol-precipitated DNA is redissolved in a solution of Na$_2$EDTA, 0.01 M and Tris, 0.1 M, pH 7.4. To this solution is added CsCl (1 g/ml) and Hoechst Dye 11258 (175 μl/10 ml). The resulting preparation is centrifuged to equilibrium at 94,000 g for 65 h at 4°C. After the buoyant density centrifugation, two bands are observed, i.e. the upper mitochondrial DNA and the lower chromosomal DNA which can be recovered using standard methods.

Discussion

Lignin degradation in *Phanerochaete chrysosporium* and *Sporotrichum pulverulentum* is a secondary metabolic event. Identification of the genes regulating the onset of secondary metabolism and secretion of extracellular proteins is of fundamental interest to the study of lignin degradation and to the understanding of mechanisms involved in the regulation of fungal metabolism. Isolation of cDNA species coding for ligninolytic enzymes opens up the possibility of producing enzymes in large quantities using bacteria. This will permit the study of such enzymes, even if they are only produced in small amounts by the ligninolytic fungus. The study of the effect of enzymes, alone or in concert, on the structure of the lignin polymer should clarify the mechanism of depolymerisation and modification of this recalcitrant molecule.

Processes using enzymes to hydrolyse lignocellulose or cellulosic material such as straw depend on a balance of a number of different enzymes to break down the substrate. The efficiency of the process is determined by this balance of enzyme activities which could be altered by manipulating the levels at which the encoding genes are expressed. In the search for preparations with good enzyme activity balances, actinomycetes are being increasingly studied and an understanding of the mechanism of fungal degradation of lignin would assist the study and manipulation of actinomycete lignocellulose breakdown.

Biological processes which utilise lignocellulosic materials are very capital intensive, and the economics of these processes are greatly influenced by the cost of pretreating materials to remove lignin. Selective removal of lignin using organisms that have been made defective in hydrolytic cellulase activity is an attractive idea. Such a process is already in operation on a pilot scale in Sweden, but there is a need to generate more efficient and versatile organisms.

In natural environments, plant material is broken down by mixed populations of organisms. The contribution of the component organisms is particularly difficult to analyse because of the complexity of the interrelationships. Isolation of the components of the degradation, using genetic methods, offers the best hope for detailed understanding of the process. Genetic methodology for analysing and manipulating processes will be an important tool for the environmental biotechnology of the future.

Note Added in Proof

Much progress has been made in understanding the biochemistry of lignin degradation since this article was written. A short summary of progress up to the end of 1983 has been published: Broda, P. & Paterson, A. 1983 Lignin biodegradation becomes biochemistry. *Nature* **306**, 737–738.

Acknowledgements

We thank the following for helpful discussions, comments on the manuscript and access to experimental results: Knut Lundquist, Department of Organic Chemistry, Chalmers University of Technology, Göteborg, Sweden and Malcolm MacDonald, Lesley Ramsey, Richard Haylock and Ute Raeder, Department of Biochemistry and Applied Molecular Biology, UMIST.

We thank the Science and Engineering Research Council, Agricultural Research Council, and British Petroleum Venture Research Unit for their part in supporting our research.

References

ANDER, P. & ERIKSSON, K.-E. 1976 The importance of phenol oxidase activity in lignin degradation by the white-rot fungus *Sporotrichum pulverulentum*. *Archives of Microbiology* **109**, 1–9.
AYERS, A. R., AYERS, S. B. & ERIKSSON, K.-E. 1978 Cellobiose oxidase, purification and partial characterisation of a haemoprotein from *Sporotrichum pulverulentum*. *European Journal of Biochemistry* **90**, 171–181.
BAVENDAMM, W. 1928 Uber des vorkommen und den nachweis von oxydasen bei holzzerstorenden pilze. *Zeitschrift fuer Pflanzenernaehrung und Bodenkunde* **38**, 257–275.

BEAUCHAMP, C. & FRIDOVICH, I. 1971 Superoxide dismutase: improved assays and an assay applicable to acrylamide gels. *Analytical Biochemistry* **44**, 276–287.

BJÖRKMAN, A. 1956. Studies on finely divided wood. Extraction of lignin with neutral solvents. *Svensk Papperstidning* **59**, 477–485.

BRUNOW, G. & LUNDQUIST, K. 1980 Comparison of a synthetic dehydrogenation polymer of coniferyl alcohol with milled wood lignin from spruce using ^{1}H NMR spectroscopy. *Paperi ja Puu* **11**, 669–672.

BRUNOW, G. & WALLIN, H. 1981 Studies concerning the preparation of synthetic lignin. *SPCI Report No. 38* Vol. 4, pp. 125–127. Stockholm: SPCI.

BUSWELL, J. A., ERIKSSON, K.-E., GUPTA, J. K., HAMP, S. G. & NORDH, I. 1982 Vanillic acid metabolism by selected soft-rot, brown-rot and white-rot fungi. *Archives of Microbiology* **131**, 366–374.

CALDWELL, E. S. & STEELINK, C. 1969 Phenoxy radical intermediates in the enzymatic degradation of lignin model compounds. *Biochimica et Biophysica Acta* **184**, 420–431.

CHANG, H. M. & ALLAN, G. G. 1971 Oxidation. In *Lignins—Occurrence, Formation, Structure and Reactions* eds. Sarkanen, K. V. & Ludwig, C. H., pp. 433–478. New York: Wiley-Interscience.

CHANG, H. M., CHEN, C. L. & KIRK, T. K. 1980 Chemistry of lignin degraded by white-rot fungi. In *Lignin Biodegradation: Microbiology, Chemistry and Potential Applications* eds. Kirk, T. K., Higuchi, T. & Chang, H. M., Vol. 1, pp. 215–230. Boca Raton: CRC Press.

CHEN, C. L., CHANG, H. M. & KIRK, T. K. 1982 Aromatic acids produced during degradation of lignin in spruce wood by *Phanerochaete chrysosporium*. *Holzforschung* **36**, 3–9.

CLARKE, L., HITZEMAN, R. & CARBON, J. 1979 Selection of specific clones from colony banks by screening with radioactive antibody. In *Methods in Enzymology* ed. Wu, R., Vol. 68, pp. 436–442. New York: Academic Press.

CONNORS, W. J. 1978 Gel chromatography of lignins, lignin model compounds and polystyrenes using Sephadex LH60. *Holzforschung* **32**, 145–147.

CRAWFORD, D. L. & CRAWFORD, R. L. 1980 Microbial degradation of lignin. *Enzyme and Microbial Technology* **2**, 11–22.

CRAWFORD, D. L., BARDER, M. J., POMETTO, A. L. & CRAWFORD, R. L. 1982 Chemistry of softwood lignin degradation by *Streptomyces viridosporus*. *Archives of Microbiology* **131**, 140–145.

CRAWFORD, D. L., POMETTO, A. L. & CRAWFORD, R. L. 1983 Lignin degradation by *Streptomyces viridosporus*: isolation and characterisation of a new polymeric lignin degradation intermediate. *Applied and Environmental Microbiology* **45**, 898–904.

CRAWFORD, R. L. 1981 *Lignin Biodegradation and Transformation*. New York: John Wiley and Sons.

CRAWFORD, R. L., ROBINSON, L. E. & CHEH, A. 1980 ^{14}C-labelled lignins as substrates for the study of lignin biodegradation. In *Lignin Biodegradation: Microbiology, Chemistry and Potential Applications* eds. Kirk, T. K., Higuchi, T. & Chang, H. M., Vol. 1, pp. 61–76. Boca Raton: CRC Press.

DAVISON, J., BRUNEL, F. & MERCHEZ, M. 1979 A new host–vector system allowing selection for foreign DNA inserts in bacteriophage λgtWES. *Gene* **8**, 69–80.

DICKERSON, A. G. & BAKER, R. C. F. 1979 The binding of enzymes to fungal β-glucans. *Journal of General Microbiology* **112**, 67–75.

EGGELING, L. & SAHM, H. 1980 Degradation of coniferyl alcohol and other lignin-related aromatic compounds by *Nocardia* sp.DSM 1069. *Archives of Microbiology* **126**, 141–148.

ELLWARDT, P.-CHR., HAIDER, K. & ERNST, L. 1981 Investigation of microbial lignin degradation by ^{13}C NMR spectroscopy of specifically ^{13}C-enriched DHP-lignin from coniferyl alcohol. *Holzforschung* **35**, 103–109.

ENOKI, A., GOLDSBY, G. P. & GOLD, M. H. 1981 β ether cleavage of the lignin model compound 4-ethoxy-3-methoxyphenylglycerol-β-guaiacyl ether and derivatives by *Phanerochaete chrysosporium*. *Archives of Microbiology* **129**, 141–145.

ERIKSSON, K.-E. 1981 Microbial degradation of cellulose and lignin. *SPCI Report No. 38* Vol. 3, pp. 360–365. Stockholm: SPCI.

ERIKSSON, K.-E. & GOODELL, E. W. 1974 Pleiotropic mutants of the wood-rotting fungus *Polysporus adustus* lacking cellulase, mannanase and xylanase. *Canadian Journal of Microbiology* **20**, 371–378.

ERIKSSON, K.-E., GRUNEWALD, A., NILSSON, T. & VALLANDER, L. 1980 A scanning electron microscopy study of the growth and attack on wood by three white-rot fungi and their cellulase-less mutants. *Holzforschung* **34**, 207–213.

ERLICH, H. A., COHEN, S. N. & MCDEVITT, H. O. 1979 Immunological detection and characterisation of products translated from cloned DNA fragments. In *Methods in Enzymology* ed. Wu, R., Vol. 68, pp. 443–453. New York: Academic Press.

FÅHRAEUS, G. & REINHAMMAR, B. 1967 Large scale production and purification of laccase from cultures of the fungus *Polysporus versicolor* and some properties of laccase A. *Acta Chemica Scandinavica* **21**, 2367–2378.

FAIX, O., LANGE, W. & SALUD, E. C. 1981 The use of HPLC for the determination of average molecular weights and molecular weight distribution of milled wood lignins from *Shorea polysperma*. *Holzforschung* **35**, 3–9.

FARMER, V. C., HENDERSON, M. E. K. & RUSSELL, J. D. 1959 Reduction of certain aromatic acids to aldehydes and alcohols by *Polystictus versicolor*. *Biochimica et Biophysica Acta* **35**, 202–211.

FENN, P. & KIRK, T. K. 1981 Relationship of nitrogen to the onset and suppression of ligninolytic activity and secondary metabolism in *Phanerochaete chrysosporium*. *Archives of Microbiology* **130**, 59–60.

FÉVRE, M. 1977 Subcellular localisation of glucanase and cellulase in *Saprolegnia monoica* Pringsheim. *Journal of General Microbiology* **103**, 287–295.

FORNEY, L. J., REDDY, C. A., TIEN, M. & AUST, S. D. 1982 The involvement of hydroxyl radical derived from hydrogen peroxide in lignin degradation by the white-rot fungus *Phanerochaete chrysosporium*. *Journal of Biological Chemistry* **257**, 11455–11462.

FREUDENBERG, K. 1956 Lignin im rahmen der polymeren naturstoffe. *Angewandete Chemie* **68**, 84–92.

FROEHNER, S. C. & ERIKSSON, K.-E. 1977 Purification and properties of *Neurospora crassa* laccase. *Journal of Bacteriology* **120**, 458–465.

FUKUZUMI, T. & KATAYAMA, Y., 1977 Bacterial degradation of a dimer related to lignin. I. Hydroxypropiovanillone and coniferyl alcohol as initial degradation products from guaiacylglycerol-β-coniferyl ether by *Pseudomonas putida*. *Mokuzai Gakkaishi* **23**, 214.

GIERER, J. 1982 The chemistry of delignification. Parts 1 & 2. *Holzforschung* **36**, 43–51 and 55–64.

GOLD, M. H., MAYFIELD, M. B., CHANG, T. M., KRISHNANGKURA, K., SHIMADA, M., ENOKI, M., ENOKI, A. & GLENN, J. K. 1982 A *Phanerochaete chrysosporium* mutant defective in lignin degradation as well as several other secondary metabolic functions. *Archives of Microbiology* **132**, 115–122.

GREEN, T. R. 1977 Significance of glucose oxidase in lignin degradation. *Nature, London* **268**, 78–80.

HAARS, A. & HÜTTERMAN, A. 1980 Function of laccase in the white-rot fungus *Fomes annosus*. *Archives of Microbiology* **125**, 233–237.

HAARS, A., MILSTEIN, O., LOHNER, S. & HÜTTERMAN, A. 1982 Rapid, sensitive and inexpensive assay for the determination of lignin degradation using different fluorescent labelled lignin derivatives. *Holzforschung* **36**, 85–91.

HAIDER, K. & TROJANOWSKI, J. 1975 Decomposition of specifically ^{14}C-labelled phenols and dehydropolymers of coniferyl alcohol as models for lignin degradation by soft and white-rot fungi. *Archives of Microbiology* **105**, 33–42.

HAIDER, K. & TROJANOWSKI, J. 1980 A comparison of the degradation of ^{14}C-labelled DHP and cornstalk lignins by micro- and macrofungi and by bacteria. In *Lignin Biodegradation: Microbiology, Chemistry and Potential Applications* eds. Kirk, T. K., Higuchi, T. & Chang, H. M., Vol. 1, pp. 111–134. Boca Raton: CRC Press.

HALL, P. L. 1980 Enzymatic transformations of lignin: 2. *Enzyme and Microbial Technology* **2**, 170–176.

HALL, P. L., GLASSER, W. & DREW, S. 1980 Enzymatic transformations of lignin. In *Lignin Biodegradation: Microbiology, Biochemistry and Potential Applications* eds. Kirk, T. K., Higuchi, T. & Chang, H. M. Vol. 2, pp. 33–50. Boca Raton: CRC Press.

HALLIWELL, B. 1982 Superoxide and superoxide-dependent hydroxyl radicals are important in oxygen toxicity. *Trends in Biochemical Sciences* **7**, 270–272.

HIGUCHI, T. 1953 Biochemical study of wood-rotting fungi. 1. Studies on the enzymes which cause Bavendamm's reaction and tyrosinase. *Journal of the Japanese Forestry Society* **35**, 77–84.

HIGUCHI, T. 1980 The biochemistry of lignification. *Wood Research* **66**, 1–16.

HIGUCHI, T. 1982 Biodegradation of lignin: biochemistry and potential applications. *Experientia* **38**, 159–166.

HIGUCHI, T. & NAKATSUBO, F. 1980 Synthesis and biodegradation of oligolignols *Kemia-Kemi* **7**, 481–488.

HOHN, B. & HINNEN, A. 1980 Cloning with cosmids in *E. coli* and yeast. In *Genetic Engineering. Principles and Methods* eds. Setlow, J. K. & Hollaender, A. Vol. 2, pp. 169–183. New York: Plenum Press.

HÜTTERMAN, A. 1978 Gel permeation chromatography of water insoluble lignins on controlled pore glass and Sepharose CL6B. *Holzforschung* **32**, 108–111.

ISHIHARA, T. 1980 The role of laccase in lignin biodegradation. In *Lignin Biodegradation: Microbiology, Biochemistry and Potential Applications* eds. Kirk, T. K., Higuchi, T. & Chang, H. M., Vol. 2, pp. 17–32. Boca Raton: CRC Press.

IWAHARA, S., NISHIHARA, T., JOMORI, T., KUWAHARA, M. & HIGUCHI, T. 1980 Enzymic oxidation of α–β unsaturated alcohols in the side-chains of lignin-related aromatic compounds. *Journal of Fermentation Technology* **58**, 183–188.

JAEGER, E., EGGELING, L. & SAHM, H. 1981 Partial purification and characterisation of a coniferyl alcohol dehydrogenase from *Rhodococcus erythropolis*. *Current Microbiology* **6**, 333–336.

KATAYAMA, Y. & FUKUZUMI, T. 1977 Enzymatic synthesis of three lignin-related dimers by an improved peroxidase–hydrogen peroxide system. *Mokuzai Gakkaishi* **24**, 664–667.

KELLEY, R. L. & REDDY, C. A. 1982 Ethylene production from α-oxo-γ-methylthio-butyric acid is a sensitive measure of ligninolytic activity by *Phanerochaete chrysosporium*. *Biochemical Journal* **206**, 423–425.

KIRK, T. K. 1981 Principles of lignin degradation by white-rot fungi. *SPCI Report No. 38* Vol. 3, pp. 66–70. Stockholm: SPCI.

KIRK, T. K. & CHANG, H. M. 1975. Decomposition of lignin by white-rot fungi. II. Characterisation of heavily degraded lignins from decayed spruce. *Holzforschung* **29**, 56–64.

KIRK, T. K., CONNORS, W. J., BLEAM, R. D., HACKETT, W. F. & ZEIKUS, J. G. 1975 Preparation and microbial decomposition of synthetic ^{14}C lignins. *Proceedings of the National Academy of Science of the United States of America* **72**, 2515–2519.

KIRK, T. K., HIGUCHI, T. & CHANG, H. M., eds. 1981 *Lignin Biodegradation: Microbiology, Biochemistry and Potential Applications*, Vols. 1 & 2. Boca Raton: CRC Press.

KOLODZLEJSKI, W., FRYE, J. & MACLEL, G. 1982 Carbon-13 Nuclear Magnetic Resonance Spectrometry with cross polarisation and magic-angle spinning for analysis of lodgepole wood. *Analytical Chemistry* **54**, 1419–1424.

KRINSKY, N. I. 1979 Biological roles of singlet oxygen. In *Singlet Oxygen* eds. Wasserman, H. N. & Murray, R. W., pp. 597–641. London: Academic Press.

KUTSUKI, H. & GOLD, M. H. 1982 Generation of hydroxyl radical and its involvement in lignin degradation by *Phanerochaete chrysosporium*. *Biochemical and Biophysical Research Communications* **109**, 320–327.

LUNDQUIST, K. 1976 Low molecular lignin hydrolysis products. *Applied Polymer Symposia* **28**, 1393–1407.

LUNDQUIST, K. 1980 NMR studies of lignins. 4. Investigation of spruce lignin by ^{1}H NMR spectroscopy. *Acta Chemica Scandinavica, Series B* **34**, 21–26.

LUNDQUIST, K. & KIRK, T. K. 1978 *De novo* synthesis and decomposition of veratryl alcohol by a lignin-degrading basidiomycete. *Phytochemistry* **17**, 1676.

LUNDQUIST, K. & KIRK, T. K. 1980 Fractionation–purification of an industrial kraft lignin. *Tappi* **63**, 80–82.

LUNDQUIST, K., OHLSSON, B. & SIMONSSON, R. 1977 Isolation of lignin by means of liquid–liquid extraction. *Svensk Papperstidning* **80**, 143–144.

MANIATIS, T., FRITSCH, E. F. & SAMBROOK, J. 1982 *Molecular Cloning—A Laboratory Manual*. New York: Cold Spring Harbor Laboratories.

MARIGO, G., RIVIERE, D. & BOUDET, A. 1981 Synthese biologique des acides ferulique et sinapique marques au ^{14}C. *Journal of Labelled Compounds and Pharmaceuticals* **18**, 695–702.

NAKAMURA, Y. & HIGUCHI, T. 1976 A new synthesis of coniferyl aldehyde and alcohol. *Wood Research* **59**, 101–105.

NAKAMURA, Y., NAKATSUBO, F. & HIGUCHI, T. 1974 Synthesis of p-coumar-, coniferyl and sinap aldehydes. *Wood Research* **56**, 1–6.

NAKATSUBO, F., REID, I. & KIRK, T. K. 1981 Involvement of singlet oxygen in the fungal degradation of lignin. *Biochemical and Biophysical Research Communications* **102**, 484–491.

ODIER, E. & MONTIES, B. 1980 Lignin degradation by non-cellulolytic bacteria. In *Colloque Cellulolyse Microbienne* pp. 101–110. Marseille: CNRS.

ODIER, E., JANIN, G. & MONTIES, B. 1981 Poplar lignin decomposition by Gram-negative aerobic bacteria. *Applied and Environmental Microbiology* **41**, 337–341.

O'FARRELL, P. H. 1975 High resolution two-dimensional electrophoresis of proteins. *Journal of Biological Chemistry* **250**, 4007–4021.

PERRIN, D. D., ARMAREGO, W. L. F. & PERRIN, D. R., eds. 1980 *Purification of Laboratory Chemicals*, 2nd edn. Oxford: Pergamon Press.

POULSON, R. 1977 Isolation, purification and fractionation of RNAs. In *The Ribonucleic Acids* eds. Stewart, P. R. & Letham, D. S., pp. 333–362. Heidelberg: Springer-Verlag.

REINHAMMAR, B. 1984 Laccase. In *Copper Proteins and Copper Enzymes* ed. Lontie, R., Vol. 3 (in press). Boca Raton: CRC Press.

SARKANEN, K. 1980 The nature of hydrolyzable linkages in lignin. *Kemia-Kemi* **7**, 489.

SCHWEERS, W. & FAIX, O. 1973 Comparative investigations on polymer models of lignins (DHPs) of different composition. *Holzforschung* **27**, 208–213.

SHIMADA, M. 1980 Stereobiochemical approach to lignin biodegradation. In *Lignin Biodegradation: Microbiology, Biochemistry and Potential Applications* eds. Kirk, T. K., Higuchi, T. & Chang, H. M., Vol. 1, pp. 195–213. Boca Raton: CRC Press.

SHIMADA, M., NAKATSUBO, F., KIRK, T. K. & HIGUCHI, T. 1981 Biosynthesis of the secondary metabolite veratryl alcohol in relation to lignin biodegradation in *Phanerochaete chrysosporium*. *Archives of Microbiology* **129**, 321–324.

SPECHT, C. A., DIRUSSO, C. C., NOVOTNY, C. P. & ULLRICH, R. C. 1982 A method for extracting high molecular weight deoxyribonucleic acid from fungi. *Analytical Biochemistry* **119**, 158–163.

TANAHASHI, M. & HIGUCHI, T. 1981 Dehydrogenative polymerisation of monolignols by peroxidase and H_2O_2 in a dialysis tube. I. Preparation of highly polymerised DHPs. *Wood Research* **67**, 29–42.

WEINSTEIN, D. A., KRISNANGKURA, K., MAYFIELD, M. B. & GOLD, M. H. 1980 Metabolism of radiolabelled β-guaiacyl ether-linked lignin dimeric compounds by *Phanerochaete chrysosporium*. *Applied and Environmental Microbiology* **39**, 535–540.

WESTERMARK, U. & ERIKSSON, K. E. 1974a Cellobiose: Quinone Oxidoreductase, a new wood-degrading enzyme from white-rot fungi. *Acta Chemica Scandinavica, Series B* **28**, 209–214.

WESTERMARK, U. & ERIKSSON, K. E. 1974b Carbohydrate-dependent enzymic quinone reduction during lignin degradation. *Acta Chemica Scandinavica, Series B* **28**, 204–208.

WHISTLER, R. L. & RICHARDS, E. L. 1970 37: Hemicelluloses. In *The Carbohydrates* eds. Pigman, W., Horton, D. & Herp, A., Vol. 11A, pp. 447–469. London: Academic Press.

WILLIAMS, B. G. & BLATTNER, F. R. 1980 Bacteriophage lambda vectors for DNA cloning. In *Genetic Engineering, Principles and Methods* eds. Setlow, J. K. & Hollaender, A., Vol. 2, pp. 201–281. New York: Plenum Press.

WOOD, D. A. 1980 Production, purification and properties of extracellular laccase of *Agaricus bisporus*. *Journal of General Microbiology* **117**, 327–328.

ZIMMERMANN, C. R., ORR, W. C., LECLERC, R. F., BARNARD, E. C. & TIMBERLAKE, W. E. 1980 Molecular cloning and selection of genes regulated in *Aspergillus* development. *Cell* **21**, 709–715.

Approaches to the Controlled Production of Novel Agricultural Composts

J. M. LYNCH,* S. H. T. HARPER, S. J. CHAPMAN AND D. A. VEAL

Agricultural & Food Research Council Letcombe Laboratory, Wantage, Oxfordshire, UK

There is no economic process in which most of the straw produced from arable crops can be utilized. However, leaving the straw on the soil can present mechanical difficulties in drilling a succeeding crop and microbial decomposition of the straw can lead to a retardation in seedling establishment and a reduction in crop yield (Lynch & Elliott 1984). In the United Kingdom, straw is therefore usually burnt in the field.

Microbial decomposition of straw can lead to the production of soil stabilizing agents. Therefore in areas where soils are subject to erosion, such as the Pacific Northwest of the United States, straw burning is discouraged. To minimize the adverse effects of microbial decomposition, such as the production of phytotoxic metabolites, and to decrease the population of organisms antagonistic to plant growth, laboratory studies have demonstrated that microbial degradation of straw can be beneficial (Lynch & Elliott 1983). In this chapter we report methods that we have used to study the decomposition of straw under controlled conditions and its effects on soils and plants.

The principal aims of our work are to increase the plant nutrient value of the straw, particularly with respect to nitrogen, to promote the production of soil-stabilizing agents and to eliminate the potential for the accumulation of phytotoxic metabolites while utilizing any plant-growth-promoting substances produced by microbial degradation. The product can therefore be regarded as a compost (compound manure). Straw from most crops is about 40% (w/w) cellulose and 35% (w/w) hemicellulose (Harper & Lynch 1981).

---

*Present address: Glasshouse Crops Research Institute, Worthing Road, Littlehampton, West Sussex, BN17 GLP, UK.

In order to understand the underlying principles of the decomposition process, especially those with defined microbial cultures, we have used pure cellulose instead of straw as the substrate in some of the studies.

Isolation of Members of N_2-Fixing Cellulolytic Communities

Isolation of single species with a particular activity from mixed populations depends on the ability of isolates to express the same characteristics in pure and mixed culture. The type of isolation procedures adopted depends on understanding the nature of interactions within the community under study.

Cellulose decomposers were isolated from composts on cellulose agar containing thiamine and ammonium as a nitrogen source (Bravery 1968). Cellulolysis was indicated both by growth and by clearing of cellulose from the agar following incubation for 21 days at 25°C. More sophisticated tests for cellulase activity were not used since they provide little indication of the potential of a microorganism to provide the cellulase function when growing in mixed culture. A general indication only was required and revealed that cellulolysis was limited to fungal isolates under aerobic conditions.

Dinitrogen fixers were isolated on N-free agar medium (Dalton & Postgate 1969) with cellobiose or glucose as C source; these sugars are products of cellulolysis that might be made available to non-cellulolytic members of a mixed population. Growth on N-free agar was restricted to obligate or facultatively anaerobic bacteria.

Selection and Testing of Potential Compost Inocula

Isolates were tested as potential compost inocula to promote the rates of either cellulolysis or N_2 fixation. No single isolate was found capable of both functions and thus pairs of isolates were compared. This identified those combinations for which substrates were made available from cellulose for N_2 fixation, the fixed N being made available to sustain cellulase production.

The screening of large numbers of isolates required a simple test. Pairs of isolates were inoculated onto sterile (autoclaved twice for 15 min at 121°C), milled (0.5 to 2.0 mm) wheat straw and moistened with 10 ml sterile N-free mineral salts solution (Dalton & Postgate 1969). The acetylene reduction test provided a non-destructive method to monitor the course of N_2 fixation but was supported by determinations of total N in the degraded straw at the end of incubation (4 weeks at 25°C). Bottles were sealed, acetylene injected to produce an atmosphere containing 10% (v/v) and ethylene determined by gas chromatography after 2 and 4 h. Ethylene was also produced in the absence of acetylene but not in significant concentrations. The extent of cellulolysis was determined as loss of weight from the straw, having made allowance for the water-soluble components. Preliminary experiments, in which the chemical components of the decomposed straw were determined using methods reported by Harper & Lynch (1981), revealed that the major losses in straw weight resulted from decomposition of cellulose and hemicellulose.

Potentially effective isolates were tested as inocula on non-sterile straw. Wheat straw (50 g) was chopped (less than 5-cm lengths) and packed into glass columns (300 × 50 mm) to occupy a volume of 0.5 litre. The columns were sealed with silicone rubber bungs carrying glass tubes for aeration and medium supply at the top and for drainage at the base. The straw was flooded with N-free mineral salts medium for 30 min and then drained before starting aeration (50 ml/min) and recycling of the medium (2 ml/min). Inoculation was carried out at flooding, adding 5×10^7 spores or viable units to the top of the column. Columns were incubated at 25°C for up to 8 weeks. At weekly intervals the course of N_2 fixation was monitored by acetylene reduction tests. The method was the same as that used to test isolates for nitrogenase with the exception that columns were flushed with a mixture of helium (0.8 atm) and oxygen (0.2 atm) to displace nitrogen gas before injecting the acetylene (50 ml). When the procedure was repeated in the absence of acetylene, the rate of accumulation of ethylene was reduced by more than 98%.

At the completion of experiments, subsamples of straw (ca. 1 g) were suspended in sterile N-free medium and a dilution series was made. Aliquots of the dilutions were spread on plates of nutrient agar medium, one-tenth strength malt agar medium and the N-free mineral salts solution supplemented with sucrose and agar. The plates of N-free agar medium were incubated in an anaerobic incubator (Don Whitley Scientific, Shipley, Yorks). Colonies were counted after 5 days at 25°C on three replicate plates of each medium. The subsamples together with the bulk of the straw were dried (100°C for 16 h) and further subsamples taken for the determination of N content by an automatic N analyser (Model 1106, Erba Science, Swindon).

Growth of Mixed Populations on Solid Substrates in a Conventional Chemostat

The conventional chemostat offers considerable advantages over batch culture techniques in studies of the microbiology of composting or other "natural" systems. For example, control of the growth rate of microorganisms to rates more typical of those in soil can influence the yield of microbial products (Lynch & Harper 1974). However, the chemostat suffers the limitation that it does not simulate the changing environmental conditions characteristic of compost. The introduction of computer control has now extended the value of the chemostat in ecological studies. Environments can be reproduced in which one or more variables are changed independently with time. The chemostat, combined with studies of mixed populations, can now provide a much better insight into the microbial interactions and population dynamics involved in the production of composts.

A problem with solid substrates in the chemostat is the difficulty in providing a feed at constant rate. With cellulose, this was achieved by ball milling the material to a fine particle size (less than 1 μm) and making a suspension which is stirred continually by a magnetic stirrer and pumped to the culture vessel.

Linked Culture Vessels

In our studies on the decomposition of straw, aerobic cellulolytic fungi and anaerobic N_2-fixing bacteria appear to co-exist at aerobic–anaerobic interfaces. To study the interaction between these communities, linked culture vessels were constructed (Fig. 1).

Vessel I was kept totally anaerobic by continual sparging with O_2-free N_2 (100 ml/min). The nitrogen was made O_2-free by passage through acidified chromous sulphate solution (Hungate 1950) prepared by the method of Skinner (1971). Washing N_2 with chromous sulphate also removed all traces of NH_3 which could otherwise invalidate work on N_2-fixing organisms (Millbank 1969). The purity of the N_2 produced was tested by bubbling through methylene blue indicator solution (Skinner 1971), which is sensitive to less than 0.05 atm of O_2. Vessel II was kept aerobic by sparging with air (1000 ml/min) and continuous stirring.

The two vessels were linked via a peristaltic pump which continuously pumped medium from vessel II to vessel I. Medium was returned from vessel I using a pressure return overflow. The air and N_2 pumped into the system were released through a single trap. The vessels were maintained at a constant temperature by a thermostatically controlled circulating water system. Values of Eh, pH and p_{O_2} can be measured using electrodes. Samples were taken from the vessels through tapped outlets. A readily measurable system was thus established where the flow of medium and the products of decomposition were controlled by a single pump.

The value of the apparatus is in the study of the transfer of products between an aerobic and an anaerobic environment. These environments in

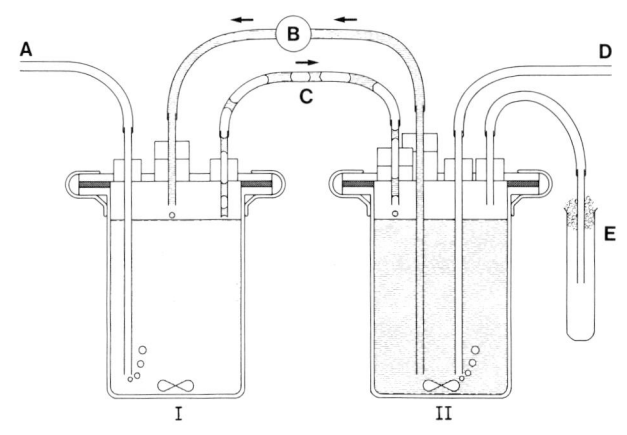

FIG. 1. Linked culture vessel. Vessel I is anaerobic; Vessel II is aerobic; (A) O_2-free N_2 supply; (B) peristaltic pump; (C) pressure return overflow; (D) air supply; (E) contamination trap.

composts may be both transient and separated by very small distances, so transfer of products between them is likely.

The Oxygen Diffusion Column

To study the effect of varying O_2 concentrations on communities containing both anaerobes and aerobes, the oxygen diffusion column was constructed. The column produced a range of O_2 concentrations, the effect of which could be simultaneously investigated.

The column was based on the design of Griffin et al. (1967) and depends solely on Fick's law of diffusion of gases to produce a range of gas concentrations. The column was checked before use to see if the O_2 concentration at the various side arms was consistent with the predictions made from Fick's law. To test the gas concentrations the outlet ports of the column were plugged with roll top Subaseal (W. Freeman, Barnsley) caps (25 mm, size 53) and left for 24 h to allow the column to equilibrate. Gas samples (5 ml) were taken from the ports and analysed for O_2, N_2 and CO_2 using gas chromatography (Hall & Dowdell 1981).

The main column (Fig. 2) was made from 3.125-mm-thick stainless steel tube (seam free), 1 m in length and 100 mm outside diameter. Both ends were screw-cut at 20 teeth per inch on a 55° thread form. Three lines of holes, 120° apart around the column and 11 cm apart down the column, were drilled at 90° to the tube wall. Stubs made from 30-mm-outside-diameter stainless steel tube with a 3.125-mm wall thickness, 37.5 mm long, were

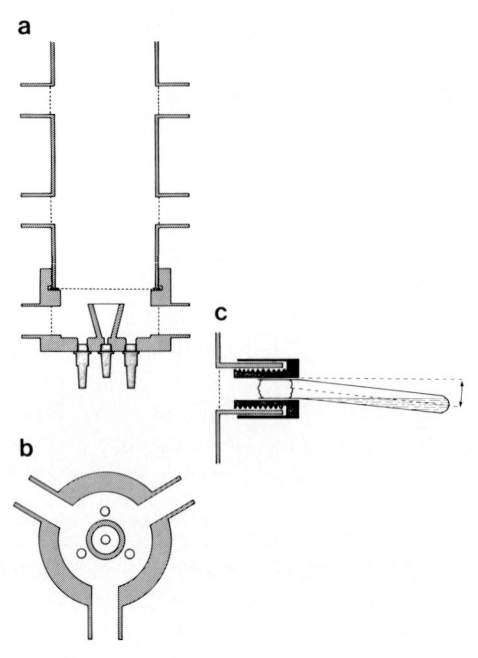

FIG. 2. The oxygen diffusion column. (a) Section; (b) plan of end cap; (c) side arm with attached culture tube.

profiled at one end to fit tightly on the column over every hole. Grooves were cut in the stubs 1.25 mm deep × 1.5 mm wide and 1.5 mm above the top of the profile, to take stainless steel mesh rings which were sprung into each groove.

The end caps were machined from solid stainless steel (125 mm diameter × 100 mm long). The internal shape was critical because various gases had to pass through the centre of each cap, but at the same time, gas from the opposite end had to pass round the cap and be expelled without producing turbulence or pressure spots. To produce an unbroken clear gas flow throughout the column, the bell mouth was machined to a 15° angle in relation to the evacuation area. Threads 15 mm long × 2.5 mm wide, with a 3.125-mm counter bore behind them, were screw-cut producing a shoulder in the end caps. Stainless steel mesh rings were produced and these were sprung into position in the counter bores. To produce a gas-tight seal, the outside diameter of the stainless steel mesh was infilled with soft solder. The main tube was tightened onto the outside of the mesh rings when the caps were applied, producing a gas-tight seal with the soft solder. An interference fit was machined into the end caps for the stubs which were pressed into place to save welding, which may have resulted in distortion of the caps. Taps were fitted to each of the three outlet holes to enable gas flow to be monitored and regulated.

To ensure that gas movement along the column was by diffusion alone, the column was packed with coarse sand (1–1.5 mm), a non-reactive porous material, and maintained at a constant temperature. The column was constructed to avoid leaks which disturb the gradient produced.

By selecting gas concentrations at each end of the column, a predictable linear concentration gradient was maintained along the column. Griffin *et al.* (1967) indicated that if the column was mounted vertically, the dense gas should diffuse upwards to avoid mass flow based on density difference. Using N_2 from a cylinder and the laboratory compressed-air supply it was found that a non-linear relationship was established between distance down the column and O_2 concentration if N_2 was introduced at the top of the column (Fig. 3). A linear relationship was established if the N_2 was introduced at the bottom of the column. It appears that the small difference in temperature between the warmer laboratory air supply and the bottled N_2 overcomes the small difference in molecular weight, and movement of the gases up and down the column is then by diffusion alone.

The column side arms were designed to take a variety of apparatus and the method to be described here was only one of a number of possibilities. Test tubes were attached horizontally to the column using a size 53 (25 mm) Subaseal cap which had had its centre removed using a 15-mm cork borer. If

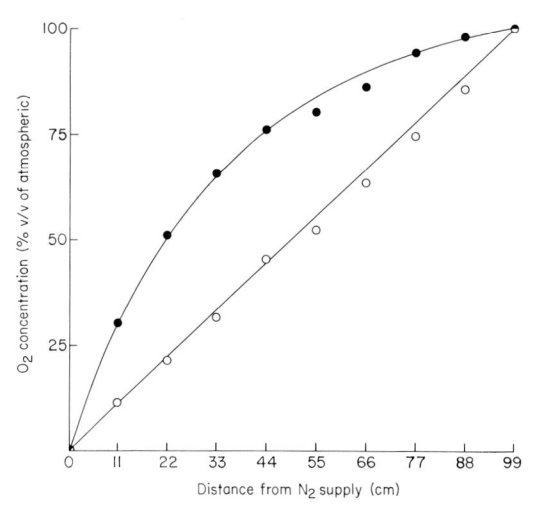

FIG. 3. Oxygen gradient production in the oxygen diffusion column. ●, N_2 introduced at top of column; ○, N_2 introduced at bottom of column.

the caps were greased, the tubes could be pushed into position producing a gas-tight seal. In order that the tubes should hold media, they were bent at 10°, 2 cm from the top (Fig. 2c). The tubes, containing media and cotton wool plugs, were autoclaved, inoculated and pushed into position. This method was far more convenient and less liable to contamination than the method using plastic cuvettes suggested by Griffin et al. (1967).

A Plant Bioassay for Effects of Microbial Degradation Products

Harper & Lynch (1980) investigated the effects of various microorganisms and microbial metabolites on barley seedling growth resulting in both inhibition and stimulation. A close correlation between longest root length and total root length necessitated only measurement of the longest root. Sand was advocated as the growth substratum as it is more homogeneous than soil. However, in the present work soft agar medium for the bioassay was used because less test substance per bioassay was needed. The agar method was also suitable for inoculation with test organisms suspected of having an effect on plant growth. Plant nutrients were incorporated into the agar medium to eliminate any stimulation due to the supply of plant nutrients in the test solution.

Hordeum vulgare (barley) seeds were selected to eliminate damaged, darkened or out-sized seeds and germinated for 36–48 h at 20°C in the dark. In the bioassay, unsterilized seeds were used since under field conditions any effects obtained from the compost would have to compete with the indigenous seed population. However, the method could easily be adapted for use with sterilized seed. The seeds were pre-germinated and then selected for similar root lengths before planting for the bioassay. This had the advantage of eliminating effects on germination and also reducing the variability in the data.

Melted medium (15 ml) containing double strength plant nutrient solution and 1.2% (w/v) agar were added to an equal volume of the fluid (test solution) from decomposed straw, which had been membrane-filtered to remove microorganisms. The medium was contained in a square (12 cm) plastic dish (Sterilin Ltd, Teddington). The dish was sloped such that the agar reached halfway across the dish. Ten seedlings were planted onto the top edge of the agar slope, and the dishes were incubated, slanted, at 20°C in the dark.

After 3 days of incubation the seedlings were carefully separated from the agar medium and the longest root of each was measured. The results were compared statistically with controls, in which 15 ml distilled water were used in place of the test solution. If one plate per test solution was used, the coefficient of variation of longest root measurements was about 12% and real differences of 11% or more could be detected at the 5% significance level by the *t* test. To detect smaller real differences more plates per test solution would be required.

Effects of Degradation Products on Soil Aggregate Stability

Microbial cell surfaces and extracellular products include polysaccharides, which are particularly effective in binding soil particles and producing aggregates which are resistant to destruction by water, such as rainfall. Natural soils can be difficult to study because of the variable effects of the organic matter. Volcanic ash, which equates with a primordial soil without organic matter, is suitable experimental material.

Volcanic ash (10 g) from Mount St. Helens, Washington State, and solution from degraded straw (10 ml) were mixed in a petri dish (9 cm diameter) and dried at 60°C. The dried "flakes" (75 mg) were added to distilled water (70 ml) in a large boiling tube (capacity 90 ml) which was then gently inverted 20 times to disperse the aggregates and challenge their stability. Care needs to be taken to standardise the amount of agitation. The aggregates were allowed to settle out for 1 min and the turbidity was then measured (Lynch 1981).

Five replicate tubes were measured from each petri dish and compared with controls using undecomposed straw suspensions. Lesser turbidity readings indicated greater stability as larger aggregates had settled out.

Discussion

Initial attempts at forcing a cellulolytic/nitrogen-fixing association between *Penicillium corylophilum* and *Clostridium butyricum* with straw as the energy substrate in glass columns produced 11.5 mg N/g straw consumed. Using 7 tonnes/ha straw which contains 3 kg N/tonne would be equivalent to returning 57 kg N/ha to the land (Lynch & Harper 1983). Improved associations and optimisation of growth conditions, particularly control of the aeration, is likely to increase the figure. Microbial degradation products of straw have greatly increased the soil aggregate stability, but inoculations have not yet proved an advantage over the native microflora (J. M. Lynch & L. F. Elliott, unpublished observations). Active plant-growth-promoting and -inhibiting fractions have been detected during straw degradation and are awaiting full characterisation. Considering the elimination of the negative value of straw by treatment (Lynch & Elliott 1983), an on-farm process may be practicable. Inevitably, any process to be economic would have to be low-grade technology and even a silage tower as a "solid substrate fermenter" or "composter" might be too sophisticated. The "clamp" in which plastic sheeting is used to cover silage in a stack could be the practical technology if our laboratory endeavours reach fruition.

Acknowledgements

We thank the mechanical workshop of the Letcombe Laboratory for the construction of the oxygen diffusion column.

References

BRAVERY, A. F. 1968 Microbiological breakdown of cellulose in the presence of alternative carbon sources. *Journal of the Science of Food and Agriculture* **19**, 133–135.

DALTON, H. & POSTGATE, J. R. 1969 Effect of oxygen on growth of *Azotobacter chroococcum* in batch and continuous cultures. *Journal of General Microbiology* **54**, 463–473.

GRIFFIN, D. M., NAIR, N. G. & BAXTER, R. I. 1967 Control of gaseous environment of organisms using diffusion column techniques. *Journal of Experimental Botany* **18**, 518–525.

HALL, K. C. & DOWDELL, R. J. 1981 An isothermal gas chromatographic method for the simultaneous estimation of oxygen, nitrous oxide and carbon dioxide content in gases in soil. *Journal of Chromatographic Science* **19**, 107–111.

HARPER, S. H. T. & LYNCH, J. M. 1980 Measurement of microbial effects on the germination and seedling growth of barley. *New Phytologist* **84**, 473–481.

HARPER, S. H. T. & LYNCH, J. M. 1981 The chemical components and decomposition of wheat straw leaves, internodes and nodes. *Journal of the Science of Food and Agriculture* **32**, 1057–1062.

HUNGATE, R. E. 1950 The anaerobic mesophilic cellulolytic bacteria. *Bacteriological Reviews* **14**, 1–49.

LYNCH, J. M. 1981 Promotion and inhibition of soil aggregate stabilization by micro-organisms. *Journal of General Microbiology* **126**, 371–375.

LYNCH, J. M. & ELLIOTT, L. F. 1983 Minimizing the potential phytotoxicity of wheat straw by microbial degradation. *Soil Biology and Biochemistry.* **15**, 221–222.

LYNCH, J. M. & ELLIOTT, L. F. 1984 Crop residues. In *Crop Establishment. Biological Requirements and Engineering Solutions* ed. Carr, M. K. V. London: Pitmans. (In press.)

LYNCH, J. M. & HARPER, S. H. T. 1974 Fungal growth rate and the formation of ethylene in soil. *Journal of General Microbiology* **85**, 91–96.

LYNCH, J. M. & HARPER, S. H. T. 1983 Straw as a substrate for co-operative nitrogen fixation. *Journal of General Microbiology.* **129**, 251–253.

MILLBANK, J. W. 1969 Nitrogen fixation in molds and yeasts—a reappraisal. *Archiv fur Mikrobiologie* **68**, 32–39.

SKINNER, F. A. 1971 Isolation of soil clostridia. In *Isolation of Anaerobes* eds. Shapton, D. A. & Board, R. G., Society for Applied Bacteriology Technical Series No. 5, pp. 57–80. London: Academic Press.

Methods for the Genetic Manipulation of *Rhizobium*

J. E. BERINGER*

Soil Microbiology Department, Rothamsted Experimental Station, Harpenden, Hertfordshire, UK

J. E. RUIZ SAINZ

Departamento de Microbiología, Facultad de Biología, Universidad de Seville, Seville, Spain

A. W. B. JOHNSTON

John Innes Institute, Norwich, Norfolk, UK

The rhizobia are gram-negative, aerobic, heterotrophic soil bacteria which are able to infect the roots of certain leguminous plants, inducing the formation of root nodules in which they fix atmospheric nitrogen. Nodulated leguminous plants routinely fix between 30 and 300 kg N/ha/year, depending on plant species and environment. The fixed N is available to the nodulated plant, some is available to companion plants and some is left in the field at harvest. Under UK agricultural conditions peas (*Pisum sativum*) and field beans (*Vicia faba*) do not need fertiliser nitrogen because they nodulate efficiently with indigenous rhizobia. Clover nodules fix large amounts of nitrogen and can contribute significant amounts to grass growing in the same sward. There is worldwide interest in increasing the input of biologically fixed N in agriculture. This might be achieved by selecting or breeding "improved" strains of *Rhizobium* and varieties of leguminous crop plants that respond particularly well to nodulation by *Rhizobium*.

There have been major advances in techniques for handling rhizobia as microorganisms for genetic experiments over the last few years. These ad-

*Present address: Unit of Molecular Genetics, The Medical School, University of Bristol, Bristol, UK.

vances have been primarily with fast-growing species of *Rhizobium*. Fortunately the main legumes cultivated in the UK are nodulated by fast-growing rhizobia, e.g., clover *Trifolium* spp. by *Rhizobium trifolii*, peas and field beans by *Rhizobium leguminosarum* and lucerne (*Medicago sativa*) by *Rhizobium meliloti*. In this chapter media and procedures used to handle and manipulate the fast-growing rhizobia are described.

Growth Conditions

Aeration and Temperature

Rhizobia are aerobic and therefore liquid cultures must be shaken or sparged with air. Most strains grow fastest at about 28°C, though true optimal temperatures at which the bacteria are not stressed may be lower. Some isolates can grow at 37°C or more, though most do not grow well above ca. 30°C. The elimination of plasmids from *Rhizobium* species (plasmid curing) has been reported at elevated temperatures (Zurkowski & Lorkiewicz 1978; Dénarié *et al*. 1981), which suggests that strains known to be unstable in their symbiotic properties should be cultured at lower temperatures.

Media

Traditionally rhizobia are cultured on a simple medium containing 0.04% (w/v) yeast extract and 1% (w/v) mannitol (Table 1). Nearly all rhizobia can grow on this medium, including some fastidious slow-growing species. This medium is unsuitable for genetic purposes because the large amount of available sugar causes the production of large amounts of polysaccharides. Also, the concentration of amino acids and other metabolic intermediates is too low to support the growth of a wide range of auxotrophic mutants.

A commonly used rich medium (TY, Table 1) was first described by Beringer (1974). This medium contains adequate nutrients for a range of auxotrophic requirements and does not induce the formation of large amounts of polysaccharides. However, rhizobia are usually sensitive to amino acids in the medium and growth abnormalities, or even the complete absence of growth, are not uncommon in rich media (Sherwood 1972; Beringer 1973). TY medium contains $CaCl_2$ to reduce the effect of amino acids on growth so that most isolates of fast-growing rhizobia grow well on it. Some strains will not grow in TY as a liquid medium, suggesting that under liquid culturing the stresses caused by amino acids are accentuated.

A number of different minimal media are available. Most contain vitamin supplements because many strains of *Rhizobium* grow best when biotin, thiamine, riboflavin or other vitamins are added. The medium SY (Table 1)

TABLE 1

Culture media

Yeast extract–mannitol medium (from Vincent *et al.* 1980)
  Mannitol, 10 g; K$_2$HPO$_4$, 0.5 g; MgSO$_4$·7H$_2$O, 0.2 g; NaCl, 0.1 g; CaCO$_3$, 4.0 g (usually omitted); yeast extract (Difco), 0.4 g; agar (when required), 15 g; distilled water, 1 litre
  Before sterilisation pH is adjusted to 6.8–7.0. Sterilise by autoclaving at 121°C for 15 min
TY medium (from Beringer 1974)
  Tryptone (Difco), 5 g; yeast extract (Difco), 3 g; CaCl$_2$·6H$_2$O, 1.3 g; agar (when required), 15 g; distilled water, 1 litre
  Sterilise by autoclaving at 121°C for 15 min without adjusting pH
SY medium (from Sherwood 1970)
  MgSO$_4$·7H$_2$O, 0.1 g; FeCl$_3$, 0.02 g (added from stock solution prepared by dissolving 20 g FeCl$_3$ in 100 ml *M* HCl and making up to 1 litre with distilled water); CaCl$_2$; 0.04 g; K$_2$HPO$_4$, 0.22 g; agar (when required) 15 g; distilled water, 1 litre
  Before sterilisation pH is adjusted to 6.8. Sterilise by autoclaving at 121°C for 15 min
Supplements
  After sterilisation the following supplements are added per litre: sodium glutamate, 1.1 g; sodium succinate, 135 mg; biotin, 1.0 mg; thiamine hydrochloride, 1.0 mg; D,L-pantothenic acid, 1.0 mg. The supplements are prepared as the following stock solutions: sodium glutamate, 44 g/litre; sodium succinate, 27 g/litre; vitamin mixture (biotin, thiamine hydrochloride and D,L pantothenic acid, 0.1 g each per litre).
  Sterilise by autoclaving at 121°C for 10 min

was developed from a minimal medium described by Sherwood (1970). It has succinate as a carbon source because polysaccharide production is not favoured, and thus the colony size is determined by the volume of bacteria not by the volume of polysaccharide that they can produce.

Media Supplements

To identify auxotrophic mutants and discriminate strains with different nutrient requirements, minimal media must be supplemented with the appropriate nutrients (Table 2).

The antibiotics commonly used for isolating resistant mutants and following the transfer of plasmids carrying drug-resistance genes (R plasmids) and transposons between *Rhizobium* strains are shown in Table 3. Concentrations given are those usually used with *R. leguminosarum*, though it should be noted that different isolates, and especially species, may require different levels.

Storage of Rhizobium

Rhizobia can be maintained for long periods (months) on slopes of yeast extract–mannitol agar medium (Table 1), especially when CaCO$_3$ is added.

TABLE 2

*Supplements for minimal media**

Stock solution[†]	Concentration (g/litre distilled water)
Adenine	1.5
L-Alanine	7.5
L-Arginine mono HCl	7.5
L-Aspartic acid	7.5
Biotin	0.01
L-Cysteine	7.5
L-Glutamic acid	7.5
L-Glycine	7.5
Guanine	1.5
L-Histidine	10
D,L-Homoserine	7.5
L-Isoleucine	7.5
L-Leucine	7.5
L-Lysine	7.5
L-Methionine	7.5
Nicotinamide	0.1
p-Aminobenzoic acid	0.1
Pantothenic acid	0.1
Pyridoxine	0.1
L-Phenylalanine	7.5
L-Proline	7.5
Riboflavin	0.1
L-Serine	7.5
Thiamine hydrochloride	0.1
L-Threonine	7.5
L-Tryptophan	7.5
L-Tyrosine	7.5
Uracil	1.5
L-Valine	7.5

* From Hopwood (1967).
[†] Sterilise by autoclaving at 115°C for 10 min except for tryptophan which is filter sterilised. Add 7.5 ml stock solutions per litre.

Cultures can be maintained for weeks on slopes of minimal medium, but it is inadvisable to store them on TY medium for more than a few days, as viability falls off rapidly. However, routine subculturing and the risk of contamination that is involved makes the maintenance of cultures on agar slopes an unsatisfactory procedure, especially if genetic purity is to be maintained.

Cultures of some, but not all, strains of *Rhizobium* can be maintained for long periods by freezing in 20% (v/v) glycerol at ca. −20°C. This technique

TABLE 3

Antibiotics used to isolate resistant mutants[*]

Antibiotic	Diluent	Stock solution (mg/ml)	Medium (µg/ml)
Chloramphenicol	Ethanol 90% (v/v)	10	50
Erythromycin	Ethanol 90% (v/v)	100	200
Gentamicin	H_2O	10	10
Kanamycin sulphate	H_2O	10	20
Nalidixic acid	NaOH[‡]	10	50
Neomycin sulphate	H_2O	10	20
Rifampicin	Methanol 95% (v/v)	1	20
Spectinomycin	H_2O	10	200
Streptomycin sulphate	H_2O	200	200
Tetracycline HCl	H_2O	5	5
Trimethoprim	Ethanol 50% (v/v)	10	100

Concentration[†]

[*] Most of these data are from Hirsch (1978).
[†] Suitable for many strains of *R. leguminosarum*, *R. phaseoli*, and *R. trifolii*, but not suitable for all strains of all *Rhizobium* species.
[‡] Made up to volume with water after dissolving in 0.02 volumes of 5 *M* NaOH.

provides a useful method for storing cultures after mutagenesis or mating so that they can be sampled repeatedly.

Lyophilisation is the best method for preserving a wide range of bacteria for long periods. It is very successful with rhizobia and should be the method of choice for preserving stock cultures. For *Rhizobium* it is usually adequate to grow cultures on agar slopes and to suspend the growth in a sucrose–peptone mixture [sucrose, 10% (w/v) and peptone, 5% (w/v); Vincent 1970] which is placed in sterile ampoules and lyophilised.

Isolation of Mutants

The isolation of mutants is important for a number of reasons. (1) Auxotrophic or antibiotic-resistance mutations or both provide a clear and unambiguous method for identifying strains. "Marked" strains of this nature are useful for ecological and genetical experiments. (2) A relatively simple way to determine whether a function is important for a particular process is to inactivate that function by mutating the gene(s) involved. (3) The isolation and manipulation of genes depends upon strains of bacteria in which the gene functions can be determined. Mutants lacking only one or a few functions are extremely useful for this purpose.

Spontaneous Mutation

This is the normal procedure for obtaining resistant mutants and other classes of mutation for which a direct selection can be made, e.g. prototrophic revertants of auxotrophic mutants.

The standard procedure is to add ca. 5×10^8 bacteria to a petri dish containing a medium that will allow only the growth of mutant bacteria. The bacteria are usually spread on the surface of agar. Because the great majority of mutants obtained by this procedure have appeared in the population before plating, it is sensible to take bacteria from only one colony (a clone) per culture from each selective medium. If more than one mutant is required, separate cultures, each derived from a small population (ca. 10 or less) of bacteria, should be plated; mutants obtained from these cultures can be assumed to be independent clones.

Induced Mutation

The frequency at which spontaneous mutation occurs is usually too low to enable mutants to be obtained unless a selective medium is available to allow them to grow while preventing the wild-type bacteria from forming colonies. Mutagenic techniques increase mutation rates. In doing this they often cause multiple mutations and thus strains obtained after mutagenesis can contain a number of mutant genes which are often not identified or noticed.

Choice of Mutagen

Not all species, or even strains of microorganisms, are equally sensitive to mutagenic treatments. Thus the choice of mutagen will be determined by the susceptibility of the strain and the willingness of the experimentor to use chemicals, such as N-methyl-N'-nitro-N-nitrosoguanidine (NTG), which are potentially harmful to man.

As with all microorganisms, mutagenic treatments are most efficient if the bacteria are allowed to replicate for a few generations to allow time for the segregation of clones carrying recessive mutations. This would appear to be particularly important with *Rhizobium* because the mucoid nature of the colonies and motility of the bacteria seem to prevent the formation of sectored colonies.

Chemical and Radiation Mutagenesis

Treatment using ultraviolet light, ionising radiation and most chemical mutagens need not be described since they are similar for most bacteria and the only major alteration to a given technique that is usually required is the time of exposure. However, not all strains of species of *Rhizobium* may be susceptible to mutagenesis by a specific mutagen (Walton & Moseley 1981). Mutagenesis by NTG (ca. 200 µg/ml) for *Rhizobium* species tends to be different from established techniques for other bacteria in that it is often performed at pH 7 with actively growing cultures. Of all the mutagens tested, NTG appears to be the most generally effective in terms of the number of mutants produced and the range of strains that are susceptible.

Transposon Mutation

Transposons are defined sequences of DNA that can integrate into other DNA sequences, sometimes at random. Some transposons carry drug-resistance genes. Therefore they can be used as mutagens because if they are introduced into a bacterium, the stable insertion into DNA within the bacterium can be obtained by selecting for drug-resistant bacteria. Transposon mutation is particularly well suited to work with *Rhizobium* because transposon-induced mutations are relatively simple to map genetically, and the DNA can be isolated and introduced into bacteriophages or plasmids (cloned) by standard techniques. The isolation of the appropriate DNA fragment for cloning is facilitated because the *Rhizobium* DNA is covalently linked to the DNA transposon. DNA carrying the transposon can be identified because it will hybridise (bind strongly to) pure radioactive single-stranded transposon DNA (Ruvkun & Ausubel 1981; Scott *et al.* 1982). The most important characteristics of rhizobia for biotechnological purposes are those involved in nodule formation and nitrogen fixation. Therefore it is genes involved in these functions that need to be transferred and manipulated to produce new, industrially important strains. Once a mutant defective in some symbiotic property has been isolated, the gene(s) involved can be isolated and transferred between bacteria by screening for the presence of the transposon (Johnston *et al.* 1978a; Buchanan-Wollaston *et al.* 1980). The most useful transposon for *Rhizobium* work to date has been Tn5, which confers kanamycin + neomycin resistance (Beringer *et al.* 1978; Kondorosi & Johnston 1981; Dénarié *et al.* 1981). Tn*1* and Tn7 (ampicillin and trimethoprim + streptomycin resistance, respectively) have been used with *Rhizobium* and *Agrobacterium* (Hernalsteens *et al.* 1978; Casadesus *et al.* 1980), but expression of sufficiently high levels of resistance and frequencies of transposition have been difficult to detect in some strains of *Rhizobium* species.

Plasmids can be transferred between *Rhizobium* and *Escherichia coli* and therefore it is possible to induce mutations of *Rhizobium* DNA in *E. coli* using techniques developed for this well-studied microorganism. Ruvkun & Ausubel (1981) have described a procedure for "directing mutation" in *Rhizobium* DNA by cloning *Rhizobium* DNA, mutating it in *E. coli*, returning the mutagenised DNA to *Rhizobium* and then forcing the introduced DNA to recombine with the corresponding DNA in the host replacing the wild-type genes. Once DNA carrying "interesting" genes is identified, this procedure can be used to induce mutations along the whole sequence of DNA and can be extended along flanking sequences of DNA when they have been identified. Theoretically it is possible to induce mutations sufficiently close to each other to have a high probability of identifying all genes which are linked and whose function can be identified.

Gene Transfer between *Rhizobium* Strains

DNA can be transferred between bacteria by transformation, transduction, conjugation and protoplast fusion. The first three procedures have been reported for most *Rhizobium* species, but gene transfer by protoplast fusion has not been so reported.

Transformation

Transformation (the addition of pure DNA to bacteria) is particularly useful for introducing plasmids into bacteria and is widely used with other bacterial species to re-introduce DNA which has been cloned into plasmids. To date, the strains of *Rhizobium* which have been genetically analysed in most detail have not been easy to transform, though Bullerjahn & Benzinger (1982) have reported that *R. leguminosarum* can be transformed at very low frequency with plasmid DNA using a procedure developed for the transformation of *Agrobacterium*. Most strains of *Rhizobium* appear to be able to conjugate with *E. coli* and exchange a wide range of plasmids. Therefore, to clone *Rhizobium* DNA it is usual to use *E. coli* which is transformable with plasmid DNA (Ruvkun & Ausubel 1981; Scott *et al.* 1982). However, because transformation has not been useful for the strains of *Rhizobium* used for a genetic analysis of symbiosis (Beringer 1980), techniques and media used are not listed in this chapter. References to papers on transformation in *Rhizobium* are given in reviews by Balassa (1963), Beringer *et al.* (1980) and Kondorosi & Johnston (1981).

Transduction

Transduction (the transfer of DNA within bacteriophages) has been used for fine-structure mapping of *R. leguminosarum* and *R. meliloti* (Buchanan-Wollaston 1979; Casadesus & Olivares 1979a) and to transfer plasmid DNA between rhizobia (Buchanan-Wollaston *et al.* 1980). The basic methods for phage preparation and infection of recipient bacteria are similar to most other published transduction procedures. For both *R. meliloti* and *R. leguminosarum*, virulent phage has been used and thus procedures were needed to inactivate phage to prevent excessive killing of the recipient population. For *R. meliloti*, antiserum was used to inactivate the phages (Casadesus & Olivares 1979b) before a non-virulent mutant was isolated (Casadesus & Olivares 1979a). With *R. leguminosarum*, the virulence of transducing phage RL38 was reduced by irradiating the phage with UV light (dose approximately 2 J/m^2/s for 5 min) to reduce the efficiency of plating approximately 10^5-fold (Buchanan-Wollaston 1979).

A "standard" medium recipe will not be given because different bacteriophages have different requirements for ions and different sensitivities to the ionic strength of the medium. For propagation of bacteriophage the medium is important. For example, with phage RL38 grown on *R. leguminosarum*, succinate was not a suitable carbon source in the minimal medium SY (Table 1) and was replaced by glucose, 0.05% (w/v) (Buchanan-Wollaston 1979).

Conjugation

Conjugation (the transfer of DNA between bacteria during pair formation) has been the most useful procedure for developing genetic studies of *Rhizobium*. In nearly all the well-documented conjugation systems in bacteria, conjugation is mediated by plasmids. For *Rhizobium* species chromosome mapping has been based primarily on chromosome mobilisation mediated by drug-resistance (R) plasmids originally identified in *Pseudomonas aeruginosa* (Beringer 1974; Meade & Signer 1977; Kondorosi *et al.* 1977; Casadesus & Olivares 1979a).

The rhizobia appear to be "typical" gram-negative bacteria in their ability to form mating pairs and exchange plasmids with other gram-negative bacteria. Of major importance to the development of *Rhizobium* genetics has been the ease with which plasmids of the P1 incompatibility group can be transferred between *Rhizobium* and *E. coli*. Most strains of *Rhizobium* can transfer plasmids such as RP4 to *E. coli* at a frequency of ca. 10^{-5}–10^{-2} per recipient; similar frequencies are observed for transfer from *E. coli* to *Rhizobium* (Datta *et al.* 1971; Beringer 1974; Mead & Signer 1977; Kondorosi *et al.* 1977).

The intergeneric transfer of plasmids has been useful for cloning experiments and for looking at the expression of *Rhizobium* genes in different genetic backgrounds. For example, Johnston *et al.* (1978b) transferred plasmids carrying *Rhizobium* genes for tryptophan biosynthesis to tryptophan-requiring mutants of *E. coli* and *P. aeruginosa* to look for gene expression in these hosts and to make tentative assignment for the function of the different *Rhizobium* genes. Lack of expression of *Rhizobium* genes in *E. coli* indicated that this was unlikely to be a suitable host to use to screen cloned *Rhizobium* DNA for gene expression.

Mating pairs induced by P1-group plasmids tend to be fragile and plasmid (or chromosome) transfer frequencies in liquid media are usually low (Jacob *et al.* 1976). Therefore conjugation experiments involving *Rhizobium* are invariably done on surfaces; either on the surface of agar in petri dishes or, more commonly, by filtering the bacteria onto the surface of a nitrocellulose membrane (usually 0.45-μm pore size) and incubating the membrane on a nutrient medium (e.g. TY) in a petri dish with the bacteria on the top surface of the membrane. Usually the bacteria are left to mate overnight. With *Rhizobium*, cultures grown on agar slopes approaching the stationary phase of growth are usually used. For *E. coli*, actively growing donors are needed to transfer plasmids to *Rhizobium*; growing or non-growing cultures can be used as recipients in crosses. Usually equal volumes of donor and recipient bacteria are mixed together each containing ca. 5×10^8 cells though, especially when *E. coli* is used as the donor species, it is preferable to have a ratio of donors to recipients of 2:1 to 10:1.

Rolfe & Holloway (1966) reported that when *P. aeruginosa* recipient cultures were incubated overnight at 43°C, they lost their ability to restrict incoming DNA. This procedure was used by Johnston *et al.* (1978a) to increase the frequency of transfer of plasmids carrying chromosomal DNA (R primes) from *R. leguminosarum* to *P. aeruginosa* about a thousandfold. Incubation of *R. leguminosarum* strain 300 overnight at 37°C (a temperature at which it does not grow) increased the frequency at which the plasmid RP4::Mu was inherited from *E. coli* by at least 10-fold (J. E. Beringer, unpublished observations), suggesting that a similar, though less dramatic, effect on restriction enzymes may occur in *Rhizobium*.

The following procedure is a "standard" method for mating rhizobia which can be adapted readily for other genera by altering the media, temperatures and times.

(1) Grow donor and recipient bacteria in TY slopes for 2 days at 28°C.
(2) Wash bacteria off slopes in ca. 5 ml sterile distilled water.
(3) Add 1 ml donor culture to 0.5 ml recipient culture and mix by shaking.
(4) Filter mixture through a nitrocellulose membrane (0.45-μm pore size).
(5) Remove membrane and place on a nutrient medium (e.g. TY) in a petri dish with the bacteria on the top surface of the membrane.
(6) Incubate overnight at 28°C.
(7) Lift off membrane and suspend the bacteria in about 5 ml sterile distilled water.
(8) Plate on suitable selective media.

Physical Analysis of the Genes that Determine Symbiotic Properties in *Rhizobium*

For a fine scale analysis of "symbiotic" genes, it is necessary to study the relevant DNA at a physical level. This requirement is made all the more important since the genes for nodulation and nitrogen fixation are not on the chromosome but on large indigenous plasmids (see reviews by Beringer *et al.* 1980; Dénarié *et al.* 1981). Although it is possible to transduce, transform and conjugate all or parts of plasmids, such methods are not suited to the mapping of the plasmid-borne genes.

A description of the techniques that are of specific relevance to the analysis of *Rhizobium* DNA follows. Techniques concerned with DNA cloning in *E. coli*, e.g. mapping with restriction endonucleases, DNA sequencing etc., have all been used for dissecting the symbiotic genes of *Rhizobium*, but the methods used are standard and are of general application to the analysis of DNA from any source.

Plasmid Isolation

Many of the techniques used to isolate relatively small plasmids, e.g. from *E. coli*, are not capable of isolating the very large symbiotic plasmids of *Rhizobium* which have molecular weights $> 200 \times 10^6$. Plasmids of such large size are fragile; if a single break occurs in one of the strands, it is impossible to distinguish plasmid from chromosomal DNA. This is because such "open circle" (OC) molecules do not migrate in agarose gels and because the OC form of plasmids has the same buoyant density as linear DNA in caesium chloride/ethidium bromide gradients. Therefore a crucial feature of all the techniques used to isolate large plasmids is that the lysis must be gentle and that subsequent handling of the lysate should be done with care.

Analytical Procedures

For many purposes it is sufficient to estimate the numbers and sizes of the plasmids in a particular strain of *Rhizobium*. Such information can be obtained by following electrophoresis of the lysate in agarose gels and estimating the sizes of the plasmids by their mobility. Although several publications have described various methods of lysis and of ways of enriching for plasmid DNA in the lysate, only one technique which has the virtues of being simple, quick and sensitive will be described.

The method first used by Eckhardt (1978) for the isolation of plasmids in *E. coli* involves the direct lysis of the cells in the agarose gels before electrophoresis of the lysate. This method keeps the manipulation of the DNA to the minimum.

Rosenberg et al. (1982) described the way in which the technique can be used to isolate very large plasmids from R. meliloti.

(1) Grow cells in TY medium to a density of 5×10^7/ml.
(2) Wash cells in 0.5 ml 0.1% (w/v) Sarkosyl in TE8 buffer (0.05-M Tris; 0.022-M EDTA; pH 8.0) to remove some of the exopolysaccharide and facilitate lysis.
(3) Suspend cells in 40 µl lysis mix, [i.e. lysozyme, 7000 units/ml; RNase, 1 unit/ml; 0.05% (w/v) bromophenol blue; 20% (w/v) Ficoll 400 in Tris-borate buffer (89-mM Tris; 2.5-mM Na$_2$EDTA; 89-mM boric acid; pH 8.2)].
(4) Transfer the suspension to the slot of a 0.7% (w/v) agarose gel in Tris-borate buffer (gel dimensions 3 mm × 140 mm × 175 mm) and leave for 10 min.
(5) Add 40 µl 0.2% (w/v) solution of sodium dodecyl sulphate (SDS) in Tris-borate buffer containing 10% (w/v) Ficoll 400. Stir the mixture gently.
(6) Seal the slots with agarose.
(7) Do electrophoresis at 8 mA for 1 h and 40 mA for 3 h.
(8) Stain the gel with ethidium bromide (0.5 µg/ml) and visualise with UV light.

This technique can be adapted for use in horizontal gels (N. J. Brewin & N. Dibb, personal communication) and has been found to be suitable for the visualisation of large plasmids in R. phaseoli, R. trifolii and R. leguminosarum. The gel is of the same constitution as that described above with dimensions 5 mm × 140 mm × 140 mm. Two parallel sets of wells are made, one set with a comb whose teeth have a width of 1 mm and the other with wider teeth (3 mm). The wide wells are placed 5 mm directly behind the thin ones and filled with 0.4% (w/v) agarose in Tris-borate buffer containing (w/v) SDS and 0.05% (w/v) bromophenol blue. The thin wells are loaded with 30 µl cell suspension and the gel is covered with Tris-borate buffer. The conditions for electrophoresis are as described above. This regime causes SDS to migrate to the wells containing the cells.

Cantrell et al. (1982) isolated plasmids from Rhizobium japonicum using the Eckhardt technique, but they preferred to use a method initially described by Currier & Nester (1976) for the isolation of large plasmids in Agrobacterium tumefaciens. This technique is fully described by Cantrell et al. (1982).

Preparative Isolation of Plasmids

Rosenberg et al. (1982) described a method for the isolation of large plasmids in R. meliloti.

(1) Grow 200 ml culture to a density of 5×10^8/ml in TY medium.
(2) Centrifuge the culture and resuspend 16 ml TE8 buffer containing 0.1% (w/v) Sarkosyl.
(3) Centrifuge and resuspend pellet in 6 ml 25% (w/v) sucrose in TE8 buffer.
(4) Divide mixture into 4×1.5 ml aliquots and to each add 0.5 ml lysozyme solution (2mg/ml).
(5) Immediately add 16 ml TE8 buffer and leave at room temperature for 15 min.
(6) Add 1.24 ml ethidium bromide (10 mg/ml) followed by 2 ml 10% (w/v) solution of Sarkosyl in TE8 buffer.
(7) Mix the lysate, add 20.2 g CsCl and dissolve.
(8) Divide each aliquot into two polycarbonate tubes and centrifuge for 16 h at 60,000 g.
(9) Withdraw the plasmid DNA band from each tube, pool the samples and recentrifuge for 4 h at 265,000 g.

Blotting of Plasmid DNA to Nitrocellulose Filters

It is possible to transfer undigested plasmid DNA from agarose gels to nitrocellulose filters (Southern blotting) as a prelude to hybridising the plasmid DNA to radioactively labelled probe DNA. Such operations have been used with the large plasmids of *Rhizobium* and have been used to show that particular plasmids contain the structural genes for nitrogenase (Hombrecher *et al.* 1981).

In order to obtain efficient transfer of the large plasmids from the agarose to the nitrocellulose it is necessary to break the plasmid DNA into small fragments. Hombrecher *et al.* (1981) did this by irradiating the gel with shortwave UV light (Mineralight Lamp, Ultra-violet Products) from a height of 5 cm for 15 min (2.7 kJ/m). The gel is then immersed for 15 min in 0.25% (v/v) HCl (Wahl *et al.* 1979) and rinsed thoroughly in distilled water. Subsequent transfer of the DNA to nitrocellulose can be done essentially as described by Southern (1975).

Vector Systems for Cloning

Rhizobium DNA can be cloned into *E. coli* vectors such as bacteriophage λ, small multicopy plasmids and cosmids. These vectors can be maintained in *E. coli*, but they cannot exist in *Rhizobium* and thus studies of the function of the cloned genes cannot be done in *Rhizobium*. In consequence a number of vectors have been developed which can exist in both *E. coli* and *Rhizobium* and which can be transferred between the two genera. These are based on the wide host range plasmids of the P and Q incompatibility groups which originated in *Pseudomonas* and members of the Enterobacteriaceae.

P-Group Plasmids

Ditta *et al.* (1980) constructed plasmid pRK290 which was derived from plasmid RK2. Plasmid pRK290 is 20 kb in size and sites for cloning with the restriction enzymes *Eco*RI and *Bgl*II. Although it has lost the transfer genes of RK2, it can be mobilised from *E. coli* to *Rhizobium* by other P-group plasmids. Ditta *et al.* (1980) have constructed a gene bank of *R. meliloti* using pRK290 as vector. Long *et al.* (1982) modified this plasmid by cloning the *cos* region of λ into pRK290 to form the cosmid LAFRI. This allows the packaging of relatively large (20 kb) regions of DNA in the cosmid, which can be transferred to *Rhizobium*.

Q-Group Plasmids

Several derivatives of the Q-group plasmid RSF1010 have been constructed. A list given by Bagdasarian *et al.* (1981) includes cosmids and vectors suitable for cloning with several different restriction enzymes.

Conclusion

The genetics of fast-growing *Rhizobium* species is well understood and there are few limitations to manipulating rhizobia genetically. It is relatively simple to transfer genes between *Rhizobium* and *E. coli*, and therefore, "modern" methods for gene cloning and manipulating DNA can be used with *E. coli* and standard *E. coli* plasmids or phages.

With *Rhizobium*, the main interest in genetic studies must be in its role in symbiotic N_2 fixation and how genetics can be used to improve the yield of crop plants. However, our present understanding of the characteristics of *Rhizobium* that are important for optimal nodule formation and nitrogen fixation is limited. The range of genetic techniques now available will facilitate studies of symbiotic characteristics and the manipulation of strains to produce rhizobia with improved properties.

References

BAGDASARIAN, M., LURZ, R., RUCKERT, B., FRANKLIN, F. C. H., BAGDASARIAN, M. J., FREY, J. & TIMMIS, K. N. 1981 Specific purpose plasmid cloning vectors. II Broad host range high copy number RSF 1010 - derived vectors and a host-vector system for gene cloning in *Pseudomonas. Gene* **16**, 237–247.

BALASSA, G. 1963 Genetic transformation of *Rhizobium:* a review of the work of *R. Balassa. Bacteriological Reviews* **27**, 228–241.

BERINGER, J. E. 1973 Genetic studies with *Rhizobium leguminosarum.* Ph.D. Thesis, University of East Anglia, Norwich.

BERINGER, J. E. 1974 R factor transfer in *Rhizobium leguminosarum. Journal of General Microbiology* **84**, 188–198.

BERINGER, J. E. 1980 The development of *Rhizobium* genetics. The fourth Fleming lecture. *Journal of General Microbiology* **116**, 1–7.

BERINGER, J. E., BEYNON, J. L., BUCHANAN-WOLLASTON, A. V. & JOHNSTON, A. W. B. 1978 Transfer of the drug resistance transposon Tn5 to *Rhizobium. Nature, London* **276**, 633–634.

BERINGER, J. E., BREWIN, N. J. & JOHNSTON, A. W. B. 1980 The genetic analysis of *Rhizobium* in relation to symbiotic nitrogen fixation. *Heredity* **45**, 161–186.

BUCHANAN-WOLLASTON, A. V. 1979 Generalized transduction in *Rhizobium leguminosarum. Journal of General Microbiology* **112**, 135–142.

BUCHANAN-WOLLASTON, A. V., BERINGER, J. E., BREWIN, N. J., HIRSCH, P. R. & JOHNSTON, A. W. B. 1980 Isolation of symbiotically defective mutants in *Rhizobium leguminosarum* by insertion of the transposon Tn5 into a transmissible plasmid. *Molecular and General Genetics* **178**, 185–190.

BULLERJAHN, G. S. & BENZINGER, R. H. 1982 Genetic transformation of *Rhizobium leguminosarum* by plasmid DNA. *Journal of Bacteriology* **150**, 421–424.

CANTRELL, M. A., HICKOK, R. E. & EVANS, H. J. 1982 Identification and characterization of plasmids in hydrogen uptake positive and hydrogen uptake negative strains of *Rhizobium japonicum. Archives of Microbiology* **131**, 102–106.

CASADESUS, J. & OLIVARES, J. 1979a Rough and fine linkage mapping of the *Rhizobium meliloti* chromosome. *Molecular and General Genetics* **174**, 203–209.

CASADESUS, J. & OLIVARES, J. 1979b General transduction in *Rhizobium meliloti* by a thermosensitive mutant of bacteriophage DF2. *Journal of Bacteriology* **139**, 316–317.

CASADESUS, J., IANEZ, E. & OLIVARES, J. 1980 Transposition of Tn1 to the *Rhizobium meliloti* genome. *Molecular and General Genetics* **180**, 405–410.

CURRIER, T. C. & NESTER, E. W. 1976 Isolation of covalently closed circular DNA of high molecular weight from bacteria. *Analytical Biochemistry* **76**, 431–441.

DATTA, N., HEDGES, R. W., SHAW, E. J., SYKES, R. B. & RICHMOND, M. H. 1971 Properties of an R factor from *Pseudomonas aeruginosa. Journal of Bacteriology* **108**, 1244–1249.

DÉNARIÉ, J., BOISTARD, P., CASSE-DELBART, F., ATHERLY, A. G., BERRY, J. O. & RUSSEL, P. 1981 Indigenous plasmids of *Rhizobium. International Review of Cytology, Supplement No. 13,* pp. 225–240.

DITTA, G., STANFIELD, S., CORBIN, D. & HELINSKI, D. R. 1980 Broad host range DNA cloning system for gram-negative bacteria: construction of a gene bank of *Rhizobium meliloti. Proceedings of the National Academy of Sciences of the United States of America* **77**, 7347–7351.

ECKHARDT, T. 1978 A rapid method for the identification of plasmid deoxy-ribonucleic acid in bacteria. *Plasmid* **1**, 584–588.

HERNALSTEENS, J. P., DEGREVE, H., VAN MONTAGU, M. & SCHELL, J. 1978 Mutagenesis by insertion of the drug resistance transposon Tn7 applied to the Ti plasmid of *Agrobacterium tumefaciens. Plasmid* **1**, 218–225.

HIRSCH, P. R. 1978 Studies of plasmids in *Rhizobium leguminosarum.* Ph.D. Thesis, University of East Anglia, Norwich.

HOMBRECHER, G., BREWIN, N. J. & JOHNSTON, A. W. B. 1981 Linkage of genes for nitrogenase and nodulation ability on plasmids in *Rhizobium leguminosarum* and *R. trifolii. Molecular and General Genetics* **182**, 133–136.

Hopwood, D. A. 1967 Genetic analysis and genome structure in *Streptomyces coelicolor*. *Bacteriological Reviews* **31**, 373–403.

Jacob, A. E., Cresswell, J. M., Hedges, R. W., Coetzee, J. N. & Beringer, J. E. 1976 Properties of plasmids constructed by the *in vitro* insertion of DNA from *Rhizobium leguminosarum* or *Proteus mirabilis* into RP4. *Molecular and General Genetics* **147**, 315–323.

Johnston, A. W. B., Beynon, J. L., Buchanan-Wollaston, A. V., Setchell, S. M., Hirsch, P. R. & Beringer, J. E. 1978a High frequency transfer of nodulating ability between strains and species of *Rhizobium*. *Nature, London* **276**, 634–636.

Johnston, A. W. B., Bibb, M. J. & Beringer, J. E. 1978b Tryptophan genes in *Rhizobium*—their organization and their transfer to other bacterial genera. *Molecular and General Genetics* **165**, 323–330.

Kondorosi, A. & Johnston, A. W. B. 1981 The genetics of *Rhizobium*. *International Review of Cytology, Supplement No. 13*, pp. 191–224.

Kondorosi, A., Kiss, G. B., Forrai, T., Vincze, E. & Banfalvi, Z. 1977 Circular linkage map of *Rhizobium meliloti* chromosome. *Nature, London* **268**, 525–527.

Long, S. R., Buikema, W. J. & Ausubel, F. M. 1982 Cloning of *Rhizobium meliloti* nodulation genes by direct complementation of Nod$^-$ mutants. *Nature, London* **298**, 485–488.

Meade, H. M. & Signer, E. R. 1977 Genetic mapping of *Rhizobium meliloti*. *Proceedings of the National Academy of Sciences of the United States of America* **74**, 2076–2078.

Rolfe, B. & Holloway, B. W. 1966 Alteration in host specificity of bacterial deoxyribonucleic acid after an increase in growth temperature of *Pseudomonas aeruginosa*. *Journal of Bacteriology* **92**, 43–48.

Rosenberg, C., Casse-Delbart, F., Dusha, I., David, M. & Boucher, C. 1982 Megaplasmids in the plant-associated bacteria *Rhizobium meliloti* and *Pseudomonas solanacearum*. *Journal of Bacteriology* **150**, 402, 406.

Ruvkun, G. B. & Ausubel, F. M. 1981 A general method for site directed mutagenesis in prokaryotes: construction of mutations in symbiotic nitrogen fixation genes of *Rhizobium meliloti*. *Nature, London* **289**, 85–88.

Scott, K. F., Hughes, J. E., Gresshoff, P. M., Beringer, J. E., Rolfe, B. G. & Shine, J. 1982 Molecular cloning of *Rhizobium trifolii* genes involved in symbiotic nitrogen fixation. *Journal of Molecular and Applied Genetics*. **1**, 315–326.

Sherwood, M. T. 1970 Improved synthetic medium for the growth of *Rhizobium*. *Journal of Applied Bacteriology* **33**, 708–713.

Sherwood, M. T. 1972 Inhibition of *Rhizobium trifolii* by yeast extracts or glycine is prevented by calcium. *Journal of General Microbiology* **71**, 351–358.

Southern, E. M. 1975 Detection of specific sequences among DNA fragments separated by gel electrophoresis. *Journal of Molecular Biology* **98**, 503–517.

Vincent, J. M. 1970 A manual for the practical study of the root nodule bacteria. *I.B.P. Handbook, No. 15*. Oxford: Blackwell.

Vincent, J. M., Nutman, P. S. & Skinner, F. A. 1980 The identification and classification of *Rhizobium* In *Identification Methods for Microbiologists*, eds. Skinner, F. A. & Lovelock, D. W., 2nd edn., pp. 49–69. London: Academic Press.

Wahl, G. M., Stern, M. & Stark, G. R. 1979 Efficient transfer of large DNA fragments from agarose gels to diazobenzyloxymethyl-paper and rapid hybridization by using dextron sulfate. *Proceedings of the National Academy of Sciences of the United States of America* **76**, 3683–3687.

Walton, D. A. & Moseley, B. E. B. 1981 Induced mutagenesis in *Rhizobium trifolii*. *Journal of General Microbiology* **124**, 191–195.

Zurkowski, W. & Lorkiewicz, Z. 1978 Effective method for the isolation of non-nodulating mutants of *Rhizobium trifolii*. *Genetic Research* **32**, 311–314.

Methods for Evaluating and Manipulating Vesicular–Arbuscular Mycorrhiza

D. S. Hayman

Soil Microbiology Department, Rothamsted Experimental Station, Harpenden, Hertfordshire, UK

Vesicular–arbuscular (VA) mycorrhizas are mutualistic symbiotic associations formed between many plant roots and certain phycomycetous fungi. They benefit plants by enhancing the uptake of nutrients, especially phosphate. These mycorrhizas are found in most soils, and the fungal symbionts can account for much of the soil microbial biomass. However, the vesicular–arbuscular mycorrhizal (VAM) fungi are not isolated with other soil micro-organisms on standard soil dilution plates because no culture medium is yet known on which they can make more than minimal growth in the absence of a living root. Consequently, alternative methods for handling them have been developed. Current techniques to monitor the distribution and activity of VAM populations in soil and their responses to various treatments are based primarily on extracting and quantifying spores, infected roots and other propagules.

Different species and strains of VAM fungi can be established in single cultures on suitable stock plants ("pot cultures") and their symbiotic efficiency evaluated. Sufficient inoculum is readily produced to screen VAM endophytes in plants grown in pots but not in large-scale field trials. Hence the relative benefits of field inoculation and of manipulating either selected VAM isolates or the indigenous VAM populations must be evaluated carefully. The identification of optimum fungus–plant–soil combinations and the development of an improved VA mycorrhizal biotechnology in specific agricultural situations are currently receiving much attention (Abbott & Robson 1982; Hayman 1982a; Tinker 1982).

Occurrence

Generally it is best to look for the large resting spores of VAM fungi in moderately fertile, arable soils growing highly susceptible crops and not treated with pesticides or large amounts of nitrogen and phosphorus fertilisers. Vesicular–arbuscular mycorrhizal root infections, on the other hand, are common in uncultivated woodland and grassland soils as well as in many arable soils. Both spores and infections are rare in recently disturbed soils such as mine spoils, in areas subject to waterlogging and in intensively cultivated soils (Mosse *et al.* 1981; Hayman 1982b).

Most plant species examined so far possess VA mycorrhizal roots. Exceptions are those in the primarily ectomycorrhizal families (e.g. Pinaceae, Fagaceae, Betulaceae), specific non-VAM endomycorrhizal groups (Orchidaceae, Ericales) and certain families which have little or no mycorrhiza of any type (e.g. Cruciferae, Chenopodiaceae; Gerdemann 1968). Some plants typically have little infection, e.g. ryegrass, whereas others are often densely infected, e.g. clover, lucerne, peas, maize, onions, citrus.

Vesicular–arbuscular mycorrhizal populations in soil decrease markedly below the top 15 cm (Redhead 1977). Their horizontal distribution varies greatly, necessitating the bulking of several sub-samples to provide material that is reasonably representative of a particular site. Samples should be collected from as close to the plant base as possible.

Identification

The VAM fungi are currently identified and classified according to the morphology of their large, thick-walled resting spores. It is not possible to distinguish between most VAM species by the anatomy of their infections in plant roots.

VAM Resting Spores

Most spores produced by VAM fungi are asexually formed chlamydospores. Some others are called azygospores because they possess a bulbous attachment, interpreted as a vestigial gametangium. Although they are placed in the Endogonaceae (order Mucorales), not all genera in this family form mycorrhizas and one, *Endogone*, forms ectomycorrhiza. There are several keys and published descriptions of VAM genera and species which should be consulted for correct identification from spore material (Mosse & Bowen 1968; Gerdemann & Trappe 1974; Becker & Hall 1976; Hall 1977; Hall & Fish 1979; Nicolson & Schenck 1979; Schenck & Smith 1982; Trappe 1982). One of the most recent (Schenck & Smith 1982) lists 38 species of Endogonaceae comprising 15 species of *Glomus* (chlamydospores with simple stalks), 12 species of *Gigaspora* (azygospores with bulbous attachments), 6 species of *Acaulospora* (sessile azygospores), 3 species of *Sclerocystis* (chlamydospores arranged symmetrically in sporocarps), and two genera, *Complexipes* and *Entrophospora*, which are not yet known to form VA mycorrhizas. Other non-mycorrhizal genera in the Endogonaceae are *Modicella* and *Glaziella*.

Examples of VAM species and genera from the Rothamsted collection are illustrated in Fig. 1.

The main diagnostic features used to identify spore types and morphological species are as follows:

(1) *Size*. VAM resting spores are the largest fungal spores known. The majority are between 100 and 250 μm in diameter, although some are nearer 50 μm and others are over 300 μm.
(2) *Shape*. Usually spherical or oval.
(3) *Colour*. Many spores are yellow-brown, others are reddish brown, greyish brown, honey, amber, pale greenish yellow or white. Colour must be determined under a stereomicroscope with incident light because under transmitted light the dense contents of some spores gives them a misleadingly dark appearance.
(4) *Cytoplasmic structure*. This is best examined with transmitted light. Mosse & Bowen (1968) differentiated between spores with reticulate (small vacuoles in network of cytoplasmic strands) and vacuolate cytoplasm (large vacuoles which coalesce with age).
(5) *Stalk attachment*. Differentiates between the three major genera: *Glomus* (simple), *Gigaspora* (bulbous) and *Acaulospora* (none).
(6) *Wall thickness and structure*. Commonly there is a thick, chitinous, brittle, coloured wall with a thin, white wall outside or inside it. In some spores the wall is simple, in others there are several layers.
(7) *Method of germination*. Either one germ tube through the main stalk (vacuolate types) or several germ tubes through the spore wall (reticulate types).
(8) *Sporocarps*. Formed in a few species only, ranging from loose clusters of spores with or without a surrounding peridium (*Glomus*) to very definite structures with spores arranged symmetrically around a central plexus (*Sclerocystis*).
(9) *Secondary spores*. These are smaller than the resting spores and have thin walls. In *Glomus* species they resemble vesicles, in *Gigaspora* they are often spiny and arranged spirally.

None of these criteria differentiates between physiological strains within any one morphological species. Strain differences have been shown in a few

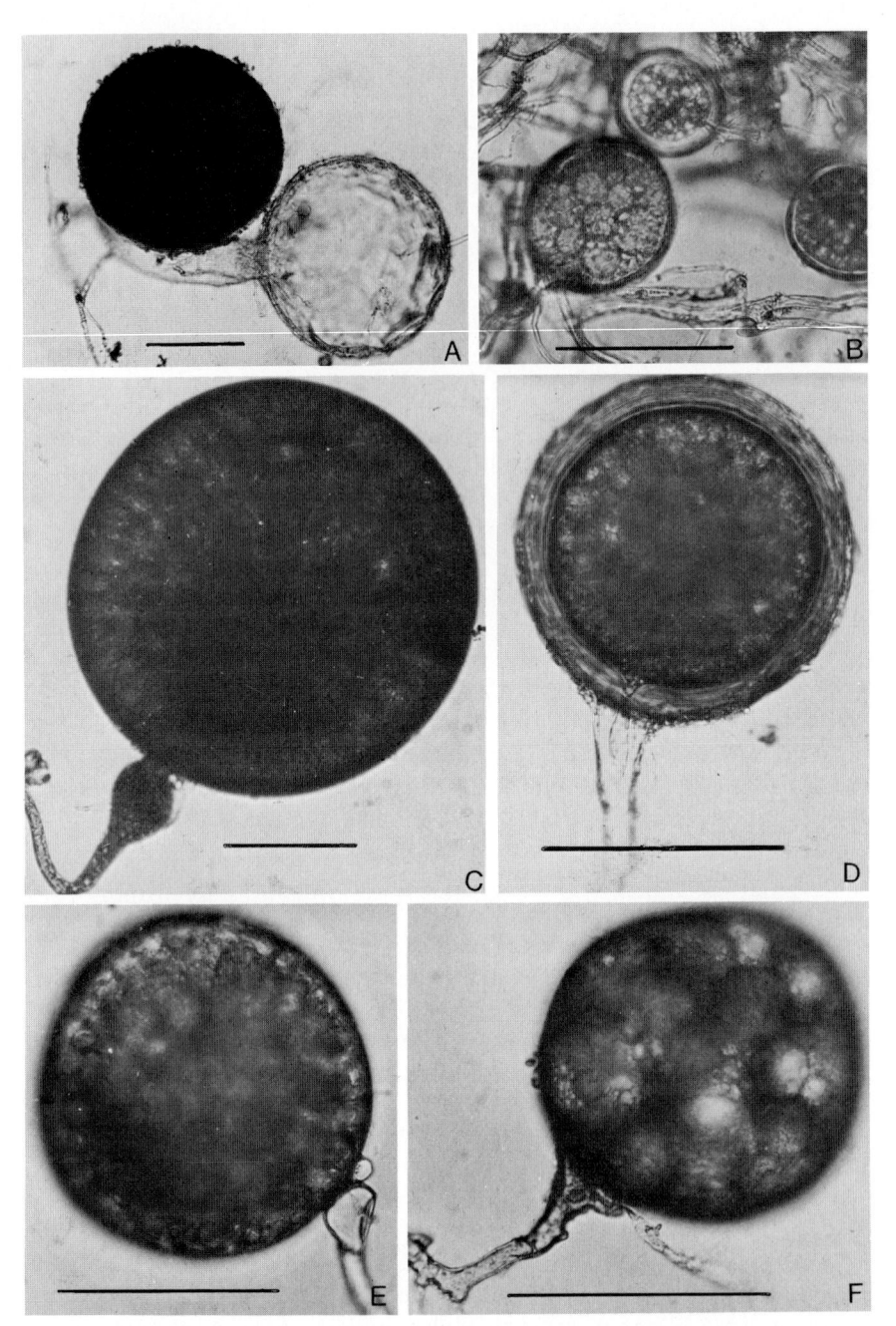

FIG. 1. Spores of vesicular–arbuscular mycorrhizal fungi. (A) *Acaulospora laevis* isolate "honey-coloured sessile"; resting spore (dark) attached to stalk of empty mother spore. (B) *Glomus fasciculatus* isolate E3. (C) *Gigaspora margarita*. (D) *Glomus macrocarpus*. (E) *Gigaspora* sp. isolate "bulbous reticulate". (F) *Glomus mosseae* isolate "yellow vacuolate". Bar = 100 μm.

plant inoculation studies but are difficult to demonstrate consistently because these endophytes cannot be cultured.

VAM Root Infections

The roots of some plants freshly washed from soil are yellow when infected with VAM fungi, e.g. maize, onion and tomato. The colour disappears after exposure to light. In addition, VAM roots tend to be more difficult to wash than non-mycorrhizal roots because their attached hyphae grow into adjacent particles of organic matter. There are no other obvious signs of VAM infection, unlike ectomycorrhizal roots of pine and beech, for example, which are shorter and fatter than non-mycorrhizal roots and branch dichotomously. Therefore differential staining is needed to prove the presence of a VAM infection.

Probably the most commonly used method is that of Phillips & Hayman (1970), which may be summarised as follows:

(1) Wash roots from soil and process either fresh or after storing in FAA fixative (formalin, 13 ml; glacial acetic acid, 5 ml; 50% (v/v) ethanol, 200 ml).
(2) Place roots in 10% (w/v) KOH in vials and heat at 90°C for ca. 1 h to clear and soften.
(3) Wash thoroughly with tap water.
(4) Rinse for about 2 min in 2% (v/v) HCl.
(5) Boil in 0.05% (w/v) Trypan blue in lactophenol for 3 min.
(6) Stand in clear lactophenol to destain.
(7) Mount in lactophenol, lactic acid or 50% (v/v) glycerol for observation under a microscope.

Only roots or root segments with a primary cortex are used, preferably the fine feeder roots. The heating time in KOH is modified according to the plant species. Roots with brown pigment are bleached (< half an hour) with H_2O_2 after heating in KOH. Other methods include boiling for 10 min in a saturated solution of chloral hydrate containing 0.01% (w/v) acid fuchsin (Gerdemann 1955). All methods require the use of a fume hood.

The diagnostic anatomical features of vesicular–arbuscular mycorrhizal infections are the vesicles (for storage) and arbuscules (for nutrient exchange), illustrated in Fig. 2. The vesicles are more often intercellular than intracellular, whereas the arbuscules form only inside the host cells, one per cell, usually filling the cell lumen but not penetrating the plasmalemma. The main hyphae are mostly intercellular and grow parallel to the root axis. Hyphal coils (peletons) are sometimes formed. Appressoria on the root surface link the mycelium inside the root to the external mycelium which

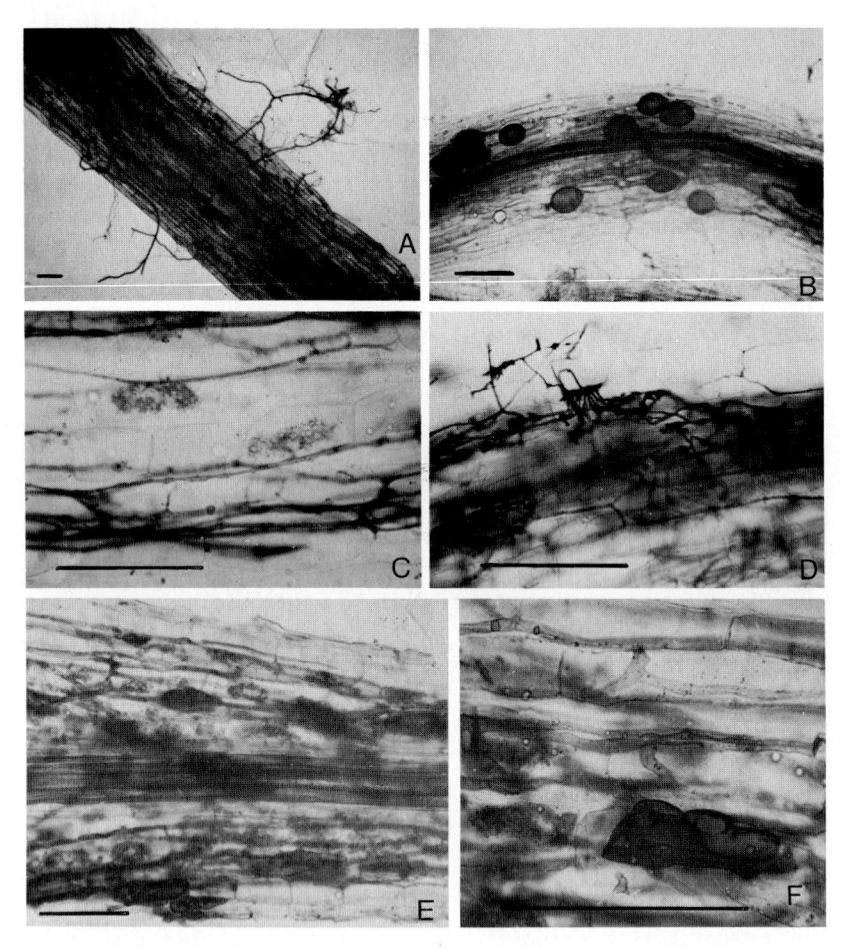

FIG. 2. Vesicular–arbuscular mycorrhizal infections. (A) Part of mycorrhizal onion root showing attached external mycelium. (B) Vesicles of *Glomus fasciculatus* E3 in white clover root. (C) Arbuscules (two granular areas, each occupying a cortical cell) of *Glomus fasciculatus* E3 in white clover. (D) Hyphae and multiple entry points of a "fine endophyte" in white clover. (E) Part of white clover root infected by an unidentified VAM endophyte indigenous to a hill grassland soil in Wales. (F) Hyphae, arbuscules and lobed vesicle of VAM endophyte in (E). Bar = 100 μm.

explores the surrounding soil. Infection is confined to the primary cortex and is absent from roots with secondary thickening. It is abnormal for the endophyte to penetrate the vascular tissues.

A few VAM endophytes can be distinguished from one another on the basis of their anatomy within the root. However, it is not usually possible to recognise any particular species from an infection, and many infections in the field are a mixture of more than one endophyte. One group, lumped together as "fine endophyte" or *Rhizophagus tenuis* (Greenall 1963), later *Glomus tenuis* (Hall 1977), is distinguished from the others by its very fine hyphae, tiny vesicles and sometimes distinctive arbuscules (Hayman 1982b). The fine hyphae are usually less than 3 μm in diameter compared with 10 to 20 μm for the others (Greenall 1963). The latter, once grouped together as "coarse endophyte" (Greenall 1963), represent most species and genera. The separate genera show some broad differences in vesicle morphology, vesicles being generally oval in *Glomus*, lobed in *Acaulospora* and rare or absent in *Gigaspora*. Abbott and Robson (1979) recognised H, Y and S junctions connecting the longitudinal intercellular hyphae and felt that such branching patterns could become a useful feature for distinguishing between certain endophytes. Unfortunately, in most instances, it is difficult to monitor on an anatomical basis the development of a selected endophyte inoculated onto a crop which is also exposed to infection by indigenous endophytes. This hinders evaluation of inoculants under field conditions.

Quantification

Populations of VAM are estimated by recovering the resting spores from soil and counting them, by examining representative root material for infection and by measuring the mycorrhizal infectivity of a particular soil.

Spore Counts

A range of techniques is available for extracting VAM resting spores from soil (Table 1). The majority involve an initial removal of the heaviest soil organic particles, sand, gravel and fine clay colloids by the wet-sieving and decanting technique described by Gerdemann & Nicolson (1963). A further separation of the spores from the bulk of the soil organic matter then follows, either by spreading out the material in a petri dish, sedimenting or centrifuging on density gradient columns (using water, gelatin, sucrose, glycerol, Percoll, etc.) or by floating in a solution with air bubbled through from below (elutriation). Also, with non-sieved soil, flotation on water and adhesion to a glass container (Sutton & Barron 1972) or collecting on filter paper (Smith & Skipper 1979) have been used; the former technique may extract many dead or empty spores and the latter is only suitable for very large spore populations that occur in pot cultures but not in field samples.

Most techniques work best with light sandy soils, reasonably well with clay soils, but poorly with highly organic soils. The approximate numbers of

TABLE 1

Techniques for extracting VAM resting spores from soil

Technique	Reference no.[*]									
	1	2	3	4	5	6	7	8	9	10
Wet-sieving and decanting	+	+	+	−	+	+	+	+	+	+
Doncaster counting dish	−	−	−	−	−	+	−	−	−	−
Sedimentation in water	+	−	−	−	−	+	+	−	−	−
Sedimentation in gelatin	−	−	+	−	−	−	−	+	−	−
Sedimentation in sucrose	−	−	−	−	−	−	+	−	−	−
Centrifugation in sucrose	−	+	−	−	−	−	+	+	−	−
Centrifugation in Percoll	−	−	−	−	−	−	−	−	+	−
Centrifugation in glycerol	−	−	−	−	−	−	−	−	−	+
Flotation in water	−	−	−	+	−	−	−	+	−	−
Flotation in glycerol	−	−	−	−	+	−	−	−	+	+
Adhesion	−	−	−	+	−	−	−	+	−	−
Bubbling/elutriation	−	−	−	−	+	−	−	−	+	−

[*] 1, Gerdemann & Nicolson (1963); 2, Ohms (1957); 3, Mosse & Jones (1968); 4, Sutton & Barron (1972); 5, Furlan & Fortin (1975); 6, Hayman & Stovold (1979); 7, Mertz *et al.* (1979); 8, Smith & Skipper (1979); 9, Furlan *et al.* (1980); 10, Kucey & McCready (1982).

spores recoverable per man–day range from 100–1570 (Ohms 1957), to a little over 5000 (Kucey & McCready 1982), to almost 10,000 (Mertz *et al.* 1979).

One version of the wet-sieving technique is as follows:

(1) Mix 50 g soil by hand in about 200 ml lukewarm water in a jug until all soil aggregates are dispersed and an even suspension obtained.
(2) Decant most of the suspension through a 710-μm sieve, supported in a funnel, into a tall, 1-litre graduated cylinder.
(3) Resuspend the residue in water and decant again through the sieve, allowing a few seconds for sedimentation, repeating this four or five times to leave sand, grit and heavy organic particles in the jug and ca. 700 ml suspension in the cylinder.
(4) Wash roots and other debris on the sieve with a jet of water from a squeeze bottle, collecting the washings in the cylinder.
(5) Resuspend the material in the cylinder by inverting several times and decant slowly through a 250-μm sieve into a second 1-litre cylinder, retaining a small volume which is resuspended in about 300 ml water and decanted to yield about 1 litre in total.
(6) Wash material on sieve and add washings to the second cylinder.
(7) Resuspend material in the second cylinder and pour most through a 105-μm sieve.

(8) Resuspend the residue in a further litre of water and pour through the 105-μm sieve.
(9) Wash material on sieve as before.
(10) Resuspend these last 2 litres and pour through a 53-μm sieve.
(11) Wash the residue on each sieve into a separate small beaker and retain for examination.
(12) Discard residues in jug and cylinders; the absence of spores in the discarded material should be verified microscopically when first using this technique.

The most convenient way to evaluate the wet-sievings is to pour aliquots into Doncaster nematode-counting dishes (Doncaster 1962) and to examine under a stereomicroscope. These concentrically grooved dishes preclude lateral movement of particles that occurs in standard petri dishes. This is probably the fastest method for yielding material in which all the spores can be counted, tentatively identified and examined for viability as judged by the extrusion of cytoplasm and vacuoles after piercing with dissecting needles (Hayman & Stovold 1979). Alternatively, the wet-sievings can be filtered through cloth with a suitable weave, e.g. Dicel (Hayman 1970), and observed under a stereomicroscope. Processing the wet-sievings through density gradients and aspirating material from the interfaces is probably slower then using Doncaster dishes if spores are to be counted in several samples. However, the density-gradient techniques probably yield more spores in less time from fewer, larger samples and thus are preferable where large batches of spores are needed for specific studies. For this purpose, Percoll and other similar media are better than sucrose for spore extraction because they have low viscosity and exert no osmotic pressure (Furlan *et al.* 1980).

As yet none of the methods is sufficiently precise to permit separation of different species. Most methods yield ca. 100–500 spores/100 g soil for field samples. The larger numbers in a few studies may reflect an abundance of small spores or the inclusion of dead spores. In pot cultures several thousand spores/100 g soil are readily extracted.

Root Infection

Several methods have been described for measuring the amount of VA mycorrhizal infection in a root system. The one chosen will depend on the degree of accuracy required, the size of the sample, the time available and the purpose for which an estimate of infection levels is needed. If the functional potential of a mycorrhizal infection is being evaluated, some cognisance of the level of infection below which the fungus is ineffective is needed. Excessive precision is laborious and often superfluous.

The commonest methods are based on the root slide technique of Nicolson (1960) in which pieces of cleared and stained root are mounted on slides and examined under a compound microscope. Infection is estimated as present or absent in individual pieces, in single microscope fields (Furlan & Fortin 1973), as a length of infected root in each piece (Hayman 1970; Biermann & Linderman 1981) or as infection intensity, i.e. length infected × proportion of cortex colonised (Hayman 1970). Usually the root segments are the same length, commonly 1 cm but sometimes 1 or 5 mm. Using these methods, a figure for percentage of root length infected of 70–80 is considered representative of a strong infection. More detailed information on the type of infection can be obtained by recording abundance of particular structures, primarily arbuscules, vesicles, hyphal coils and entry points.

A more subjective but faster estimate of infection can be obtained by spreading out the root system in a petri dish and determining the proportion of stained fungal material using a stereomicroscope. This also gives some idea of the amount of external mycelium, although some is lost by washing, etc.

Ambler & Young (1977) used a line-intersect method for determining both percentage and total mycorrhizal root length. Giovannetti & Mosse (1980) described a similar method in which the roots are spread over a square dish with a grid pattern. Percentage infection is based on recording under a stereomicroscope the presence or absence of infection at 100 or 200 points where roots and lines intersect. If the lines are exactly $\frac{1}{2}$ in. apart, and the roots are cut into 1-cm pieces, the number of intersections also gives the total root length in centimetres (Marsh 1971). Total mycorrhizal root length can then be calculated, which may be more meaningful than percentage root length mycorrhizal in some studies, especially on nutrient uptake. Sub-samples should be looked at under a compound microscope to verify identification of the endophyte, especially for plants inoculated in non-sterile soils.

A major problem with the aforegoing methods is that a two-dimensional measurement (length) is being made on three-dimensional material. This gives an overestimate of fungal biomass, as indicated by the much smaller figures obtained (ca. 10%) when transverse sections are examined and the proportion of root tissue or individual cells colonised are observed (Strzemska 1975). Unfortunately, this last method is very time consuming. Recently, the technique of morphometric cytology has been applied to squashed mycorrhizal roots, recording presence or absence of endophyte structures under dots on a symmetrical grid. Using a formula to convert area to volume determinations, this method gives figures of 10–20% infection (Toth & Toth 1982).

A chemical measurement of the amount of fungal tissue present can be made colorimetrically after converting the fungal chitin to glucosamine (Hepper 1977). This method is objective and the values obtained (4–17%) are of the same order as the visual and volumetric methods of Strzemska

(1975) and Toth & Toth (1982). However, it will not differentiate between mycorrhizal and other fungi in roots and so is chiefly applicable to experimental systems with controlled inoculations rather than mixed infections in field samples. Bethlenfalvay *et al.* (1981) compared the chitin assay method with standard histological methods and found that chitin assays were best if there was much infection (> 60%), but similar at low levels of infection. An alternative colorimetric technique is based on hot water extracts of the yellow pigment characteristic of some VA mycorrhizas, e.g. *Glomus fasciculatus* on onion roots (Becker and Gerdemann 1977). A third colourimetric method is noted by Herrera and Ferrer (1980) who measured the blue stain in the fungus after eluting the colour from stained mycorrhizal roots.

All the methods referred to so far involve destruction of the sample and time-consuming chemical procedures. However, a rapid non-destructive method was reported by Ames *et al.* (1982) who discovered that arbuscules in fresh pieces of VA mycorrhizal root autofluoresce under UV light. Although vesicles and hyphae did not autofluoresce, the values obtained for percentage root length infected in young roots using UV or clearing–staining techniques agreed well. Moreover, autofluorescence continued even in completely collapsed, non-functional arbuscules.

How much infection is functional at any one time is not revealed by standard clearing and staining methods, but may be gauged by cytochemical techniques which stain certain enzymes in the fungus (Macdonald & Lewis 1978).

In order to standardise comparisons between mycorrhizal studies, it is recommended that percentage root length infected be determined in addition to any other measurements that workers may favour. This should be done by the gridline intersect method as described by Giovannetti & Mosse (1980), provided infections in representative root segments are also examined at high magnification to check the proportion of mycorrhizal to other fungi. This check is particularly important in field studies and surveys to evaluate the variability of endophytes in relation to fungal species, host plant, soil amendments, etc.

In experiments on plant growth, infection should be measured periodically because early infection is probably more relevant to nutrient uptake than is the final level of infection. Also the amount of external mycelium is probably more directly correlated with nutrient uptake than is internal infection (Hayman 1982b), but is more difficult to measure.

Soil Infectivity

The numbers of spores and the levels of VAM root infection in soil are not always related. Furthermore, spores can sometimes be too few to account for soil infectivity as measured by the potential for seedlings to become mycorrhizal in a particular soil. Thus, in soils containing endophyte species which form no spores or under conditions where the indigenous endophytes do not readily sporulate, alternative propagules occur. These include clumps of mycelium and fragments of infected but moribund root (Hayman 1982b).

Porter (1979) estimated the number of VAM propagules in soil by the "most probable number" technique. This involves mixing a non-sterile soil with sterilised soil in a series of dilutions (e.g. 2- or 10-fold) until seedlings sown in the soil mixture no longer become infected. Propagule numbers estimated by this method were found to agree with spore numbers in one soil but not in another. One problem is that preparation of a dilution series can break up propagules such that one infected root piece, for example, can become several. The use of seedlings as bait plants in undiluted soil (Hayman & Stovold 1979) gives a measure of soil infectivity but not actual numbers of propagules.

Isolation

Because VAM spores are so large, they can be collected from wet-sievings of soil in a petri dish under a stereomicroscope, using a capillary micropipette, hair from a camel-hair brush or (for the largest spores) fine watchmakers' forceps. The material on the coarse sieves (> 250 μm pore diameter) may contain sporocarps. Spores of morphologically uniform types are transferred to watch glasses of tap water, re-examined under a stereomicroscope, doubtful spores and debris are discarded, and the remaining spores placed on moist filter paper in petri dishes and stored at 2–4°C for inoculation onto stock plants.

Roots from the coarse sievings can be used for inoculating seedlings after staining sub-samples to verify the presence of VAM. If enough seedlings are inoculated with small pieces (0.1–1 cm) of root, at least some plants should develop single infections. Alternatively, a portion of the wet-sievings is used directly as inoculum. Thus spores (25 is a suitable number), root fragments or crude sievings are obtained and placed in contact with the roots of young seedlings of onion, clover, strawberry, sorghum or other suitable host plant raised in a sterilised growth medium. Then the inoculated seedlings are replanted in pots of sterilised soil. After some 2–4 months, new resting spores are generally formed around the infected roots. Uniformity of spore type indicates the isolation and establishment of a single morphological species. If there is more than one morphological type, spores of each should be picked out and used for re-isolating the different endophytes. This is usually much easier than collecting spores from the original field samples.

An indirect method is to sow seeds of highly susceptible plants in non-sterile soil and remove after a few weeks. These seedlings act as bait plants for VAM fungi native to the test soil (Gilmore 1968; Hayman & Stovold 1979) and at least some should develop single infections. Their roots are then washed and the seedlings transplanted to pots of sterilised soil and checked, as before, for subsequent spore production. Native seedlings likely to have become naturally infected can also be used similarly to bait plants.

Cultures

When single VAM endophytes have been isolated, they must be established and maintained on one or more suitable host plant species. Usually this is done in open pot cultures, but for specific purposes aseptic, monoxenic cultures may be obtained.

Pot Cultures

The choice of stock plants for the establishment, maintenance and multiplication of VAM fungi in open pot cultures is wide because these fungi show little host specificity. Commonly used stock plants are onion, strawberry, white clover, black pepper, citrus, sorghum, *Nardus stricta*, *Coprosma robusta* and *Stylosanthes* spp. according to availability, growth requirements and glasshouse conditions. Maize is useful for building up large quantities of inoculum quickly but is limited for pot cultures by not being perennial.

The growth medium is usually a soil of moderately low fertility, mixed with sand or other material to improve its physical structure, and sterilised to remove indigenous mycorrhizas. Sterilisation by γ-irradiation (0.8 Mrad) is preferred, otherwise steaming can be used. Some endophytes prefer acidic soils, others neutral or alkaline soils, and they are also influenced by other soil factors. Hence several soils should be tested for their ability to support VAM proliferation and sporulation. These stock cultures normally provide an abundance of spores for a year or so, after which time new stock cultures should be established with inoculum from the old culture.

Pot cultures are the standard way of maintaining VAM collections around the world. Inevitably, however, there are some drawbacks. These include the risk of cross-contamination between adjacent pots of endophytes, the large amounts of bench space required, the impurity of inoculum containing other microorganisms and the lack of genetic stability under these conditions. The existence of different physiological strains within a single morphological species has been reported (Daniels & Duff 1978; Gildon & Tinker 1981), but such strain differences are difficult to accommodate in a culture collection based on morphological criteria. Also, it is conceivable that pot cultures may induce the selection of some strains favoured by the particular growth conditions.

Monoxenic Cultures

Spores can be surface-sterilised in 2% (w/v) chloramine T containing 200 mg streptomycin/litre and a trace of detergent (Mosse & Phillips 1971) in watch glasses for 15 min. They are then transferred by capillary pipettes through three changes of sterile water. Seedlings raised aseptically on agar in flasks or test tubes are inoculated by placing 10–15 surface-sterilised spores to germinate near the roots growing on the agar surface. Seeds of clover (*Trifolium repens* and *Trifolium parviflorum*), the most commonly used plant in these studies, are surface-sterilised by 10–15 min in concentrated H_2SO_4.

Mosse & Phillips (1971) described a range of media for synthesising pure two-membered cultures of host plant and VAM fungus. They discussed the negative effects of phosphate on infection and found optimum levels of ca. 100 mg P/litre as $CaHPO_4$. With 265 mg P/litre infection occurred only when the medium lacked N. Calcium phytate, DNA and inositol greatly stimulated growth of the fungal mycelium external to the root.

Although monoxenic VAM cultures permit physiological relationships to be studied without interference from rhizosphere organisms and soil, there have been surprisingly few applications of these methods. Successful syntheses include *Glomus mosseae* on *Trifolium repens* (Pearson & Tinker 1975), *Glomus fasciculatus* on *Bouteloua gracilis* (Allen *et al*. 1979), *Glomus mosseae* and *Glomus caledonius* on both *Trifolium parviflorum* and *Trifolium repens* (Hepper 1981), *Glomus caledonius* on *Trifolium parviflorum* (Macdonald 1981) and *Gigaspora margarita* and *Glomus mosseae* on *Trifolium repens* (St. John *et al*. 1981). Attempts at producing two-membered cultures with callus tissues have not been successful, probably in part because of the tendency of the endophyte not to colonise meristematic tissue, although VAM fungi can be established in root organ cultures (cf. references cited by Hepper 1981).

The medium used by Hepper (1981) contains as macronutrients (mg/litre) the following: KNO_3, 303; $Ca(NO_3)_2 \cdot 4H_2O$, 2040; $MgSO_4 \cdot 7H_2O$, 368; KH_2PO_4, 44; FeNaEDTA, 46; pH adjusted to 6.8 before sterilising at 121°C for 15 min; micronutrients (mg/litre) are $MnSO_4 \cdot 4H_2O$, 2.23; $CuSO_4 \cdot 5H_2O$, 0.24; $ZnSO_4 \cdot 7H_2O$, 0.29; H_3BO_3, 1.86; $(NH_4)_6Mo_7O_{24} \cdot 4H_2O$, 0.035. For solid media 1.2% (w/v) agar (Difco Bacto) was added; for liquid media macronutrients and micronutrients were used at 10% of the stated concentrations. Some micronutrients inhibit spore germination and therefore were added after the fungus had become established. Plants were supported either on agar slopes or on strips of chromatography paper or microscope slides suspended in liquid media.

Genetics

The improvement of bacterial and fungal strains for commercial purposes by genome modification is an aspect of biotechnology that may have great potential for VA mycorrhizal fungi. Although as yet the technical difficulties are considerable and our present knowledge of the life cycles and genetics of VAM fungi is sparse, some assessment of the possibilities will be attempted here.

Vesicular–arbuscular mycorrhizal fungi have typical phycomycetous, coenocytic hyphae that lack regular septa. Of their two types of resting spore, chlamydospores are formed asexually on somatic hyphae and azygospores are formed parthenogenically on one or both of two gametangium-like organs. Both spore types are multinucleate. Sexual reproduction is unknown and has been shown only in one genus of the Endogonaceae, viz. *Endogone*. This contrasts with a number of other members of the Mucorales where gametangial copulation and zygospore formation is well documented in cultural studies. This is believed to involve plasmogamy and karyogamy followed by meiosis, the short-lived zygote being the only diploid stage. However, even with these fungi, zygospore formation is believed to be rare in nature.

It is clear from the aforegoing that attention should be directed towards non-sexual mechanisms for genetic variation and recombination. One is mutation followed by multiplication of the mutant nuclei among the wild types. Another is anastomosis between dissimilar strains followed by nuclear migration and maintenance of both types of nuclei in the hybrid mycelium. Both mechanisms lead to heterokaryosis, which is almost certainly the normal nuclear state of any individual VAM isolate. Heterokaryosis allows rapid adaptation to environmental changes, but is a problem in maintaining strain stability in pot cultures. However, it provides a wide gene pool for the selection of strains for specific purposes.

The parasexual cycle is well known as a source of mitotic recombination during somatic nuclear divisions and occurs in several of the Fungi Imperfecti. It has been detected in Ascomycetes and Basidiomycetes but not in Phycomycetes. However, the possibility of its occurrence in VAM fungi should not be totally dismissed, especially in view of their uncertain taxonomic and phylogenetic relationships.

Another possibility is the transmission of non-nuclear genetic factors called "plasmons" (Burnett 1976). This can occur after hyphal fusion, with somatic segregation of these cytoplasmic factors from two different fungi leading to the formation of a heteroplasmon. An example of this is the stable maintenance of DNA-containing mitochondria from two different sources in yeast.

The production of new recombinants may be accelerated by direct mutagenesis and by genetic engineering. Whether these techniques, currently attracting much attention with prokaryotes and some culturable fungi, can be adapted for use with VAM fungi is a subject of great interest and speculation. Protoplasts of VAM fungi might provide suitable material for studies in this hitherto neglected area. Although the creation of VAM strains tailor-made for specific conditions may be a long way off, it is clear that less empirical, more controlled methods for defining and manipulating strains should be adopted as far as possible in current symbiotic investigations.

Inoculum Production

Specific endophytes should be inoculated onto suitable host plants and grown in an appropriate soil (see under *Pot Cultures*). Inoculum produced by this means is usually concentrated by wet-sieving the soil–root mass from each pot to produce a mixture of infected root pieces, resting spores, mycelium and much organic matter.

Inoculum should be produced on plant species not closely related to the crop being tested to minimise the possible development of host-specific

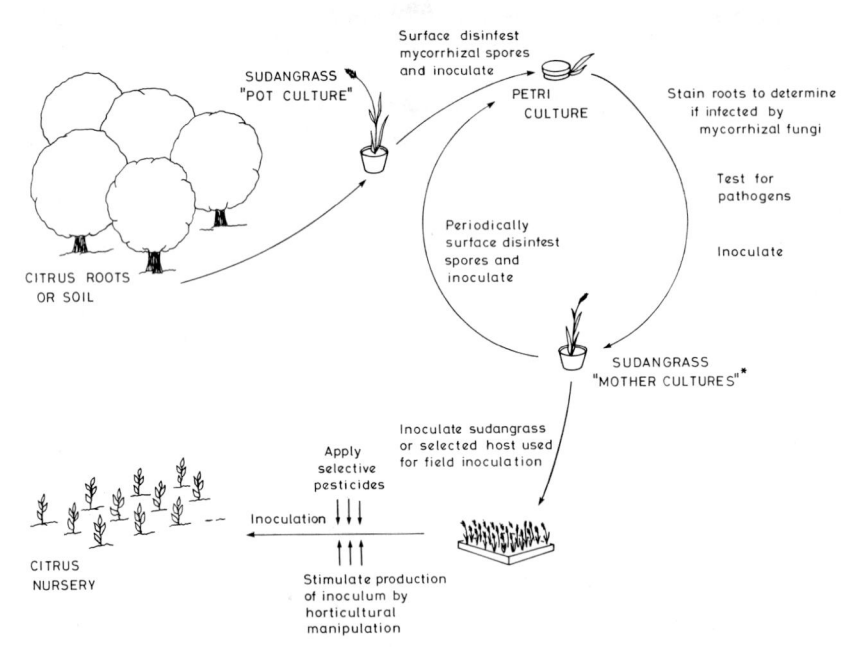

FIG. 3. Proposed scheme for commercial production of mycorrhizal inoculum, redrawn from Menge *et al.* (1977) with permission. * "Mother cultures" must be on a crop species not related to the crop being inoculated in the field.

pathogens in the open pot cultures. In addition, the inoculum can be treated with pesticides not harmful to mycorrhizal fungi in order to kill nematodes and other pests which may be present. These principles were summarised by Menge et al. (1977) and are reproduced in Fig. 3.

One VAM species, *Glomus epigaeus*, forms loose sporocarps on the soil surface. These sporocarps can be scraped off to give much purer inoculum than is possible with other VAM species which sporulate underground. An average of 4.5 sporocarps/pot culture/month were obtained on sudangrass in California, and each sporocarp contained ca. 1×10^5 to 7.5×10^6 spores (Daniels & Menge 1981). This spore-producing capacity, coupled with successful storage (up to 83% germination after 3 months) and symbiotic efficiency on a wide range of host plants suggested that *Glomus epigaeus* may have considerable commercial potential.

Another way of producing fairly pure inoculum, namely infected roots with attached mycelium but no soil, is to raise inoculated plants in nutrient film technique (NFT) cultures (Elmes & Mosse 1980). The longevity of such roots is uncertain, however, and modifications of the technique such as incorporating plants in peat blocks in the NFT solutions are being developed (Warner et al. 1982). This development, using nursery-grown lettuces, facilitates the production of large amounts of inoculum, a factor which has hitherto limited the size and number of VAM field experiments.

Field Inoculation Methods

Various methods exist for introducing VAM inoculum into a field-grown crop, but most are restricted to small-scale usage, mainly because of the difficulties of producing large quantities of relatively pure inoculum (see under *Inoculum Production*). To achieve good infection it is essential that inoculum be concentrated near the seed, not broadcast thinly over a wide area. The most commonly used methods for field inoculations will be summarised. However, in view of the variety of methods available, the form of inoculum to be used needs to be considered. Mixed inocula may be better in some instances, leaving the plant and soil to select the best endophyte *in situ*.

Pre-Inoculated Transplants

Seedlings are raised in sterilised soil supplied with selected VAM fungi in containers and planted out when mycorrhizal infection is well established. This helps the introduced endophytes to compete with endophytes indigenous to the planting site. This method is most feasible for crops which are normally transplanted. It is also useful for screening endophytes in the field before attempting large-scale inoculation procedures.

Coated Seeds

Seeds are coated with an adhesive, e.g. methyl cellulose, to which inoculum is expected to stick and subsequently infect the emerging radicles. Unfortunately, because of their large size, it is much more difficult to attach VAM propagules than bacteria in this way. However, it has proved satisfactory for large-seeded crops such as citrus in field nurseries (Hattingh & Gerdemann 1975).

Pellets

Rather than coat seeds with VAM inoculum, it seems technically more feasible to incorporate seeds into the inoculum to form multiseeded pellets. These pellets, ca. 1 cm in diameter, consist of soil inoculum from pot cultures stabilised with clay (Powell 1979; Hayman et al. 1981), or peat inoculum. Good infection has been achieved by this means. Hall & Kelson (1981) have described a system for producing some 5000 VAM-infected soil pellets per man-day with seeds attached by gum arabic.

Inoculation in Furrows

Placing inoculum under or beside seeds sown in a furrow (Owusu-Bennoah & Mosse 1979; Hayman et al. 1981) is probably the most common and effective method of inoculating plants in small field plots. It usually gives good results, but is impractical on a large scale because some 2–3 tonnes inoculum/ha would be needed (Owusu-Bennoah & Mosse 1979).

Fluid Drilling

Incorporation of seeds and inoculum into a uniform suspension in a viscous fluid, e.g. 4% (w/v) methyl cellulose, and applying in furrows as a slurry is another means of placing seeds and inoculum in proximity. This has proved successful under field conditions (Hayman et al. 1981), but germination of the seeds must be rigorously controlled to prevent damage to the emerging radicles. The advantage of this technique is that the inoculum is less bulky (wet-sievings representing about one-seventh of the original soil inoculum used in the previous inoculation in furrows method) and does not increase appreciably the volume of material drilled.

Pre-Cropping

Populations of VAM propagules can be raised *in situ* by growing strongly mycorrhizal host plants and leaving the infected roots and associated spores in the soil to infect the next crop. Hence, where mycorrhizal benefits look possible, judicious crop rotations might be considered as a means of manipulating the native mycorrhizas. These can be increased in bulk if they are effective, otherwise efficient introduced endophytes can be built up on the pre-crop. The top soil could be used as crude inoculum elsewhere.

Endophyte Screening

Some means of evaluating and selecting the most efficient VAM endophytes from a stock culture collection is a prerequisite for manageable field trials of a few elite fungal strains. The standard method is to grow plants with different endophytes and uninoculated controls in pots of sterilised low-phosphate soil in the glasshouse. Symbiotic efficiency is judged on the basis of enhanced plant dry matter production after about 10 weeks. This will reveal large differences between endophytes, allowing poor ones to be rejected, but overall shoot growth is not always correlated with fruit or grain yield. Also, selection should not be made on the basis of one soil and one host plant because of differences in endophyte specificity (Mosse 1973) in terms of the degree of positive interaction between plant and fungal endophyte and, more importantly, endophyte preferences for particular soils and susceptibility to various soil amendments.

Serial harvesting may show trends at an early stage of crop development. Measurement of the rate of early infection can be useful because this is sometimes directly related to early growth responses. However, there are endophytes which infect well but do not stimulate growth appreciably.

Another parameter reflecting endophyte efficiency is rate of phosphate uptake. Consequently, analysis of P concentrations in plant tissues could point to the most actively functioning endophyte before increases in growth become obvious.

Other features such as persistence, survival and ability to infect in competition with indigenous endophytes and the soil microflora also need to be evaluated. Unfortunately, it is difficult to monitor the fate of introduced inoculants infecting roots alongside native VAM fungi because in most instances they cannot be distinguished anatomically. Immunofluorescence techniques might be developed for monitoring purposes.

Additional characteristics to look for when screening endophytes are pH optima, performance at realistic levels of soil P and tolerance to pesticides. These characteristics vary considerably between species.

Finally, those endophytes which produce the most external mycelium may be the most efficient by reason of their increased absorbing surface, but more important may be how well this mycelium is distributed through the soil away from the root. This is because they function primarily by absorbing more of the soluble soil phosphate, not by solubilising extra amounts of inert soil phosphate.

Conclusions

Vesicular–arbuscular mycorrhizas can greatly enhance plant growth in certain situations. These situations must be identified as clearly as possible if we wish to manipulate VAM fungi to advantage by cultivation treatments and inoculation programmes and evaluate potential benefits. In essence, benefits are determined by endophyte efficiency, soil fertility and plant species. Positive responses to VAM are most likely in low-phosphate soils with mycorrhiza-dependent plants inoculated with rapidly infecting and functioning fungal endophytes. Marginal (e.g. hill grasslands), reclaimed (e.g. coal tips) and fumigated lands are currently being evaluated in this context. In addition, VA mycorrhizas are being studied in infertile and normal arable soils. The ultimate evaluation should be made in relation to plants given adequate phosphate fertiliser because if mycorrhizal inoculation is to be considered commercially, perhaps in conjunction with reduced fertiliser input, it must emulate results with standard farming practices. Field trials are necessary for the final evaluation because pot trials are not always a good guide to final yields (Hayman 1982a; Tinker 1982).

It is difficult to study and manipulate VA mycorrhizas because VA fungi cannot be grown in pure culture. Therefore methods for identifying species, measuring spore populations and root infections, determining their distribution and activity in soil and assessing the contribution of indigenous mycorrhizas to crop growth assume considerable importance. In view of the widespread interest in VA mycorrhiza that has arisen during the last decade, it is likely that the methodology, understanding and biotechnological application of this mutualistic symbiosis will continue to improve in the coming years.

Acknowledgements

M. Tavares and C. Grace are thanked for preparing and photographing material for parts of Figs. 1 and 2.

References

Abbott, L. K. & Robson, A. D. 1979 A quantitative study of the spores and anatomy of mycorrhizas formed by a species of *Glomus*, with reference to its taxonomy. *Australian Journal of Botany* **27**, 363–375.

Abbott, L. K. & Robson, A. D. 1982 The role of vesicular–arbuscular mycorrhizal fungi in agriculture and the selection of fungi for inoculation. *Australian Journal of Agricultural Research* **33**, 389–408.

Allen, M. F., Moore, T. S. & Christensen, M. 1979 Growth of vesicular–arbuscular mycorrhizal and nonmycorrhizal *Bouteloua gracilis* in a defined medium. *Mycologia* **71**, 666–669.

Ambler, J. R. & Young, J. L. 1977 Techniques for determining root length infected by vesicular–arbuscular mycorrhizae. *Soil Science Society of America Journal* **41**, 551–556.

Ames, R. N., Ingham, E. R. & Reid, C. P. P. 1982 Ultraviolet-induced autofluorescence of arbuscular mycorrhizal root infections: an alternative to clearing and staining methods for assessing infections. *Canadian Journal of Microbiology* **28**, 351–355.

Becker, W. N. & Gerdemann, J. W. 1977 Colorimetric quantification of vesicular–arbuscular mycorrhizal infection in onion. *New Phytologist* **78**, 289–295.

Becker, W. N. & Hall, I. R. 1976 *Gigaspora margarita*, a new species in the Endogonaceae. *Mycotaxon* **4**, 155–160.

Bethlenfalvay, G. J., Pacovsky, R. S. & Brown, M. S. 1981 Measurement of mycorrhizal infection in soybeans. *Soil Science Society of America Journal* **45**, 871–875.

Biermann, B. & Linderman, R. G. 1981 Quantifying vesicular–arbuscular mycorrhizae: a proposed method towards standardization. *New Phytologist* **87**, 63–67.

Burnett, J. H. 1976 *Fundamentals of Mycology*, 2nd ed., 673 pp. London: Arnold.

Daniels, B. A. & Duff, D. M. 1978 Variation in germination and spore morphology among four isolates of *Glomus mosseae*. *Mycologia* **70**, 1261–1267.

Daniels, B. A. & Menge, J. A. 1981 Evaluation of the commercial potential of the vesicular–arbuscular mycorrhizal fungus, *Glomus epigaeus*. *New Phytologist* **87**, 345–354.

Doncaster, C. C. 1962 A counting dish for nematodes. *Nematologica* **7**, 334–337.

Elmes, R. & Mosse, B. 1980 Vesicular–arbuscular mycorrhiza: nutrient film technique. *Rothamsted Report for 1979*, Part 1, p. 188.

Furlan, V. & Fortin, J. A. 1973 Formation of endomycorrhizae by *Endogone calospora* on *Allium cepa* under three temperature regimes. *Naturaliste Canadien* **100**, 467–477.

Furlan, V. & Fortin, J. A. 1975 A flotation-bubbling system for collecting Endogonaceous spores from sieved soil. *Naturaliste Canadien* **102**, 663–667.

Furlan, V., Bartschi, H. & Fortin, J. A. 1980 Media for density gradient extraction of endomycorrhizal spores. *Transactions of the British Mycological Society* **75**, 336–338.

Gerdemann, J. W. 1955 Relation of a large soil-borne spore to phycomycetous mycorrhizal infections. *Mycologia* **47**, 619–632.

Gerdemann, J. W. 1968 Vesicular–arbuscular mycorrhiza and plant growth. *Annual Review of Phytopathology* **6**, 397–418.

Gerdemann, J. W. & Nicolson, T. H. 1963 Spores of mycorrhizal *Endogone* species extracted from soil by wet-sieving and decanting. *Transactions of the British Mycological Society* **46**, 235–244.

Gerdemann, J. W. & Trappe, J. M. 1974 The Endogonaceae in the Pacific Northwest. *Mycologia Memoir* No. 5, 76 pp.

Gildon, A. & Tinker, P. B. 1981 A heavy metal-tolerant strain of a mycorrhizal fungus. *Transactions of the British Mycological Society* **77**, 648–649.

GILMORE, A. E. 1968 Phycomycetous mycorrhizal organisms collected by open-pot culture methods. *Hilgardia* **39**, 87–105.
GIOVANNETTI, M. & MOSSE, B. 1980 An evaluation of techniques for measuring vesicular–arbuscular mycorrhizal infection in roots. *New Phytologist* **84**, 489–500.
GREENALL, J. M. 1963 The mycorrhizal endophytes of *Griselinia littoralis* (Cornaceae). *New Zealand Journal of Botany* **1**, 389–400.
HALL, I. R. 1977 Species and mycorrhizal infections of New Zealand Endogonaceae. *Transactions of the British Mycological Society* **68**, 341–356.
HALL, I. R. & FISH, B. J. 1979 A key to the Endogonaceae. *Transactions of the British Mycological Society* **73**, 261–270.
HALL, I. R. & KELSON, A. 1981 An improved technique for the production of endomycorrhizal infested soil pellets. *New Zealand Journal of Agricultural Research* **24**, 221–222.
HATTINGH, M. J. & GERDEMANN, J. W. 1975 Inoculation of Brazilian sour orange seed with an endomycorrhizal fungus. *Phytopathology* **65**, 1013–1016.
HAYMAN, D. S. 1970 *Endogone* spore numbers in soil and vesicular–arbuscular mycorrhiza in wheat as influenced by season and soil treatment. *Transactions of the British Mycological Society* **54**, 53–63.
HAYMAN, D. S. 1982a Practical aspects of vesicular–arbuscular mycorrhiza. In *Advances in Agricultural Microbiology* ed. Subba Rao, N. S., pp. 325–373. New Delhi: Oxford & IBH Publ. Co.
HAYMAN, D. S. 1982b Influence of soils and fertility on activity and survival of vesicular–arbuscular mycorrhizal fungi. *Phytopathology* **72**, 1119–1125.
HAYMAN, D. S., MORRIS, E. J. & PAGE, R. J. 1981 Methods for inoculating field crops with mycorrhizal fungi. *Annals of Applied Biology* **99**, 247–253.
HAYMAN, D. S. & STOVOLD, G. E. 1979 Spore populations and infectivity of vesicular–arbuscular mycorrhizal fungi in New South Wales. *Australian Journal of Botany* **27**, 227–233.
HEPPER, C. M. 1977 A colorimetric method for estimating vesicular–arbuscular mycorrhizal infection in roots. *Soil Biology and Biochemistry* **9**, 15–18.
HEPPER, C. M. 1981 Techniques for studying the infection of plants by vesicular–arbuscular mycorrhizal fungi under axenic conditions. *New Phytologist* **88**, 641–647.
HERRERA, R. A. & FERRER, R. L. 1980 Vesicular–arbuscular mycorrhiza in Cuba. In *Tropical Mycorrhiza Research* ed. Mikola, P., pp. 156–162. Oxford: University Press.
KUCEY, R. M. N. & MCCREADY, R. G. L. 1982 Isolation of vesicular–arbuscular mycorrhizal spores: a rapid method for the removal of organic detritus from wet-sieved soil samples. *Canadian Journal of Microbiology* **28**, 363–365.
MACDONALD, R. M. 1981 Routine production of axenic vesicular–arbuscular mycorrhizas. *New Phytologist* **89**, 87–93.
MACDONALD, R. M. & LEWIS, M. 1978 The occurrence of some acid phosphatases and dehydrogenases in the vesicular–arbuscular mycorrhizal fungus *Glomus mosseae*. *New Phytologist* **80**, 135–141.
MARSH, B. A' B. 1971. Measurement of length in random arrangements of lines. *Journal of Applied Ecology* **8**, 265–267.
MENGE, J. A., LEMBRIGHT, H. & JOHNSON, E. L. V. 1977 Utilization of mycorrhizal fungi in citrus nurseries. *Proceedings of the International Society of Citriculture* **1**, 129–132.
MERTZ, S. M., HEITHAUS, J. J. & BUSH, R. L. 1979 Mass production of axenic spores of the endomycorrhizal fungus *Gigaspora margarita*. *Transactions of the British Mycological Society* **72**, 167–169.
MOSSE, B. 1973 Advances in the study of vesicular–arbuscular mycorrhiza. *Annual Review of Phytopathology* **11**, 171–196.
MOSSE, B. & BOWEN, G. D. 1968. A key to the recognition of some *Endogone* spore types. *Transactions of the British Mycological Society* **51**, 469–483.

MOSSE, B. & JONES, G. W. 1968 Separation of *Endogone* spores from organic soil debris by differential sedimentation on gelatin columns. *Transactions of the British Mycological Society* **51**, 604–608.
MOSSE, B. & PHILLIPS, J. M. 1971 The influence of phosphate and other nutrients on the development of vesicular–arbuscular mycorrhiza in culture. *Journal of General Microbiology* **69**, 157–166.
MOSSE, B., STRIBLEY, D. & LE TACON, F. 1981 Ecology of mycorrhizae and mycorrhizal fungi. *Advances in Microbial Ecology* **5**, 137–210.
NICOLSON, T. H. 1960 Mycorrhiza in the Gramineae. II. Development in different habitats, particularly sand dunes. *Transactions of the British Mycological Society* **43**, 132–145.
NICOLSON, T. H. & SCHENCK, N. C. 1979 Endogonaceous mycorrhizal endophytes in Florida. *Mycologia* **71**, 178–198.
OHMS, R. E. 1957 A flotation method for collecting spores of a phycomycetous mycorrhizal parasite from soil. *Phytopathology* **47**, 751–752.
OWUSU-BENNOAH, E. & MOSSE, B. 1979 Plant growth responses to vesicular–arbuscular mycorrhiza. XI. Field inoculation responses in barley, lucerne and onion. *New Phytologist* **83**, 671–679.
PEARSON, V. & TINKER, P. B. 1975 Measurement of phosphorus fluxes in the external hyphae of endomycorrhizas. In *Endomycorrhizas* eds. Sanders, F. E., Mosse, B. & Tinker, P. B., pp. 277–287. London: Academic Press.
PHILLIPS, J. M. & HAYMAN, D. S. 1970 Improved procedures for clearing roots and staining parasitic and vesicular–arbuscular mycorrhizal fungi for rapid assessment of infection. *Transactions of the British Mycological Society* **55**, 158–161.
PORTER, W. M. 1979 The "most probable number" method for enumerating infective propagules of vesicular–arbuscular mycorrhizal fungi in soil. *Australian Journal of Soil Research* **17**, 515–519.
POWELL, C. LL. 1979 Inoculation of white clover and ryegrass seed with mycorrhizal fungi. *New Phytologist* **83**, 81–85.
REDHEAD, J. F. 1977 Endotrophic mycorrhizas in Nigeria: species of the *Endogonaceae* and their distribution. *Transactions of the British Mycological Society* **69**, 275–280.
SCHENCK, N. C. & SMITH, G. S. 1982 Additional new and unreported species of mycorrhizal fungi (Endogonaceae) from Florida. *Mycologia* **74**, 77–92.
SMITH, G. W. & SKIPPER, H. D. 1979 Comparison of methods to extract spores of vesicular–arbuscular mycorrhizal fungi. *Soil Science Society of America Journal* **43**, 722–725.
ST. JOHN, T. V., HAYS, R. I. & REID, C. P. P. 1981 A new method for producing pure vesicular–arbuscular mycorrhiza—host cultures without specialized media. *New Phytologist* **89**, 81–86.
STRZEMSKA, J. 1975 Occurrence and intensity of mycorrhiza and deformation of roots without mycorrhiza in cultivated plants. In *Endomycorrhizas* eds. Sanders, F. E., Mosse, B. & Tinker, P. B., pp. 537–543. London: Academic Press.
SUTTON, J. C. & BARRON, G. L. 1972 Population dynamics of *Endogone* spores in soil. *Canadian Journal of Botany* **50**, 1909–1914.
TINKER, P. B. 1982 Mycorrhizas: the present position. In *Whither Soil Research*, Transactions of the 12th International Congress of Soil Science, New Delhi, 8–16 February 1982, Vol. 5, pp. 150–166. New Delhi: Indian Society of Soil Science.
TOTH, R. & TOTH, D. 1982 Quantifying vesicular–arbuscular mycorrhizae using a morphometric technique. *Mycologia* **74**, 182–187.
TRAPPE, J. M. 1982 Synoptic keys to the genera of zygomycetous mycorrhizal fungi. *Phytopathology* **72**, 1102–1108.
WARNER, A., GEE, P. & FYSON, A. 1982 Vesicular-arbuscular mycorrhiza: The production of inoculum in nutrient film culture. *Rothamsted Report for 1981*, Part 1, p. 211.

Single- and Multi-Stage Fermenters for Treatment of Agricultural Wastes

P. N. HOBSON, R. SUMMERS AND C. HARRIES

Microbial Biochemistry Department, Rowett Research Institute, Bucksburn, Aberdeen, UK

Anaerobic digestion of farm wastes is not a new concept. Digestion of crop residues and animal excreta was used, on the continent of Europe for instance, during and after the last war. The reason for digestion then was to obtain a fuel on which tractors and other machinery could be operated and thus save scarce petrol and diesel oil. The digesters were not very efficient, and since the object was to obtain one usable fuel, overall energy balances did not matter, providing a non-oil fuel was used to heat the digesters. Thus, some digesters were heated by coal boilers.

With the return of plentiful and cheap oil, anaerobic digestion of farm wastes was more or less forgotten for some years, except in the late 1950s with the Gobar gas scheme in India. This was to provide fuel and a better fertilizer than the ashes that remained if dried cow dung was used for fuel.

In the 1960s, problems of pollution were becoming apparent, particularly from the intensive units that seemed to be the future style of farming; animals were kept without bedding and the excreta were collected in the form of a slurry of faeces and urine. Our work was at that time directed towards using anaerobic digestion as a method of reducing pollution from such units. A large intensive pig unit, for instance, smells badly and can produce the pollution equivalent [in terms of biochemical oxygen demand (BOD), etc.] of a medium-sized town.

The rise in prices, and uncertainty in supply, of oil-based fuels in the 1970s added a major impetus to research on anaerobic digestion, i.e. the production of "alternative" fuels. Although this aspect has tended to overshadow others, pollution control is still overall a reason for using anaerobic digestion on farms. Farm pollution control is given higher priority on the continent of Europe than in Britain; here pollution control by anaerobic digestion is more thought of for municipalities and factories.

Anaerobic Digestion

"Anaerobic digestion" is a process carried out under non-sterile conditions by a consortium of naturally occurring anaerobic bacteria. This consortium can be developed from bacteria in, for instance, faecal wastes, or organic matter decaying anaerobically, by providing suitable conditions of anaerobiosis, temperature and feedstock. Once developed, the consortium is very stable and the majority of digesters are run as continuous cultures for periods of years. Such digesters settle down to a steady state in so far as such can be obtained with a mixed culture carrying out a complex series of reactions on an inhomogenous substrate.

Detailed descriptions of starting digesters, the reactions and bacteria, construction, feedstocks used, etc. are given in numerous papers and reviews (e.g. Hobson *et al*. 1974, 1981; McInerney & Bryant 1981). Only a summary of the reactions in digesters will be given here.

Organic matter in the feedstock (including the pollutants) is converted to "biogas", a mixture of CH_4 (50–75%) and CO_2. The methanogenic bacteria can utilise only a limited number of substrates, principally $H_2 + CO_2$ and acetic acid. Higher fatty acids are converted to CH_4 via degradation to acetic acid and H_2 with, according to chain length, propionic acid which can also be converted to CH_4 (CO_2 is formed in some of these reactions).

Higher fatty acids are derived from bacterial hydrolysis of lipids in the feedstock, but the lower fatty acids, H_2 and CO_2 are largely products of sugar fermentations. In pure culture the fermentative bacteria form many products, but in the mixed culture, because H_2 is removed in CH_4 formation, the sugar fermentations are biased towards production of acetic acid rather than more reduced acids and ethanol. The sugars may be contained in the feedstock, e.g. in food processing, but are more generally derived from polymeric carbohydrates by hydrolysis by the digester bacteria, e.g. in farm animal wastes. The feedstock contains only residues of the animal feeds undigested in the gut, with a little spilt feed and possibly some bedding material, e.g. straw and sawdust. The remainder of faecal material is largely intestinal bacteria, with some intestinal epithelial cells and secretions.

The maximum growth rates of the methanogenic bacteria are low and methanogenesis is the rate-limiting reaction in digestion of dissolved sugars. However, the vegetable feedstuffs of animals have been subject to gastric and intestinal digestion, or to microbial digestion in rumen or caecum, or both. Therefore, what remains is lignified and ordered cellulose and hemicellulose which has resisted degradation in the animal. Such material is only partially degradable by bacteria in anaerobic digesters and what is hydrolysable is degraded only slowly. The final stages take 10, 20 or more days in an anaerobic digester, and thus fibre degradation becomes the rate-limiting step in the conversion of the feedstock to methane. The rate and extent of hydrolysis and subsequent fermentation of fibrous feedstocks can be increased by chemical or physical treatments, e.g. alkali or autoclaving, which "loosen" the fibre structure.

Types of Digester

Single-Stage Systems

A number of types of apparatus can be used with a pure culture of a microorganism growing on substrates in solution. The type of apparatus which can be used for anaerobic digestion is limited by the particular feedstock and by economic considerations. Basically, anaerobic digestion has to be a low-cost process. The value of pollution control varies. A municipality, faced with the possibility of disease, will value reduction in pollution quite highly, as will a factory faced with pollution-control regulations and the cost of discharging waste to the sewers; but for most farmers there is no monetary return from reducing pollution. However, even for municipalities and factories pollution control must be as cheap as possible otherwise it will add too much to rates or cost of product. The cost of production of biogas as fuel must be commensurate with that of oil, gas or electricity. The refinements of an industrial fermenter producing a high-value product cannot be economically sustained for a digester, and, fortunately, are not necessary. Another point to consider, particularly with farm digesters, is that they have to operate outside in all climatic conditions, and have to have minimum attention from relatively unskilled operators. Finally, the feedstocks for farm, and some other, digesters are slurries with variously sized suspended solids which can settle and block pipes and tanks. The slurries are also likely to contain miscellaneous debris: stone, wood, plastic bags, etc. Digesters have to be rugged, and the type of pumps and other ancillary equipment must be able to deal with the conditions and the feedstocks.

Anaerobic digesters, like other fermenters, can be run at temperatures suitable for mesophilic (about 25–44°C) or thermophilic (about 50–65°C) bacteria. Since the bacterial reactions are anaerobic and slow, metabolic heat generation in digesters is low and in both ranges heat must be applied to the digester. Because of the additional heating required, and also because of some instabilities in digestion of particular feedstocks, thermophilic digestion is seldom used. Thus, while the following discussion could apply to mesophilic or thermophilic digestion, in practice most, if not all, full-scale digesters are run in the lower range, usually ca. 35°C.

The simplest type of continuous-flow digester is the single-stage, stirred tank. In this, feedstock is added as continuously as possible (generally, for practical reasons, this means every hour or so) and an equal volume of digested material is removed. Since there is no necessity to force air into solution, mixing is only sufficient to provide thermal and microbial homogeneity in the tank contents, and is now usually done by gently sparging some of the biogas through the tank at intervals. Heating is usually by an internal hot water system (e.g. Hobson *et al.* 1981) for some details of construction of digesters). The biogas produced is collected either in a floating roof of the digester tank, or is led from a fixed-roof tank to a separate gas holder. The retention time of the feedstock in the tank is dependent on fibre degradation rate and is determined by experiment.

For farm slurries with high fibre content (e.g. cattle slurry), a retention time of 20 days is necessary, thus requiring a large but feasible digester capacity. However, with dissolved substrates in a factory effluent produced at hundreds or thousands of gallons an hour, a retention time of some 5 days imposed by the growth rate of the methanogenic bacteria could lead to a tank digester of impracticable size or a large number of tanks running in parallel.

In such circumstances, to overcome the inherent disadvantage of the low growth rate of the methanogens, recourse is made to retaining in some way the methanogenic bacteria for longer than the hydraulic retention time. This can be done in the "contact" digester by separating the bacteria from the outflow of a stirred tank and returning it to the tank. However, problems arise in separating an active and gassing bacterial mass from the outflow, and other systems are being tested and used. In the "upflow sludge-blanket" digester, the bacteria grow as a granular mass which is held in the digester tank by a combination of gravity settling, baffles, and upwards movement of the feedstock pumped in at the bottom. A similar system is the "fluidised-bed" digester where the bacteria grow attached to small particles, e.g. glass beads a few millimetres in diameter. The particles are kept floating in the tank by the upwards flow of feed liquid. In the fixed-film digester (or anaerobic filter), the bacteria grow on the surface of a fixed, inert, support matrix of stones, plastic, unglazed porcelain, etc. in blocks, tubes or other shapes. The feed liquid passes over the filter matrix in an upwards or downwards direction. In all these cases, of course, some of the bacterial mass breaks off the support or passes the baffles and escapes from the digester; thus the bacteria have a finite retention time in the tank.

Another type of digester is the "tubular" or "plug-flow" system in which feedstock flows unmixed along a tube at a regular rate. Since, in theory, an input slug of feedstock is not mixed with preceeding feed, no inoculation can take place and inoculation must be provided by feedback of some of the bacteria-containing effluent. Theoretical behaviour in a tubular digester cannot be achieved with a dissolved feedstock and theory becomes less approachable with a gas-producing slurry containing large amounts of suspended solids. However, farm waste digesters on this principle have been built on an experimental scale, mainly because they can offer some structural advantages over a small stirred tank, even though their mode of operation is a compromise between a theoretical tubular reactor and long stirred tank.

Two-Phase Systems

Any of the single-stage types of digester can be combined with a similar or different type. However, with farm animal wastes certain combinations can be used only as a "two-phase" system, not a "two-stage" system. These two systems are different in concept. In the two-phase system an attempt is made to separate the fermentative and methanogenic steps of anaerobic digestion.

While such a phase separation would be easy to achieve in a fermenter system with a sterile feedstock which could be inoculated with the appropriate bacteria, it is extremely difficult with a non-sterile feedstock such as animal excreta which already contains the inoculum. Attempts at obtaining two-phase digestions and some results obtained have been reviewed elsewhere (e.g. Hobson *et al.* 1981), but one particular farm two-phase system described by Colleran *et al.* (1982) will be briefly mentioned here as it could be an alternative to the two-stage systems discussed in the next section (p. 124).

During the establishment of a mesophilic digester flora, production of acids must take place before methanogenesis can start. Acids accumulate and are then utilised as the methanogenic flora develops during the first few weeks after starting a digester. However, if the digester is overloaded, acid production reduces the pH sufficiently to prevent methanogenic bacteria from multiplying; at ca. 35°C, the decrease in pH usually continues until hydrolysis and fermentation cease and the whole digestion goes "sour" (e.g. Hobson & Shaw 1973). When farm animal wastes are kept in storage tanks at ambient temperature, the solids settle and digestion proceeds with slow breakdown of solids and fermentation predominating over methanogenesis. Since the reactions are slow it is possible to terminate the process before the pH becomes too acidic. On the other hand, since maximum degradation of fibre polysaccharides takes some 15 days for pig excreta and 15–25 days for cattle excreta at 35°C (Van Velsen 1977; Bousfield *et al.* 1979; Summers &

Bousfield 1980), to obtain reasonable degradation of solids at less than 20°C requires a very long retention time in a storage tank.

However, it is this slow, storage tank stage that has been used in a type of two-phase digestion of farm wastes by Colleran *et al.* (1982). Pig faecal slurry was allowed to stand at ambient temperature in collecting tanks. Acids accumulated to 11,700 mg/litre during a storage period of 2–3 weeks. The supernatant liquor was then run off and passed through an anaerobic filter with a liquid retention time of 3 days running at 28–30°C where some 90% of the acids was converted to biogas. The large ammonia content gives farm slurries considerable buffering capacity and the slurry in the first (collecting) tank decreased only to pH 6.8 at which value the activities of the bacteria in the filter were not inhibited. Therefore, as the acids were utilised the pH of the filter contents increased to ca. 7.4. As these workers reported the concentration of acids in the "raw" pig slurry to be unusually large, i.e. 9200 mg/litre, some degradation of solids had presumably taken place in the piggery before the slurry was collected for the acidification stage. The additional acidification in the first-phase tank represents conversion of only ca. 7% of the solids present, considering the degradable solids as hexose polymers. Hydrogen and CH_4 produced in the acidification stage would have been lost.

If the two phases of acidification and methanogenesis are not completely separated, then a two-phase digester becomes an inefficient two-stage digester.

Multi-Stage Systems

In a multi-stage digester the reaction vessels are connected in series, the output from the first becoming the input to the second, and so on. There is no feedback from the final output, or from any of the intermediate vessels, to the input, but it would be possible to add fresh substrate to each of the vessels.

Physically a two-stage digester resembles a two-phase digester. The distinction lies in that in the two- or multi-stage system all the reactions of digestion—hydrolysis (if required), fermentation of primary substrate, methanogenesis—take place in each vessel.

A multi-stage digester can, in theory, consist of any type of digester or mixture of types. In the case of feedstock with completely dissolved primary substrate, two or more sludge-bed or filter digesters could be connected. With particulate substrates, a first-stage filter is impossible, and a first-stage tank followed by a second-stage filter must operate as a two-phase digester. It might be noted that the titles of some papers suggest that various types of fixed-film digesters can treat, say, "pig waste". However, the texts show that what is being treated is not normal pig waste slurry, but separated slurry from which much of the particulate matter has been removed either mechanically or by gravity. Also the slurry is often much diluted with washing water or other extraneous liquid.

Multi-stage "mechanical" digesters for whole slurries from piggeries, cattle or poultry houses can be only of either the stirred-tank type or the tubular type. There is also one other system involving a stirred-tank digester which will be considered later (see p. 131).

It should be noted that a multi-stage digester cannot give greater feedstock breakdown or gas production than a single-stage digester run at a correspondingly longer retention time. However, in theory, it gives equal digestion at a shorter retention time than a single-stage digester. Thus there are the possibilities of having a smaller digester plant or improving the effluent from an existing digester running at a retention time which does not allow complete digestion of the feedstock.

Theoretical Aspects of Stirred-Tank Digestion

Some of the following theoretical aspects of multi-stage digesters have been previously discussed by Hobson *et al.* (1981).

In a single-stage, continuous-flow, stirred-tank, piggery waste digester, Hobson & McDonald (1980) showed that the conversion of fermentation acid to biogas could be described by the mathematical model that Herbert *et al.* (1956) applied to a single-stage, continuous culture of one bacterium on one limiting substrate. This model presupposes that the feed to the culture is sterile. Farm waste feedstocks contain many bacteria, but the concentration of bacteria which form the degradative and methanogenic flora of the digester is extremely small (Hobson & Shaw 1974). Therefore as far as digestion is concerned, the feed could be assumed to be sterile. The residual acid concentration in steady-state running was that required to keep the bacteria growing at the rate imposed by the retention time of the digester. As retention time was increased, residual acid concentration decreased. All experiments show that the amount of degradation of the solids in the feedstock of farm waste, stirred-tank digesters approaches asymptotically some maximum value as the retention time of the digester increases. This maximum is actually only a small part of the organic solids in the feedstock slurry; the remainder is non-biodegradable under anaerobic conditions.

Thus, the effluent from a stirred-tank digester running at optimum, or near optimum, retention time for degradation of feedstock and production of biogas should contain little degradable substrate. The theoretical tubular digester is designed to maximise substrate utilisation and again the residual substrate concentrations in the effluent will be very small.

The addition of a second stage to such digesters would seem, purely from observations, to be able to offer little in the way of further gas production. Consideration of experimental results suggests that some additional digestion could be obtained only if the first-stage digester is running at less than the optimum retention time, and theory supports this pragmatic suggestion.

Single-Stage, Stirred-Tank Digesters

The theoretical model used by Hobson & McDonald (1980) was originally applied to substrates in solution (Herbert *et al.* 1956). In a digester, the acids and the H_2 and CO_2 used by the methanogens are also in solution. The feedstock solids are not in solution, but the same theoretical treatment can describe solids breakdown in farm waste digesters if it is assumed that the degradable portion of the solids can be divided into two substrates with different rates of degradation. Each maximum specific growth rate (μ_{max}) value then becomes a maximum growth rate of the bacteria on one particular solid and the K_s value (substrate concentration at which half maximum growth rate is obtained) becomes a weight concentration related to surface area of the solid available for attack by bacteria (Hobson 1983).

The undegradable solids are not altered in concentration as they pass through the digester, but each part of the degradable solids is reduced to a concentration $\bar{S}$ such that

$$\bar{S} = DK_s/(\mu_{max} - D) \tag{1}$$

where $\bar{S}$ is the steady-state residual substrate concentration and D is the dilution rate. The amount of solids degraded is then $(S_R - \bar{S})$ where S_R is the input concentration of degradable solids. The solids are converted to acetic acid, H_2 and CO_2. The fermentation acid, plus acid in the feedstock, is converted to CH_4, leaving a residual acid concentration described by Eq. (1). Hydrogen is similarly reduced in concentration by conversion to CH_4, and CO_2 is in excess.

The experimentally determined gasification of solids in a piggery waste digester running at different retention times (at 35°C and pH 7.2) with a feedstock containing 35 g total solids (TS)/litre (of which 70% is organic material) and 3.7 g/litre acetic acid, can be explained by assuming the feedstock to have 14.7 g/litre degradable solids of which 3.7 g is slowly degraded with K_s 0.782 g/litre and μ_{max} 0.139/day, and 11 g is rapidly degraded with K_s 0.50 g/litre and μ_{max} 0.36/day. Figure 1 shows the experimental and computed results. The residual acids in the digester can be described by a methanogenesis reaction with K_s 0.85 g/litre and μ_{max} 0.40/day (Fig. 2). The conversion of H_2 to CH_4 has K_s 2 µg/litre and μ_{max} 0.35/day. Residual H_2 concentrations cannot easily, if at all, be experimentally verified as they are very small.

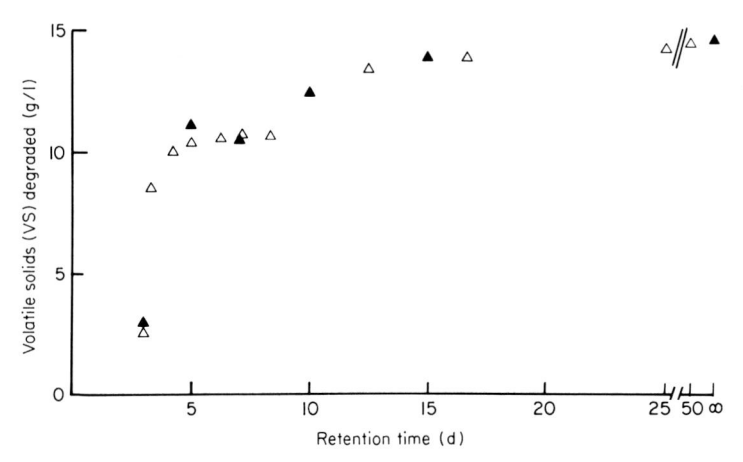

FIG. 1. Volatile solids (VS) degradation at 35°C in a piggery waste anaerobic digester. ▲, Experimental results; △, computed results.

This theoretical treatment can also describe the degradation of cattle waste solids in mesophilic and thermophilic digestions, although the values for K_s and μ_{max}, and the proportions of slowly degraded and rapidly degraded solids, differ from those of the piggery waste digestion. This would be expected as the proportions of fine and coarse solids differ in the wastes

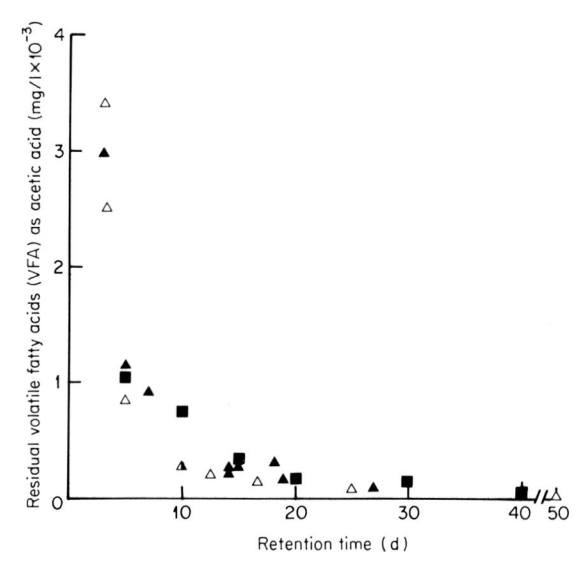

FIG. 2. Residual volatile fatty acids (VFA) during degradation of animal waste at 35°C in anaerobic digesters. ▲, Experimental results, piggery waste; ■, experimental results, cattle waste; △, computed results.

from the different animals, and with cattle the types of faecal solids vary with the feed. The constants for acid degradation should, however, be the same for all digesters running at the same temperature.

The piggery waste digestions in Fig. 1 contained non-inhibitory concentrations of ammonia. Ammonia inhibitions and inhibitions caused by large feedstock solids concentrations can be described in the model by functions which reduce μ_{max}. The effects of temperature of digestion in the mesophilic and thermophilic ranges can be described by modifying μ_{max} values in accordance with the equations of Ratkowsky et al. (1982).

With a single-stage digester run at 10-day retention time with piggery waste, it can be seen from Fig. 1 that some 85% of the degradable solids have been converted to gas and at 15-day retention time some 94% have been degraded. A retention time of 10–15 days is the normal for piggery waste digestions running at about 35°C with feedstock containing about 4–5% (w/v) TS.

Multi-Stage, Stirred-Tank Digesters

The kinetic equations used to describe a pure-culture, dissolved-substrate chemostat can thus describe a stirred-tank anaerobic digestion, as the various reactions are carried out by different groups of bacteria. It might be expected that the development of these equations to multi-stage chemostats would also be applicable to multi-stage anaerobic digesters. The following discussion is based on the paper of Herbert (1964).

If two stirred-tank fermenters of volume V_1 and V_2 are connected in series with a flow (F) of feedstock passing through, then the dilution rate in the first tank (D_1) will be F/V_1 and the dilution rate in the second tank (D_2) will be F/V_2. If the concentration of bacteria in the first tank is x_1 and in the second tank is x_2, then the rate of increase of bacteria in the second tank = growth rate + input rate − outflow rate; that is

$$dx_2/dt = \mu_2 x_2 + D_2 x_1 - D_2 x_2$$

where μ_2 is the growth rate of the bacteria in the second tank. At steady-state conditions $dx_2/dt = 0$, and it then follows that $\mu_2 = D_2 - D_2 (\bar{x}_1/\bar{x}_2)$, where $\bar{x}_1$ is the concentration of bacteria coming from the first tank (also in steady state), and $\bar{x}_2$ is the concentration of bacteria in the second tank. Therefore μ_2 is always less than D_2; also,

$$\bar{x}_2 = D_2 \bar{x}_1/(D_2 - \mu_2)$$

Thus as long as there is substrate and bacteria supplied from the first tank, $\bar{x}_2$ must be greater than zero and no washout will occur however great is D_2. It is thus theoretically possible to run a second-stage fermentation at a dilution rate greater than the maximum growth rate of the fermenter bacteria.

The growth of bacteria and substrate utilisation in the first fermenter is described by the equations in the previous section (*Single-stage, stirred-tank digesters*). Therefore if $\bar{S}_1$ is the residual substrate concentration from the first fermenter running in steady state, then by also equating the rate of change of substrate in the second tank to zero it can be shown that $D_2\bar{S}_1 - (\mu_2 \bar{x}_2/Y) - D_2\bar{S}_2 = 0$, where Y is the bacterial yield coefficient such that $\bar{x}_2 = Y(S_R - \bar{S}_2)$, where S_R is the input substrate concentration to the first fermenter where $\bar{x}_1 = Y(S_R - \bar{S}_1)$.

If the growth rate of the bacteria in the second tank is related to the residual substrate as in a single-stage fermenter, then $\mu_2 = \mu_{max} \bar{S}_2/(K_s + \bar{S}_2)$ where μ_{max} and K_s are as in the first fermenter.

From the above equations a value for $\bar{S}_2$ can be found:

$$\bar{S}_2 = \{-(D_2\bar{S}_1 - \mu_{max}S_R - D_2K_s) \pm [(D_2\bar{S}_1 - \mu_{max}S_R - D_2K_s)^2 - 4 D_2\bar{S}_1 K_s(\mu_{max} - D_2)]^{0.5}\}/2(\mu_{max} - D_2) \qquad (2)$$

The analysis can be applied to more than two tanks in series. It would seem, however, that in the case of anaerobic digestion, economics and practical matters such as difficulties of transfer of sludges would preclude the use of more than two tanks in a farm multi-stage digester. Also from theoretical considerations there seems little reason for having more than two stages.

The multi-stage fermenter can work only if bacteria are growing in the first stage. If the flow rate of feedstock is such that the washout dilution rate of bacteria in the first fermenter is exceeded, then no growth can occur in the first or subsequent stages, unless a subsequent fermenter volume exceeds the critical "washout" volume, when this would become the first stage of a new series; this configuration is not likely.

If all the fermenters are of the same size, washout will occur at the dilution rate equal to that of a single fermenter of this size. Thus the washout dilution rate for a chain of equal size fermenters will be overall much less than that of a single fermenter. To cite Herbert (1964), if washout of a culture occurs at $D = 0.5/h$, then a 1-litre fermenter or a chain of 10 1-litre fermenters will wash out at a flow rate of 0.5 litre/h. A single 10-litre fermenter will wash out at $D = 0.5/h$, but in this case the flow rate is 5 litres/h.

There is also the possibility of supply of feedstock to each stage of a multi-stage tank fermenter. Again, economics and practical difficulties appear to prohibit the use of such a farm digestion system. In theory, however, as long as the first fermenter is run at a dilution rate less than the washout rate, then the feedstock supply to the second fermenter can be as high as desired and washout will not occur in the second fermenter. Therefore this type of fermenter can be operated at higher total feedstock flow rates than the single-stream, two-stage fermenter. The second tank could also be run at a different temperature from the first tank. In this case, of course, μ_{max} values for the bacteria in the two stages would differ. Unlike Eq. (1), Eq. (2)

contains S_R; thus the initial input substrate concentration affects the residual substrate concentration in the second stage ($\tilde{S}_2$). Increasing the input substrate tends to reduce the residual substrate. In the first stage, as in the single-stage digester, the residual substrate is independent of input substrate concentration.

Tubular Digesters

At its simplest, the tubular digester can be considered as a batch culture in space as well as time, and the increase in bacterial growth over the length of the fermenter will be $(x_1 e^{\mu t} - x_1)$ where x_1 is the input concentration of bacteria from the stirred-tank fermenter, and t is the residence time in the tubular fermenter (volume divided by flow rate) (Fig. 3). Substrate utilisation is $(x_1 e^{\mu t} - x_1)/Y$ where Y is the bacterial yield coefficient.

This simple treatment holds only if μ is constant and therefore is applicable only when either the substrate concentration is much above K_s, or t is so short that there is little change in substrate concentration and thus in μ. In practice the substrate concentration from the stirred tank will be usually less than K_s. A tubular fermenter (with bacterial input) can also be considered as a multi-stage tank fermenter with an infinite number of tanks of very small volume, which Powell & Lowe (1964) considered theoretically and mathematically. The detailed mathematical treatment is not easy, but in theory a tubular fermenter of sufficient length should reduce substrate concentrations to very small values because very low growth rates on very small substrate concentrations are possible as the bacteria pass down the tube. In practice, however, a theoretical tubular fermentation cannot be obtained and a tubular fermenter, especially with a gassing reaction, has some characteristics of a stirred tank. With feedstocks such as piggery waste, settlement of solids in the tube provides an additional complication.

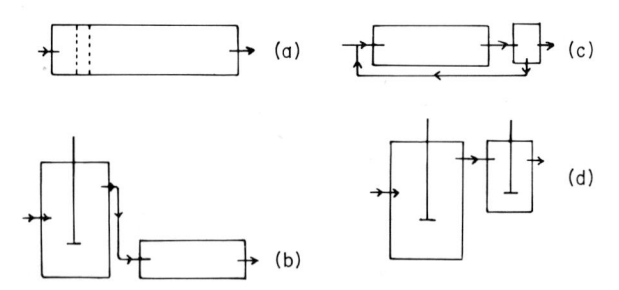

FIG. 3. (a) Tubular digester. Arrows indicate flow and dotted lines the theoretical slug of input passing through the tube. (b) Tubular digester fed from a single-stage, stirred-tank digester. (c) Tubular digester with feedback. (d) Two-stage stirred tank digester.

Some Theoretical and Practical Results from Two-Stage Digesters

Two Stages Run at Different Temperatures

A Low-Temperature Second Stage

One type of two-stage system which is in general use (although not usually considered as such) is a stirred-tank digester plus an anaerobic lagoon or storage tank. This system is, in fact, a necessity for a farm digester where digester effluent is being spread on land as fertiliser. Practical considerations of optimum use of nutrients by plants and the undesirability of spreading liquid slurry onto wet land should preclude slurry spreading on every day of the year and some storage of digested slurry generally is provided. Although it may be possible to provide a volume sufficient only for storage of a comparatively few days of digester output, ideally the storage facilities should be for some 200-day output. Such storage tanks or lagoons will probably be gradually filled by the digester effluent and then emptied over a short period for slurry spreading. Therefore the retention time of the contents will not be uniform. But it is possible, where a number of storage tanks or lagoons are used, for a tank to be filled comparatively quickly and then left for a long time before being emptied.

Slow evolution of gas is visible in such storage tanks or lagoons, and solids slowly circulate with temporary gas flotation or convection currents caused by temperature or wind effects. To a first approximation a lagoon could then be considered as a stirred-tank second stage to a digester, with a very long retention time and running at low temperature. Lagoon storage for 200 days effluent from a 13.5-m^3 stirred-tank, piggery waste digester running at 10 days retention time and 35°C was used in experimental plant. The results from the 13.5-m^3 digester were very similar to those from the 100-litre digester (Mills 1977; Hobson *et al.* 1980; Fig. 1) and could be described by the same model. At 12°C, which was about the maximum temperature attained by the lagoon for much of the 200-day storage, the μ_{max} for acid degradation is reduced from 0.4/day to 0.037/day; there was no ammonia inhibition (Hobson 1983). In Eq. (2), for μ_{max} 0.037/day, D_2 0.005/day, $\bar{S}_1$ 283 mg acid/litre (see Fig. 2), and S_R nominally 5 g/litre, the value computed for $\bar{S}_2$ is 6.3 mg/litre. With S_R 16.5 g/litre [the acids produced from degradation of 3.5% (w/v) total solids input slurry with associated acid at 10 days retention time], $\bar{S}_2$ is 2.2 mg/litre. The residual degradable solids (S_1) from the 10-day retention first-stage digester would be 2.2 g/litre (Fig. 1) which theoretically would be reduced to 0.16 g/litre in the lagoon, the μ_{max} values for solids breakdown being reduced by temperature change.

In practice it is difficult to confirm these results as large undegraded solids gradually settle and compact, making sampling difficult. In the case quoted the average BOD of the digester output was 4250 mg/litre which would be a measure of residual acids and easily biodegradable residues, mainly small particles. The total chemical oxygen demand (COD) of the lagoon liquid, consisting of dissolved material and fine solids, was 300 mg/litre (Mills 1977). These figures cannot be exactly equated with the theoretical calculations, but do show that a decrease in residual biodegradable substrates had occurred.

A Low-Temperature First Stage

The two-phase system of a storage tank plus a filter previously described (p. 123) can be regarded as a first-stage, low-temperature, stirred-tank reactor, although mixing of solids in the first stage is only by convection and gas flotation. Theoretical consideration of the temperature effects on μ_{max} values for solids breakdown shows that at ambient temperatures such as those in Britain, extended retention times are needed for any digestion of solids to take place. Calculations for piggery waste show that at 13°C, washout retention time for solids degradation will be 25 days (Hobson 1983); Van Velsen (1977) found no gas production or solids breakdown in piggery waste in a stirred-tank digester running at 20 days retention time and 13°C. At 15 and 20°C predicted washout retention times for piggery waste are 17 and 8 days, respectively. The first stage of the two-phase plant of Colleran *et al.* (1982) required 2–3 weeks at ambient (unspecified) temperature for some degradation of solids.

Short retention times in either first or second stage require higher than ambient temperatures in temperate climates.

Two Stages Run at the Same Temperature (35°C)

While a storage lagoon after a digester, as used with domestic sewage digesters, is the only two-stage system in present use on farms, future farm digesters might incorporate two mesophilic stages if the extra complexity of the apparatus proved worthwhile. The following experiments are part of a continuing programme of investigation of two-stage farm waste digestion.

Methods

The digester is a stainless steel fermenter, with continuous mechanical stirring, heated or cooled by an external water jacket, and similar in design to the 100-liter digester used in experiments described previously (e.g. Bousfield *et al.* 1974; Summers & Bousfield 1980). The first stage is of 150-litre capacity fed by a Lina-Flow pump (Wilkinson Rubber Linatex Ltd., Stanhope Road, Camberley, Surrey) from a continuously stirred feed vessel. Effluent from this vessel overflows to a second similar vessel which can be operated at 75- or 25-litre capacity. The second vessel overflows to a third, unheated vessel large enough for effluent to be collected, mixed, the volume recorded and sampled for analysis every 2–3 days. The two digester stages have individual systems for liquid sampling. All three vessels have an interconnecting head space; gas is led off to a gas meter and can be sampled for analysis. Analytical methods were described by Summers & Bousfield (1980).

The 150-litre vessel had previously been used for poultry waste digestion (Bousfield *et al.* 1979; Hobson *et al.* 1980). It was adapted to piggery waste by using the previous digestion as seed culture and feeding piggery waste at ca. 20 days retention.

A second inoculum from a cattle waste digester was given in the 20-week interval between the first and second phases of the experiment (Table 1) after a mechanical failure had allowed water into the digester. The digester had been running on piggery waste for some 17 weeks before the first results were obtained; the digester has now been running for 90 weeks. Digestion temperature was 35°C in both stages.

The waste came originally from the same piggery as that used in previous

TABLE 1

Acid degradation in piggery slurry two-stage digesters

Phase of experiment	Retention time in stage 1 digester* (days)	Total solids (%)	Slurry input Acetic acid (mg/litre)	NH_3N (mg/litre)	Digesters Stage 1 acid† (mg/litre)	Stage 2 acid† (mg/litre)
1	14.9	1.1	5172	2155	336 (228)	150 (15)
2	16.4	2.0	2196	1624	144 (153)	<120 (17)
3	14.4	1.4	8070	2810	1015 (628)‡	1208 (55)‡
4	18.4	2.3	2320	1433	160 (136)	<120 (13)
5	15.2	2.2	2423	1838	247 (173)	<120 (19)
6	16.3	4.0	2363	809	262 (153)	<120 (15)
7	15.6	4.1	4215	1278	277 (162)	<120 (8)

* Retention in stage 2 = half that in stage 1. Values are average of four to six determinations over ca. 4-week periods with pump nominally set to nearest day or half day rates.
† Figures in parentheses are computed values.
‡ For input NH_3N of 3100 mg/litre computed values for acid become: stage 1, 2335 mg/litre; stage 2, 432 mg/litre (see text).

experiments (Summers & Bousfield 1980), but a change in the pig drinkers and piggery waste system made only low-solids waste available. This was used at first (phases 1–5) but then there was a change of source of slurry to the Rowett piggery to give a thicker slurry of a more usual consistency.

It was decided that the first-stage retention time would be varied during the experiment and the second stage retention time would be half that of the first stage. A long retention time was used in these initial tests because experiments (e.g. Van Velsen 1977; Summers & Bousfield 1980) had shown that the limits of digestion of solids seemed to be approached at 15–20 days retention time, and the model explaining these results was based on a degradation at infinite retention time only a little greater than at 15 days. Theory then predicted that degradation in a two-stage system could not exceed this infinite retention time limit, i.e. at long retention times there would be little more degradation in a two-stage than in a one-stage system. Low-solids waste was used for the initial experiments on acid digestion as these provided a known amount of acid as feed substrate concentration (S_R). If fermentable solids were present, then S_R would be "acid in feedstock + acid from solids degradation". Substrate acid in the second stage would not be residual acid from the first stage ($\bar{S}_1$), but ($\bar{S}_1$) + acid from solids degradation in the second stage.

It was found that as neither the titrimetric nor gas chromatographic methods available could accurately determine less than ca. 120 mg acid/litre, it was not possible to prove conclusively that the second stage behaved as predicted theoretically. Nevertheless, the results (Table 1) show that decreased, and probably theoretical, acid concentrations can be obtained with a second stage and that fermentation of solids in the first and second stages does not affect appreciably the results. Slurry samples with a large average NH_3N content (Table 1, phase 3), individually had widely varying concentrations, on two occasions being > 3000 mg/litre. At NH_3N concentrations > 1800 mg/litre, inhibition of acid degradation increases rapidly with increase in ammonia; the figures in parentheses show how fluctuations in ammonia could affect residual VFA concentrations. The results suggest also that large ammonia concentrations inhibit acid digestion in the second stage more than in the first stage. Concentrations of NH_3N were the same in the slurry input and in both digestion stages.

Solids Degradation, COD and BOD

In the phases of the experiment with low-solids waste, 1–1.5% (w/v) TS represents mainly salts in solution, some faecal bacteria and a little fibrous material. In theory there would be little solids degradation, and this was so in practice within limits of experimental error. Gas production was from acids in the feedstock.

In the phases with higher solids waste it was expected that degradation could be detected. However, two problems have arisen. The first, the regular movement of slurry from input tank through stages 1 and 2 to output, could also be a problem in large-scale plants. Some hold-ups of solids on the overflow weirs between tanks occurred at intervals and this led to problems in solids determinations. Secondly, the piggery waste contains more sawdust bedding than that from the other piggery used in earlier experiments. This may be a cause of the solids hold-ups on the weirs, but it also means that the organic solids appear less digestible than before.

Analysis of the input TS in phase 7 (Table 1) showed that organic matter and fibre content were greater than had been reported by Summers & Bousfield (1980) for the type of slurry used in phases 1–5, i.e. volatile solids, 80% of TS; neutral detergent fibre, 64% of TS, compared with 70 and 46%, respectively; acid detergent fibre and lignin content was also greater in phase 7 samples. From analyses, and assuming lignin was undigested, it was calculated that in the stage 1 digester there was a loss of 15% of the TS as cellulose and hemicellulose plus 4% as non-fibre material (based on input TS), and in stage 2, 4.5% of TS as hemicellulose (none as cellulose) plus 5% as other material. Similar calculations from analyses had previously been found to agree closely with gravimetric determinations of solids digestion in a single-stage digester (Summers & Bousfield 1980). Determinations of COD and BOD (Table 2) are consistent with these figures. Values for COD_{total}, BOD_{total} and BOD_{acid} show that material other than acid is degraded, e.g. decrease in $COD_{non-acid}$ is 32.1% (20.2% in stage 1, 11.9% in stage 2).

TABLE 2

BOD and COD values in piggery slurry two-stage digester (phase 7)[*]

	Slurry input (mg/litre)	Digesters Stage 1 (mg/litre)	Stage 2 (mg/litre)
Acid	4215	277	<120
BOD_{acid} (calculated)	4400	295	0[†]
BOD_{total}	11,354	2344	2239
$BOD_{non-acid}$[‡]	6954	2049	2239
COD_{total}	57,375	42,571	35,876
$COD_{non-acid}$[‡]	52,975	42,276	35,876

[*] All values are averages of four measurements during phase 7 (Table 1).
[†] Value taken as zero because small amount of acid remaining was less than sensitivity of detection methods, i.e. 120 mg/litre.
[‡] BOD_{total} (or COD_{total}) − BOD_{acid}.

A model using the two-stage theory already described and giving a loss near to that observed in practice can be made if some changes are made in the constants used previously, although more results would be needed to properly verify the model. Some organic matter is more slowly degraded, but the proportions and amounts of slowly and rapidly degraded solids are altered, i.e. K_s 0.782 g/litre, μ_{max} 0.130/day, degradable solids 7 g/litre for slow degradation; and K_s 0.50 g/litre, μ_{max} 0.36/day, 3 g/litre for rapid degradation (compare data on p. 126). The remaining solids in a 4.1% TS input are undegradable. This gives at 16 days retention, computed values of 22% degradation of TS in the first stage and 2% in the second stage.

Bacterial Counts

Over 40 sets of counts of strictly anaerobic bacteria were done during the periods covered in Table 1, i.e. "total" fermentative bacteria on a multi-sugar medium, amylolytic bacteria (starch hydrolysis appears to be a common property of digester bacteria: Hobson & Shaw, 1974; unpublished data) cellulolytic and hemicellulolytic bacteria, and methanogenic bacteria growing on H_2 plus CO_2. The methods used were slight modifications of those applied by Hobson & Shaw (1974).

Counts for total fermentative bacteria were ca. 10^7/ml, for amylolytic bacteria were 10^6/ml and for methanogenic bacteria were 10^5–10^7/ml. However, in all cases counts in the stage 2 digester were less than in stage 1. Table 3 gives some examples.

The slurry input to piggery waste digesters contains large numbers of viable bacteria, but very small numbers of the cellulolytic and other bacteria which form the active digester flora (Hobson & Shaw 1974). Death of many of the original faecal bacteria in the digester, particularly in the second stage, may, in part at least, account for the decrease in counts for total and amylolytic bacteria. Low substrate concentrations may account for death of methanogenic and possibly other bacteria in the second stage. Theory predicts that bacteria will grow in the second stage if substrate is present.

TABLE 3

Counts of bacteria in piggery slurry in two-stage digesters at 35°C (phase 7)

Groups of bacteria	Stage 1 (bacteria/ml)	Stage 2 (bacteria/ml)
"Total" fermentative bacteria	3.2×10^7	2.1×10^7
	1.7×10^7	1.3×10^7
Amylolytic bacteria	9.8×10^6	9.2×10^6
	5.4×10^6	3.4×10^6
Methanogenic bacteria	2.3×10^7	1.3×10^6

Only once in the period up to phase 6 (Table 1) were cellulolytic bacteria detected, i.e. only 6×10^2/ml in stage 1. During phase 6 1×10^4 cellulolytic bacteria/ml were detected in stage 1, but none in stage 2. In phase 7 very small numbers of cellulolytic bacteria were only occasionally cultured. A small cellulolytic population present may be almost entirely attached to fibre particles and not dislodged by the isolation techniques. Therefore counts may be an underestimate of the population. Hemicellulolytic bacteria were cultured from both stages in phase 7, the only phase examined for these organisms. Counts of lactate-fermenting and lipolytic bacteria showed the same pattern as the others; i.e. lower counts were obtained in stage 2 than in stage 1.

Conclusion

A two-stage digester system for farm wastes can take a number of forms, but two stages operated as stirred tanks at the same temperature (35°C) should give most rapid and efficient operation; more than two stages would offer little, if any, advantage.

The experiments described here are only part of a long-term investigation of two-stage mesophilic digestion and obviously a large amount of work remains to be done. However, at the longer retention times tested, the digestion of acid supports the suggestion that the behaviour of a two-stage digester is the same as the theoretical performance of a two-stage, pure-culture fermenter supplied with dissolved substrate. Although mechanical problems have arisen in investigation of solids degradation, so far as can be shown at present the digestion of solids, as determined by COD removal, appears to be described by the extension of this culture theory to solids digestion with more than one substrate. However, bacterial counts in the second stage digester suggest that death of bacteria at low dilution rates and low substrate concentrations predominates over growth.

Theory and practice suggest that substrates not degraded at extended retention times in a single-stage digester will not be degraded in a two-stage system. Further experiments will show whether good digestion can be obtained with two stages running at very much shorter retention times than the optimum for a single-stage digester. The main use for a second digester stage for farm wastes may lie in polishing digester effluent to make the liquid suitable for discharge to a water course.

Acknowledgements

We would like to thank Mr. A. Richardson and his assistants in our department for the bacterial counts.

References

BOUSFIELD, S., HOBSON, P. N. & SUMMERS, R. 1974 Pilot-plant high-rate digestion of piggery and silage wastes. *Journal of Applied Bacteriology* **37**, 11.

BOUSFIELD, S., HOBSON, P. N. & SUMMERS, R. 1979 A note on anaerobic digestion of cattle and poultry wastes. *Agricultural Wastes* **1**, 161–163.

COLLERAN, E., BARRY, M., WILKIE, A. & NEWELL, P. J. 1982 Anaerobic digestion of agricultural wastes using the upflow anaerobic filter design. *Process Biochemistry* **17**, 2, 12–17.

HERBERT, D. 1964 Multi-stage continuous culture. In *Continuous Cultivation of Microorganisms* ed. Malek, I. *et al.* pp. 23–44. Prague: Academia.

HERBERT, D., ELSWORTH, R. & TELLING, R. C. 1956 The continuous culture of bacteria: a theoretical and experimental study. *Journal of General Microbiology* **14**, 601–622.

HOBSON, P. N. 1983 The kinetics of anaerobic digestion of farm wastes. *Journal of Chemical Technology and Biotechnology* **33B**, 1–20.

HOBSON, P. N., BOUSFIELD, S. & SUMMERS, R. 1974 The anaerobic digestion of organic matter. *Critical Reviews in Environmental Control* **4**, 131–191.

HOBSON, P. N., BOUSFIELD, S. & SUMMERS, R. 1981 *Methane Production from Agricultural and Domestic Wastes*. London: Applied Science Publishers.

HOBSON, P. N., BOUSFIELD, S., SUMMERS, R. & MILLS, P. J. 1980 Anaerobic digestion of piggery and poultry wastes. In *Anaerobic Digestion* eds. Stafford, D. A., Wheatley, B. I. & Hughes, D. E., pp. 237–253. London: Applied Science Publishers.

HOBSON, P. N. & MCDONALD, I. 1980 Methane production from acids in piggery-waste digesters. *Journal of Chemical Technology and Biotechnology* **30**, 405–408.

HOBSON, P. N. & SHAW, B. G. 1973 The anaerobic digestion of waste from an intensive pig unit. *Water Research* **7**, 437–449.

HOBSON, P. N. & SHAW, B. G. 1974 The bacterial population of piggery waste anaerobic digesters. *Water Research* **8**, 507–516.

MCINERNEY, M. J. & BRYANT, M. P. 1981 Basic principles of bioconversion in anaerobic digestion and methanogenesis. In *Biomass Conversion Processes for Energy and Fuels* eds. Sofer, S. S. & Zaborsky, O. R., pp. 277–296. New York: Plenum.

MILLS, P. J. 1977 Comparison of an anaerobic digester and an aeration system treating piggery waste from the same source. In *Food Fertilizer and Agricultural Residues* ed. Loehr, R. C., pp. 415–422. Ann Arbor: Ann Arbor Science Publishers.

POWELL, E. O. & LOWE, J. R. 1964 Theory of multi-stage continuous cultures. In *Continuous Cultivation of Microorganisms* eds. Malek, I. *et al.* pp. 45–53. Prague: Academia.

RATKOWSKY, D. A., OLLEY, J., MCMEEKIN, T. A. & BALL, A. 1982 Relationship between temperature and growth rate of bacterial cultures. *Journal of Bacteriology* **149**, 1–5.

SUMMERS, R. & BOUSFIELD, S. 1980 A detaile study of piggery waste digestion. *Agricultural Wastes* **2**, 61–78.

VAN VELSEN, A. F. M. 1977 Anaerobic digestion of piggery waste. 1. The influence of detention time and manure concentration. *Netherlands Journal of Agricultural Science* **25**, 151–169.

Methanogenesis in the Anaerobic Treatment of Food-Processing Wastes

B. H. Kirsop, M. G. Hilton, G. E. Powell and D. B. Archer

Agricultural Research Council, Food Research Institute, Norwich, UK

Disposal of industrial waste organic matter may involve either treatment on site or discharge via the sewerage system for treatment by the local authority. Because of the cost of the latter approach, there is increasing interest in treatment on site and, especially, in anaerobic treatment. The two principal advantages of this are that the expensive and energy-intensive process of aeration is eliminated and that a useful fuel, biogas, is obtained. Biogas contains approximately 70% CH_4 and 30% CO_2 and may be used in conventional burners with little difficulty.

In spite of its basic attraction, the use of anaerobic digestion for methanogenesis has not become widespread, principally because the conversion of organic matter to CH_4 is often slow, but also because the process has a poor reputation for reliability. Slowness results in increased capital cost and poor reliability makes for substantial difficulties when waste utilisation is an integral part of a production process. Detailed consideration of the potentialities of the process, however, has suggested that the difficulties reflect the shortage of detailed knowledge of the microbiology of methanogenesis. It is believed that, with proper investigation, a rapid process can be developed, and that the improved knowledge of the physiology of anaerobic digestion will allow reliable control protocols to be developed.

Methane is formed by very strictly anaerobic methanogenic bacteria, which are a group of morphologically diverse organisms unified by physiological and other characteristics and now known as members of one of the three recently proposed "urkingdoms" of microorganisms, the archaebacteria (Woese & Fox, 1977; Woese *et al.* 1978). The initial proposal that the archaebacteria should be separated from the eubacteria was based on studies of the highly conserved 16S ribosomal RNA of these organisms (Fox *et al.* 1977), but it has subsequently been shown that the archaebacteria—which includes also extreme halophiles, acidophiles and thermophiles—are characterised by the presence of unusual RNA molecules in the translation apparatus, by the absence of teichoic acids in the cell walls and by the presence of phytanyl, ether-linked, glycerol compounds as lipids, which replace the fatty acid glycerol esters characteristic of eubacteria. Analysis of 16S RNA has allowed methanogens to be classified into orders, families, genera and species, and the hierarchy established (Balch *et al.* 1979) is in agreement with arrangements based on more traditional characteristics. Recent reviews may be consulted for full description of the taxomony, morphology and physiology of the methanogens which now include 12 genera (Zeikus 1977; Balch *et al.* 1979; Wolfe 1979, 1982; Kirsop 1984).

In nature and in waste utilisation plants methanogens occur in mixed population; they form the final link in the chain of events which converts organic matter anaerobically to CH_4 and CO_2. They are able to utilise relatively few substances for methane formation. In anaerobic digesters and in nature the significant substrates are almost always acetate and H_2/CO_2 although formate, methanol and methylamines are also methanogenic substrates which can be produced from appropriate wastes. Usually, approximately 70% of the CH_4 generated arises from acetate (Jeris & McCarty 1965; Smith & Mah 1966). The reaction is an aceticlastic one, with CH_4 arising principally from the methyl group of acetate (Barker 1956; Smith & Mah 1978).

The breakdown of organic matter by fermentative bacteria leads to the formation of acetate, H_2 and CO_2, which are used as energy sources by methanogens, and to the production of other fatty acids, organic acids and a range of compounds which are inaccessible to the CH_4-producing bacteria. Methanogenic populations therefore need to contain, in addition to fermentative bacteria, organisms which are able to metabolise these non-methanogenic end products to yield substrates used by methanogens. Organisms able to do this are collectively known as acetogens. The organisms responsible are relatively little studied, primarily because they usually exist in syntrophic association with other bacteria and cannot be cultured as single organisms.

The growth rate of methanogens utilising acetate is relatively low, primarily because the energy yield is low. Generation times are long, often at least 24 h and in many cases of the order of several days. It is therefore not surprising that the rate-limiting step in methanogenesis from soluble wastes is often the conversion of acetate to CH_4 and CO_2 (Archer 1983). Unsatisfactory performance of a digester is usually accompanied by the accumulation of acetate as a result of the inability of the population to metabolise this material at an adequate rate. Accordingly, the study of acetate metabolism by pure and by mixed methanogenic populations is of particular significance in the area of waste utilisation and we are carrying out such a study. Methanogens are widely recognised to be difficult to handle and it is hoped that a detailed description of some aspects of their manipulation will be helpful to others contemplating investigations.

Choice of Experimental Fermenter System

The choice of experimental systems for studying industrial methanogenesis requires difficult decisions to be made regarding the extent to which factors affecting the performance of anaerobic digesters on the larger scale should be incorporated in the laboratory system. In the present work, importance was attached to the fact that large scale fermenters contain solids and that close associations develop between the bacterial population and the surfaces both of the solids and of the vessel. Accordingly, the present study was carried out in fermenters containing finely divided calcium carbonate. Early in the study it was demonstrated that the methanogenic activity was associated with the chalk particles; if the solids were allowed to sediment and supernatant liquid was transferred to another vessel CH_4 was not formed, whereas transfer of solids led to the initiation of a new fermentation. This is in agreement with the work of Barker (1936) and Baresi *et al.* (1978).

Another important consideration is that the rate at which waste is supplied to industrial waste utilisation units varies with fluctuations in production rate. Much of the difficulty experienced in using anaerobic systems for waste treatment arises either when shock loads are added to the reactor or when waste production ceases for a period of time. Initially, therefore, a study was made of the influence on CH_4 generation of intermittent supply of acetate, the principal source of CH_4 and the rate-limiting substrate; this was supplied as the sole carbon source, intermittently, and in large concentrations. The intervals were of variable length, so that on some occasions the population was without supplies of carbon for relatively long periods; on others, further nutrient was provided before that previously added had been fully utilised. The study was carried out in batch fermenters using stirred and unstirred conditions. Results from the stirred enrichment have been published (Powell *et al.* 1983).

The data showed that large concentrations of acetate do not inhibit methanogenesis, unlike reports in some of the earlier literature. In some cases ca. $0.16\text{-}M$ acetate concentrations were attained without any indication that CH_4 generation was inhibited. It may be that, in previous studies, inhibition of methanogenesis resulted either from changes in pH associated with the accumulation of acetate or from inhibitory effects of cations present at large concentrations. Calcium, used in these studies as the cation associated with acetate, is relatively non-toxic (McCarty & McKinney 1961) and is also removed from solution as carbonate by combination with CO_2 liberated when acetate is utilised for CH_4 generation. Furthermore, the presence of chalk serves to regulate the pH at levels suitable for the growth of methanogens.

Procedures for Experimental Fermenter Systems

Enrichment Cultures

Sewage sludge (40 ml) from an anaerobic digester at a municipal sewage works is added to 200 ml acetate enrichment medium containing 2% (w/v) calcium acetate (see Appendix) in a 250-ml flask equipped with a gassing head. After incubation for 18 days at 37°C, and when gas production has ceased, the solids in the flask are transferred to 800 ml of the same medium contained in a Duran bottle (capacity 1 litre) fitted with a fermentation lock containing 1% (w/v) sodium metabisulphite solution. This is then incubated at room temperature for 14 weeks.

Fermenter Systems

An unstirred, fed batch culture is established by adding the whole second enrichment culture to a 10-litre flask, together with 8.7 litres fermenter medium (see Appendix) containing 1% (w/v) $CaCO_3$ and 300 ml $Na_2S \cdot 9H_2O$ solution (1%, w/v) as reductant; pH is adjusted to 7.0 with $Ca(OH)_2$. At intervals, 400 ml fermenter maintenance medium (see Appendix) is added.

A 1-litre stirred batch culture is initiated by adding 1 litre mixed solids and liquid of a 188-day-old 10-litre unstirred batch culture to a 1-litre fermenter (500 series Mark II, LH Engineering, Stoke Poges, Buckinghamshire) which is stirred and controlled at 35°C. The pH is maintained at 6.0–6.5 without control. At intervals, 40 ml fermenter maintenance medium are added after withdrawing an appropriate volume to ensure that the total volume remains constant.

A 1-litre stirred continuous culture is initiated in a fermenter as specified above. Sewage sludge (100 ml) is added to 900 ml acetate enrichment medium (2%, w/v, calcium acetate) and 30 ml $Na_2S \cdot 9H_2O$ solution (1%, w/v) as reductant; pH is adjusted to 7.2. After incubation for 3 days at 35°C without stirring and when 0.5 l/day gas is produced, stirring (100 rev/min) and medium flow [125 ml/day acetate enrichment medium containing 1% (w/v) calcium acetate, medium pH 7.5–8.0] are begun. At the stirring speed used, the chalk particles sediment such that the aqueous phase occupies the top 25% of the fermenter volume while the remainder contains a particulate phase of $CaCO_3$ dispersed in liquid.

Additions to and Removals from Fermenters

The fermenters and the medium to be added are flushed with O_2-free N_2 throughout the operations. In all batch fermentations, medium is transferred by using the pressure of O_2-free N_2 to pass it into the fermenter. Medium was added to continuous fermentations via a peristaltic pump unit (Model MHRE 2, Watson Marlow Ltd., Falmouth, Cornwall).

Samples are removed from 10- and 1-litre batch cultures by using gas pressure to force liquid through small tubes inserted into the fermenter to a standard depth. In the case of continuous fermentation, standard devices of the fermenter are used.

Gas Collection and Measurement

Gas released from the continuously fed fermenters may be separated from liquid effluent by a simple T-piece. Liquid passes into a U-tube which prevents access of gas to the effluent-collecting flask; the volume of liquid collection per day is recorded.

Collection of biogas over water acidified with sulphuric acid to pH 1–2 is unsuitable because of losses caused by solubility of CO_2 in the fluid. Losses are greatly reduced by using NaH_2PO_4 solution (4 M), acidified to pH 2 with H_2SO_4. Gas is collected in gas towers (Fig. 1) and the volume recorded either daily or, in the case of the 1-litre stirred fed batch fermenter, automatically at more frequent intervals by time-lapse photography. Calculations of true gas volume at STP are made by adjusting for the decrease in pressure which arises from supporting a column of liquid of varying height. For the routine assay of CH_4 and other gases a katharometer method is used. Gas samples are removed from the collecting towers by syringe and injected via a gas sampling loop onto a a Pye series 104 Katharometer containing a stainless steel column, 76.2 cm × 6.35 mm diameter, packed with Porapak Q (Waters Associates Inc, Mass; USA) 80–100 mesh; a copper delay column, 213.4 cm × 3.18 mm diameter, empty; and a stainless steel column, 274.3 cm × 6.35 mm diameter, packed with Molecular Sieve 5A, 30–60 mesh (Phase Separations Ltd; Clywd). The columns are operated at ambient temperature with argon as the gas phase, flow rate 50 ml/min. The filaments are also operated at ambient temperature with a bridge current of 90 mA.

FIG. 1. Gas collection from the fed batch calcium acetate enrichment. Gas levels are recorded by time-lapse photography.

Methanogenesis in Fed Batch Fermentations

Fed batch fermentations were continued for nearly 2 years, throughout which time CH_4 formation varied in a characteristic manner. The rate rose relatively quickly from very low levels, when acetate was added to an inactive culture, until a peak of activity was reached; this peak, which on occasion was prolonged for a few days, was followed by a decline in rate to a low level (Fig. 2). The shape of the individual curves varied with circumstance; in particular the pattern was sharper in the stirred than in the unstirred fermenters (Fig. 2). The response to added acetate occurred less rapidly as the period between the end of the previous cycle and the addition of further nutrient was increased (Fig. 2) (Powell *et al.* 1983). The data as a whole have been analysed (Powell 1983) to give kinetic information (Fig. 3) and this has revealed that the overall pattern consists of a sequence of four distinct phases. These will be discussed separately.

Phase 1 is that in which the rate of CH_4 production rises rapidly, to increase severalfold in less than a day. In view of the lengthy generation times of the

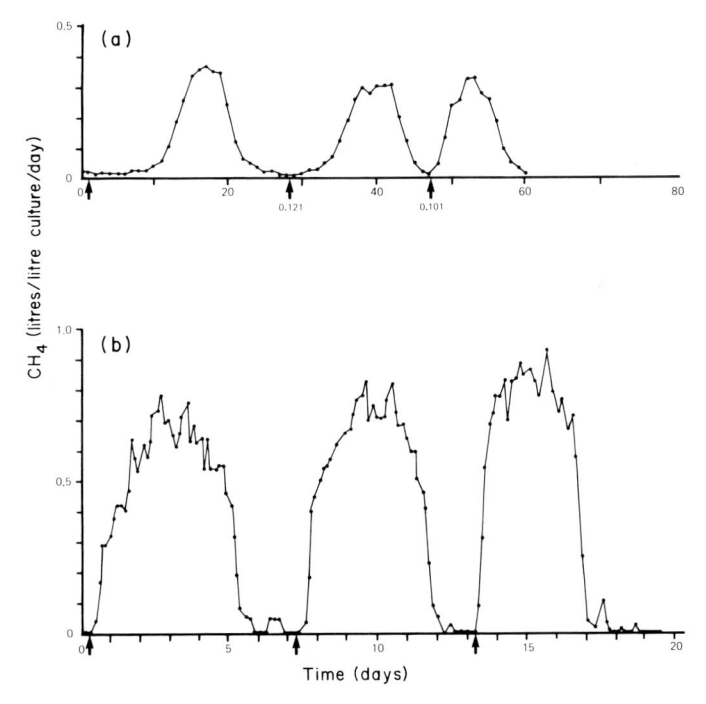

FIG. 2. Rate of methane production from three consecutive cycles in a 10-litre fed batch fermentation. (a) Unstirred; (b) stirred; (▲) Addition of calcium acetate solution at 0.126 mol/litre except where indicated. Note the rapid response of the stirred culture (b) to the addition and depletion of acetate.

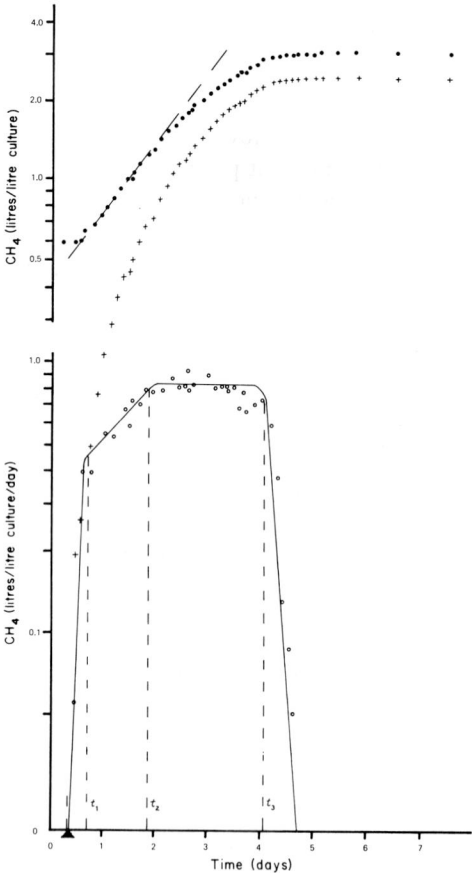

FIG. 3. The kinetics of methanogenesis for a typical cycle from the stirred fed batch calcium acetate enrichment culture. ○, Rate of CH_4 production, ●, accumulated CH_4 since feed time t_0 (uncorrected), +, corrected accumulated CH_4. The logarithmic vertical scales result in phases of exponentially increasing quantities following a linear plot. The kinetically distinct phases begin at the times t_0, t_1, t_2 and t_3; t_4 is the time at which the rate of CH_4 production has fallen to 0.2 litre/litre culture/day. A comparison of the curves (●) and (+) shows that the error introduced by the "zeroing effect" completely masks the early exponential phase 2 (see Powell 1983).

methanogens which are able to utilise acetate, this phase cannot represent cell multiplication; rather it is believed to represent the return to activity of a viable population which has been starved of energy source.

Phase 2 is one of exponential increase in the rate of methane production, during which the specific rate of increase is compatible with the known doubling times of acetate-utilising methanogens. It is concluded that phase 2 is a period in which growth of methanogens is occurring exponentially and is tightly coupled to CH_4 production.

Phase 3 is characterised by a constant rate of CH_4 formation, the level being that reached at the end of phase 2. It clearly represents a particularly interesting phase of methanogenesis, for it suggests that CH_4 production may have become uncoupled from growth. Detailed microbiological information is lacking but on the view that phase 2 represents a period of exponential growth in which methane production is directly coupled with biomass production, it seems that the phase 3 represents either (1) a period in which cell growth and methanogenesis continues but in which cell death also occurs at a balancing rate, such that there is no net increase in cell mass and there is thus a constant rate of methanogenesis by a constant population or (2) a period in which methanogenesis and energy production continue at unchanged rate uncoupled from biomass production. Further work is necessary to determine whether one of these explanations is correct or whether some other reason must be sought.

Phase 4 is a phase of rapid decline in methanogenesis occurring as acetate concentration decreases.

A period characterised by a very slow rate of CH_4 production follows during which the acetate concentration is reduced to very low levels. When this period was relatively short, the addition of more acetate led to an early increase in the rate of CH_4 generation, and phase 2 commenced at the rate close to that found in phase 3 of the previous cycle. This suggests that the methanogens had not lost viability to a substantial degree in the period between cycles. When this period was extended to a number of days, however, phase 2 began at a lower rate of CH_4 formation than in phase 3 of the previous cycle suggesting that the size of the methanogenic population had decreased very considerably.

The foregoing interpretation of events is derived from study of CH_4 generation rates and their variation in response to changed conditions. It will clearly be necessary to confirm this interpretation by direct measurement of the methanogens present in various circumstances. This has not so far proved possible because of difficulties in enumerating methanogens by conventional techniques. Two major problems are that the known acetate-utilising methanogens occur in masses (*Methanosarcina barkeri*) or in long filaments (*Methanothrix soehngenii*) so that colony-forming units are a poor guide to numbers of organisms present. The second difficulty is that these organisms can be difficult to grow on solid media. The procedures listed below allow a wide range of methanogens to be cultured on solid media and have allowed two methanogens, *Methanosarcina barkeri* and a species of *Methanobacterium*, to be isolated from one of the acetate-fed batch fermenters described (Archer 1984). Whilst it seems possible that plating efficiency will improve sufficiently to allow organisms growing as single cells to be enumerated, there seems little doubt that the species growing in masses or filaments must be quantified by indirect means; here the measurement of one of the coenzymes in methanogens, such as F_{420}, will be appropriate (Keltjens & Vogels 1981).

With regard to the composition of the microbial population utilising acetate, it is interesting that the supply of acetate as sole carbon source for prolonged periods (ca. 15 months) did not result in a population containing only acetate-utilising species. Only one of 14 organisms isolated from the uncleared 10-litre batch fermenter, *Methanosarcina barkeri* (Fig. 4) was able to utilise acetate. It is believed that another acetate-utilising methanogen was present, for the characteristic filaments of *Methanothrix soehngenii* were seen under the microscope, but we were unable to isolate this organism. The remaining isolates included one sulphate-reducing bacterium and another methanogen, *Methanobacterium* sp. As neither these nor the other 11 isolates were able to grow in monoculture in the modified acetate enrichment medium, containing acetate as sole carbon and energy source, it is assumed that they are dependent on the acetate-utilising methanogen(s) for essential nutrients. The population present is clearly a mixture of interdependent organisms. Although there was no evidence that acetate-utilising organisms were

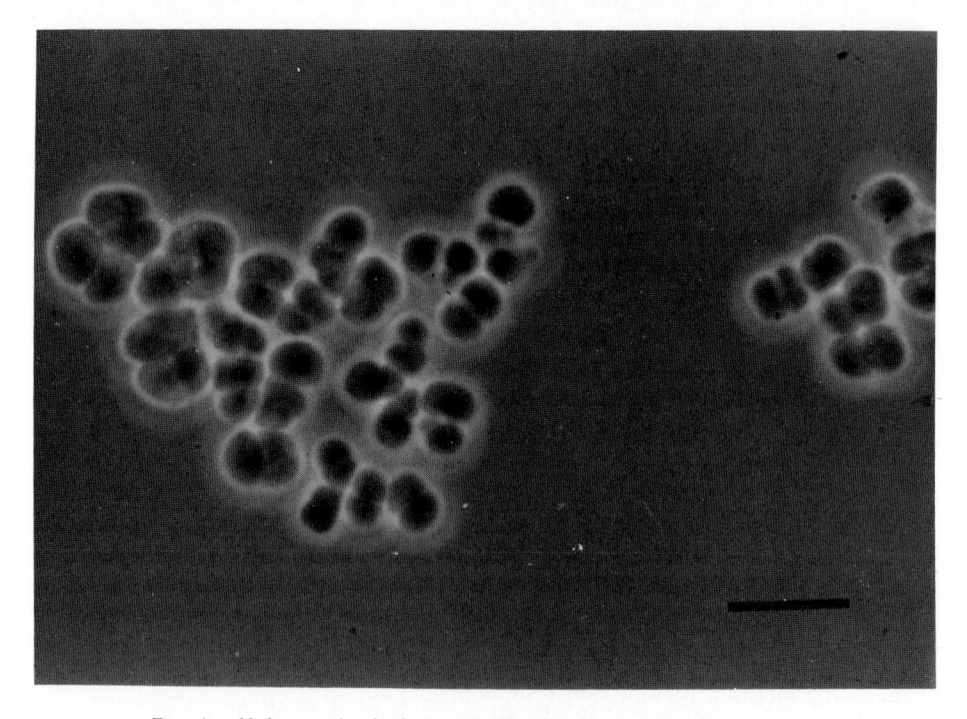

FIG. 4. *Methanosarcina barkeri* strain FR-1 (phase contrast). Bar = 10 μm.

affected by the other bacteria present using the batch culture described here, it seems likely that such a situation occurred in continuous culture supplied with acetate medium. The rate of CH_4 generation increased after initial inoculation with sewage sludge to a high level but subsequently declined gradually over a period of some months. Increasing the concentration of all of the medium constituents or addition of yeast extract failed to restore the rate of gas generation. However, the addition to the culture of a further quantity of anaerobic digester sludge led to a marked increase in rate, even when this material was contained in a dialysis tube which prevented the microorganisms present from contaminating the culture. Addition of a mixture of vitamins (Balch *et al.* 1979) increase the rate of methanogenesis threefold. These findings are interpreted as indicating that organisms producing a nutrient, probably a vitamin, which is stimulatory to methanogenesis were gradually lost from the continuously fed enrichment.

While this suggestion must be regarded as speculative, there is no doubt that the overall rates of waste utilisation and CH_4 production depend on the extent to which the nutritional requirements of the methanogens and non-methanogens are satisfied by constituents of the waste and by primary or secondary metabolites produced by one species and used by another. The precise population established in a waste utilisation plant greatly influences the overall methanogenic activity of that population.

Isolation of Methanogenic Bacteria

Isolation Media

The growth requirements of methanogenic bacteria have been summarised elsewhere (Balch *et al.* 1979; Mah & Smith 1981). Mesophilic and thermophilic species are known; with few exceptions a pH value close to neutrality is required and all species require a low redox potential of about −300 mV for growth. Although some species have specific organic requirements for growth, many are autotrophic. Nearly all methanogens utilise $H_2 + CO_2$ for growth and many utilise formate (Table 1). In addition to $H_2 + CO_2$, *Methanosarcina barkeri* and *Methanosarcina mazei* can utilise acetate, methanol and methylamines as sole carbon and energy sources (Weimer & Zeikus 1978; Balch *et al.* 1979; Mah 1980). Acetate but not $H_2 + CO_2$ is used by *Methanothrix soehngenii* (Zehnder *et al.* 1980; Huser *et al.* 1982). As the nutritional requirements of methanogenic bacteria are not known in all cases, it is common practice to add an undefined nutrient extract of the inoculum habitat to the medium for isolation from complex mixtures of bacteria (Mah & Smith 1981). Methanogenic bacteria are unaffected by antibiotics which

TABLE 1

Diagnostic properties of methanogenic bacteria[*]

Genus	Species	Morphology	Methanogenic substrates[†]	NaCl requirement[‡]	G + C (mol%)
Methanobacterium	formicicum	Short–long rods; filaments	$H_2 + CO_2$; formate	–	40.7
	bryantii	Short–long rods	$H_2 + CO_2$	–	32.7
	thermoautotrophicum	Short–long rods; filaments	$H_2 + CO_2$	–	52
Methanobrevibacter	ruminantium	Short rods	$H_2 + CO_2$; formate	–	30.6
	smithii	Short rods	$H_2 + CO_2$; formate	–	32.0
	arboriphilus	Short rods	$H_2 + CO_2$	–	27.5
Methanococcus	vannielii	Cocci	$H_2 + CO_2$; formate	–	31.1
	voltae	Cocci	$H_2 + CO_2$; formate	+	30.7
	maripaludis	Cocci	$H_2 + CO_2$; formate	+	34.0
	deltae	Cocci	$H_2 + CO_2$; formate	+	40.5
	jannaschii	Cocci	$H_2 + CO_2$	+	31.0
	thermolithotrophicus	Cocci	$H_2 + CO_2$; formate	+	31.3
Methanomicrobium	mobile	Short rods	$H_2 + CO_2$; formate	–	48.8
	paynteri	Short rods	$H_2 + CO_2$	–	44.9
Methanogenium	cariaci	Irregular cocci	$H_2 + CO_2$; formate	+	51.6
	marisnigri	Irregular cocci	$H_2 + CO_2$; formate	+	61.2
	olentangyi	Irregular cocci	$H_2 + CO_2$	–	54.4
	thermophilicum	Irregular cocci	$H_2 + CO_2$; formate	+	59
Methanospirillum	hungatei	Spirillum	$H_2 + CO_2$; formate	–	45
Methanoplanus	limicola	Plate shaped	$H_2 + CO_2$; formate	+	47.5
Methanothermus	fervidus	Short rods	$H_2 + CO_2$	–	33
Methanothrix	soehngenii	Thick rods; filaments	Acetate	–	51.9
Methanolobus	tindarius	Irregular cocci	Methanol; methylamines	+	45.9
Methanococcoides	methylutens	Irregular cocci	Methanol; methylamines	+	42
Methanosarcina	barkeri	Pseudosarcina	H_2/CO_2; methanol	–	38.8
	mazei	Pseudosarcina; cocci	Acetate; methylamines	–	42.2

[*] Modified from Balch *et al.* (1979); references for descriptions of isolations made since 1979 can be obtained from DSM catalogue.
[†] Organic compounds can be stimulatory or required for the growth of some species.
[‡] NaCl may also be stimulatory.

inhibit eubacterial cell wall synthesis (Hilpert *et al.* 1981), and such antibiotics have been used successfully to reduce eubacterial growth during isolations of methanogenic bacteria (Mah & Smith 1981). The following media and procedures for the isolation and maintenance of methanogenic bacteria from an acetate enrichment culture of sewage sludge organisms (Archer & King 1983) have also been found to be useful in the culture of methanogenic bacteria from culture collections and from natural environments.

Defined Medium

A defined medium is described in the Appendix. Cefoxitin (0.1 g/litre), a cefamycin antibiotic (Merck, Sharpe & Dohme Ltd., UK) is added for isolation purposes but not for growth of pure cultures.

The defined medium supports the aceticlastic growth of *Methanosarcina barkeri* and can be modified slightly for the growth of $H_2 + CO_2$-utilising methanogens in the following way: replace TES buffer (see Appendix) with 25 ml/litre Na_2CO_3 solution (8%, w/v) and then proceed as described except that the medium should be boiled, cooled and dispensed under deoxygenated $N_2 + CO_2$ (4:1 by volume); sodium acetate can be omitted except for those methanogens with an acetate requirement for cell carbon synthesis.

Following inoculation the medium should be flushed with $H_2 + CO_2$ (4:1 by volume) and pressurised to 2 atm. All additions, flushings and samplings after sterilisation should be done using syringes and needles by passage through the butyl rubber septa.

Complex Medium

The composition of this medium is that of the basal medium supplemented with yeast extract (5 g/litre), sodium formate (2 g/litre) and K_2HPO_4 (0.9 g/litre); the sodium acetate concentration is reduced to 2 g/litre and the TES buffer is replaced by the addition of Na_2CO_3 as described for the use of the defined basal medium with a $H_2 + CO_2$ headspace. Cysteine sulphide without additional Na_2S solution is the reductant and the medium is dispensed under $N_2 + CO_2$ (4:1 by volume) into Hungate tubes with 0.5 cm deep butyl rubber septa before sterilisation. Before use, the headspace gas is replaced with $H_2 + CO_2$ (4:1 by volume) at 2 atm pressure. Further nutrient additions such as trypticase or sludge and rumen extracts may be necessary for the growth of the more fastidious methanogens.

Solid Media

Liquid medium is solidified with Difco agar (1%, w/v) for use in petri dishes. Roll tubes (Hungate 1969) with 2% (w/v) agar were not used in this work but have been used with these media formulations for other purposes. Some methanogens do not grow well on agar and polysilicate plates must be used (Stetter *et al.* 1981).

General Technique

Plastic, vented petri dishes are pre-reduced by storage in an anaerobic cabinet (Forma, model 1024, USA). Pre-reduced medium is cooled to 52°C and poured into the petri dishes within the anaerobic cabinet and allowed to solidify and dry for 30 min. Plates should be stored in a sealed container; storage of plates within the cabinet should be avoided because at pH 6.8 sulphide within the medium is lost as H_2S. Inoculation of the solid medium from liquid cultures is normally done by streaking the inoculum across the surface of the medium using sealed Pasteur pipettes. Alternatively, 10–20 µl of liquid cultures may be used to inoculate the solid medium without spreading.

Plates are incubated in a stainless steel pressure vessel similar in design to that described by Balch *et al.* (1979). The gas within the pressure vessel (i.e. the gas within the anaerobic cabinet: $N_2 + CO_2 + H_2$, 7:2:1 by volume) is removed by evacuation and replaced with $H_2 + CO_2$ (4:1 by volume) at 2 atm for growth of methanogens on $H_2 + CO_2$.

The pressure vessels contain a palladium catalyst to ensure a reduced atmosphere in the presence of H_2. For growth of methanogens on acetate-containing medium in the absence of H_2 reduction of the medium can be a problem. Although both methanogens isolated from the acetate enrichment culture were obtained in the presence of H_2, we have since confirmed that reducing conditions may be maintained under N_2 by introduction of H_2S gas (Jones *et al.* 1983).

Plates should be examined in the anaerobic cabinet twice weekly. Samples of bacterial colonies may be transferred to liquid medium by picking from the centre of individual colonies with a finely drawn out Pasteur pipette. Colonies of methanogens may be recognised by fluoresence under UV light (Edwards and McBride, 1975) but may be surrounded by satellite colonies of non-methanogens (Ward *et al.* 1978). Therefore purity of a culture should be checked microscopically after growth in complex medium not containing antibiotics, and repeated transfers from solid to liquid media are required to establish purity. Microscopic recognition of methanogenic bacteria is aided by their epifluoresence when excited by light of 420 nm (Mink & Dugan 1977; Doddema & Vogels 1978).

Storage of Pure Cultures

Most laboratories, including that of the Deutsche Sammlung von Mikroorganismen (Hippe 1984), store methanogenic bacteria by freezing in liquid N_2. Cryopreservants such as 10% (v/v) glycerol or 5% (v/v) dimethylsulphoxide have been used but not routinely in this laboratory. Wheaton pre-scored cryules of 2-ml volume (Jencons Ltd., UK) are sterilised and flushed with deoxygenated $N_2 + CO_2$ (4:1, by volume). To a methanogenic culture in mid-exponential phase of growth is added a 100-fold dilution of Ti^{3+} solution (Zehnder & Wuhrmann 1976). A 0.5-ml volume of the culture is transferred to the cryule which is then sealed in the flame as the gassing canula is removed. The cultures are then frozen at $-20°C$ before transfer to liquid N_2. Cultures are thawed at room temperature and fresh medium is inoculated with 10% by volume of the stored culture. Anaerobiosis is maintained either by opening the cryule within an anaerobic cabinet, or by flushing with $N_2:CO_2$ as soon as the cryule top is snapped off. Estimations of loss of viability have not been made, but no cultures have yet been lost using this procedure.

Other methods of storage have been examined and proved to be less reliable. Storage of cultures in growth tubes at 4 or $-20°C$ is unreliable and no detailed studies were made. Some species appear to retain viability when stored as colonies on agar slopes (P. A. Pinn, personal communication). Hippe (1984) has reported the successful freeze-drying of methanogens.

Characterisation of Methanogenic Bacteria

The properties of methanogenic bacteria have been fully documented elsewhere (Balch *et al.* 1979; Mah & Smith 1981). Preliminary identification and characterisation of methanogenic bacteria may be achieved using the simple techniques to be described, but definitive descriptions require more sophisticated techniques (Balch *et al.* 1979).

Production of CH_4 is, of course, a diagnostic property of methanogens and is readily assayed by GLC. In this laboratory small concentrations of CH_4 are estimated using a Pye 104 gas chromatograph equipped with a flame ionisation detector and a 150×0.6 cm glass column packed with Porapack QS, 80–100 mesh (Waters Associates, USA). The oven and detector are both operated at room temperature. Argon is used as the carrier gas at a flow rate of 20 ml/min. Amounts of gas are calculated by comparision with peak heights obtained from standard mixtures of CH_4 in N_2 (Phase Separation Ltd., UK). Fluoresence is also a diagnostically valuable property of methanogenic bacteria but, as reported by Mah & Smith (1981), we have found fluoresence intensity to be dependent upon species and culture conditions.

Assignation to a genus can sometimes be made on the basis of cell morphology in the light microscope. Electron microscopy can reveal other features to aid in identification (Zeikus 1977; Mah & Smith 1981). Nutritional requirements of isolated methanogens can be used as a guide to species designation. Autotrophy can be determined in the basal medium provided that acetate and cysteine are omitted and the final concentration of sodium sulphide is raised to 2 mM by addition of 20 ml 2.5% (w/v) Na$_2$S·9H$_2$O to a 1 litre medium (Archer & King 1983). The ability of the isolate to use a variety of carbon and energy sources is readily determined in this medium. Specific organic requirements, such as the need for acetate as a carbon source, cysteine, other amino acids and 2-mercaptoethanesulphonate (coenzyme M) are documented for some methanogens (Balch *et al.* 1979). A large concentration of NaCl is required by saltwater isolates, a growth medium for which has been described by Balch *et al.* (1979). Notwithstanding the problems of growing the more fastidious methanogens, most known species can be partly characterised by use of a microscope and simple growth tests. However, it is recognised that many species may remain undiscovered because of their complex requirements for growth.

Conclusion

Methanogenesis from acetate is a major, but slow, route in the anaerobic digestion of wastes. Calcium acetate enrichment cultures enable aceticlastic methanogenesis to be studied although it is recognised that the large concentrations of acetate in the enrichments are not found in well-functioning anaerobic digesters. Methanogenic bacteria were isolated from the unstirred fed batch enrichment and the kinetics of methanogenesis from the stirred fed batch enrichment were studied in detail. Interpretation of the gas kinetics from the unstirred enrichment is severely hindered by the effect of substrate diffusion through chalk, with which the methanogenic bacteria are closely associated. The chalk particle size was much smaller in the stirred enrichment and substrate diffusion was not thought to be a limitation in interpreting the gas kinetics.

In each of the calcium acetate enrichments described, *Methanosarcina barkeri* was the predominant acetate-utilising methanogen present. This organism was isolated from the unstirred enrichment but its presence in the stirred fed batch and continuously fed enrichments was confirmed by microscopy. It is expected that *Methanosarcina barkeri* should predominate in the enrichments over the other known mesophilic aceticlastic methanogen, *Methanothrix soehngenii*, because *M. barkeri* has the greater specific growth rate (Smith *et al.* 1978). *Methanothrix soehngenii* would only be expected to predominate when the acetate level is low because the K_s (acetate) is approximately one-tenth that of *M. barkeri* (Smith *et al.* 1978). In all three calcium acetate enrichments described, the bacteria were associated with the chalk layer. No chalk was removed from the unstirred enrichment and therefore, despite its low specific growth rate, it is not surprising that *M. soehngenii* was detected microscopically. It was thought not to be a significant utiliser of acetate until the acetate concentration was small in the fed batch system. Evidence from the kinetic studies indicated that death of *M. barkeri* occurred owing to the small concentrations of acetate at the end of a batch fed cycle in the stirred enrichment. This is probably also the case in the unstirred enrichment and was found to occur in axenic culture (Archer & King 1984). The non-methanogenic bacteria present in the enrichments presumably metabolised the dead *M. barkeri* cells, with a H_2-utilising organism serving as the terminal electron acceptor in the enrichment. A *Methanobacterium* sp. isolated from the unstirred and the stirred enrichments (J. E. Harris, unpublished data) probably participates in interspecies H_2 transfer.

Analysis of the kinetics of methanogenesis from the stirred enrichment revealed four distinct phases, three of which have parallels in conventional pure culture studies: reactivation of a starved population (phase 1), exponential growth (phase 2) and starvation (phase 4). Phase 3 is characterised by a constant rate of methanogenesis. An exponentially increasing production of CH_4 has not been observed previously in acetate enrichment cultures because of the zeroing error in gas volumes (Powell 1983). Plotting rates of methanogenesis instead of accumulated gas volumes circumvents the zeroing error and allows the transition from exponentially increasing rate of methanogenesis to a constant rate to be observed. This transition occurs when the level of acetate is still high, but other essential nutrients may be limiting. Understanding the requirements for exponential growth of bacteria in mixed methanogenic population is essential to ensure optimum performance of a digester, and is important in the industrial context.

Appendix

Media for Enrichment Cultures

Acetate Enrichment Medium

$CaCO_3$, 100 g; $Ca(CH_3COO)_2$, 20 g; NH_4Cl, 1 g; $K_2HPO_4 \cdot 3H_2O$, 0.4 g; $MgCl_2 \cdot 6H_2O$, 0.21 g; resazurin, 0.001 g; trace metals solution (see under *Isolation media*), 10 ml; glass-distilled water to 1 litre. Add to the autoclaved medium 30 ml filter-sterilised solution containing 1% (w/v) $Na_2S \cdot 9H_2O$, 5% (w/v) Na_2CO_3.

Fermenter Medium

NH_4Cl, 1 g; $K_2HPO_4 \cdot 3H_2O$, 0.4 g; $MgCl_2 \cdot 6H_2O$, 0.21 g; resazurin, 0.001 g; trace metal solution, 10 ml; $Ca(CH_3COO)_2$, 10 g; glass-distilled water to 1 water to 1 litre. Add to the autoclaved medium 30 ml filter-sterilised solution of 1% (w/v) $Na_2S \cdot 9H_2O$.

Fermenter Maintenance Medium

$Ca(CH_3COO)_2$, 250 g; NH_4Cl, 1 g; $K_2HPO_4 \cdot 3H_2O$, 0.4 g; $MgCl_2 \cdot 6H_2O$, 0.21 g; trace metals solution, 10 ml; glass-distilled water to 1 litre.

Preparation of Fermenter Media

Sterilisation is at 121°C for 15 min for volumes of less than 250 ml, 30 min for volumes of 0.25–1 litre, 60 min for volumes of 1–5 litres and 90 min for volumes of 5–20 litres. Directly after sterilisation the medium is cooled at room temperature, whilst O_2-free N_2 is passed through (Hungate 1969). Large volumes of media used in continuous cultures are continuously flushed with this gas. Reducing agent ($Na_2S \cdot 9H_2O$) is sterilised by filtration, boiled to deoxygenate and cooled under O_2-free N_2. All tubing is O_2-impermeable butyl rubber (Sterilin Ltd., Teddington, Middlesex).

Isolation Media

Defined Medium

Basal medium: mineral solution, 75 ml; trace metals solution, 10 ml; $FeSO_4$ solution, 1 ml; resazurin solution (0.1%), 1 ml; TES buffer, 100 ml; anhydrous sodium acetate, 4.1 g; glass-distilled water, to 1 litre.

Mineral solution: NaCl, 12 g; $(NH_4)_2SO_4$, 12 g; KH_2PO_4, 6 g; $CaCl_2 \cdot 6H_2O$, 2.4 g; $MgSO_4 \cdot 7H_2O$, 2.5 g; glass-distilled water, 1 litre.

Trace metals solution (modified from Balch *et al.* 1979): nitrilotriacetic acid, 1.5 g; $MgSO_4 \cdot 7H_2O$, 3 g; $MnSO_4 \cdot 2H_2O$, 0.5 g; NaCl, 1.0 g; $FeSO_4 \cdot 7H_2O$, 0.1 g; $CoSO_4$, 0.1 g; $CaCl_2 \cdot 2H_2O$, 0.1 g; $ZnSO_4$, 0.1 g; $CuSO_4 \cdot 5H_2O$, 0.01 g; $AlK(SO_4)_2$, 0.01 g; H_3BO_3, 0.01 g; $Na_2MoO_4 \cdot 2H_2O$, 0.01 g; $NiCl_2 \cdot 6H_2O$, 0.01 g; Na_2SeO_3, 0.01 g; glass-distilled water, 1 litre.

$FeSO_4$ solution: $FeSO_4$, 1% (w/v) in HCl, 1% (v/v).

TES buffer: *N*-tris(hydroxymethyl)methyl-2-aminoethane sulphonic acid (0.5 M, Sigma) buffered to pH 6.8 with 1-M NaOH.

Cysteine–sulphide solution: cysteine hydrochloride, 2.5 g; $Na_2S \cdot 9H_2O$, 2.5 g; distilled water, 100 ml. Dissolve cysteine hydrochloride in water at pH 10.0, add sodium sulphide solution, boil, cool under deoxygenated N_2, dispense under N_2 in stoppered tubes; sterilise at 121°C for 15 min.

Boil and cool the medium under deoxygenated N_2 (Hungate 1969). Add 15 ml cysteine–sulphide solution and 5 ml 2.5% (w/v) $Na_2S \cdot 9H_2O$. Check that the pH is 6.8 and measure Eh (ca. −200 mV). Dispense the medium under N_2 into serum bottles, stoppered with 1.5 cm deep butyl rubber septa (Bellco Inc. USA) and sterilise at 121°C for 15 min. A Ti^{3+} solution

(Zehnder & Wuhrman 1976) can be used to lower the Eh to approximately −400 mV, but is not used routinely. Medium containing Ti^{3+} supports aceticlastic growth of *Methanosarcina barkeri* in the absence of H_2 but some precipitation in the medium occurs.

Acknowledgements

We are grateful for excellent technical assistance by M. Tatton, H. Gravenall and R. Makins.

References

ARCHER, D. B. 1983 The microbiological basis of process control in methanogenic fermentation of soluble wastes. *Enzyme and Microbial Technology* 5, 162–170.

ARCHER, D. B. 1984 Hydrogen-using bacteria in a methanogenic acetate enrichment culture. *Journal of Applied Bacteriology* 56, 125–129.

ARCHER, D. B. & KING, N. R. 1983 A novel ultrastructural feature of a gas-vacuolated *Methanosarcina*. *FEMS Microbiology Letters* 16, 217–223.

ARCHER, D. B. & KING, N. R. 1984 Isolation of gas vesicles from *Methanosarcina barkeri*. *Journal of General Microbiology* 130, 167–172.

BALCH, W. E., FOX, G. E., MAGRUM, L. J., WOESE, C. R. & WOLFE, R. S. 1979 Methanogens: re-evaluation of a unique biological group. *Microbiological Reviews* 43, 260–296.

BARESI, L., MAH, R. A., WARD, D. M. & KAPLAN, I. R. 1978 Methanogenesis from acetate: enrichment studies. *Applied and Environmental Microbiology* 36, 186–197.

BARKER, H. A. 1936 Studies upon the methane producing bacteria. *Archiv fuer Mikrobiologie* 7, 420–438.

BARKER, H. A. 1956 *Bacterial Fermentations*. New York: Wiley.

DODDEMA, H. J. & VOGELS, G. D. 1978 Improved identification of methanogenic bacteria by fluorescence. *Applied and Environmental Microbiology* 36, 752–754.

EDWARDS, T. & MCBRIDE, B. C. 1975 New method for the isolation and identification of methanogenic bacteria. *Applied Microbiology* 29, 540–545.

FOX, G. E., MAGRUM, L. J., BALCH, W. E., WOLFE, R. S. & WOESE, C. R. 1977 Classification of methanogenic bacteria by 16S ribosomal RNA characterization. *Proceedings of the National Academy of Sciences of the United States of America* 74, 4537–4541.

HILPERT, R., WINTER, J., HAMMES, W. & KANDLER, O. 1981 The sensitivity of *Archaebacteria* to antibiotics. *Zentralblatt fuer Bakteriologie, Mikrobiologie und Hygiene, Abteilung I. Originale* C2, 11–20.

HIPPE, H. 1984 Maintenance of methanogenic bacteria. In *Maintenance of Microorganisms. A Manual of Laboratory Methods* eds. Kirsop, B. E. & Snell, J. J. S., pp. 69–81. London: Academic Press.

HUNGATE, R. E. 1969 A roll tube method for cultivation of strict anaerobes. In *Methods in Microbiology* eds. Norris, J. R. & Ribbons, D. W., Vol. 3B, pp. 117–132. New York: Academic Press.

HUSER, B. A., WUHRMANN, K. & ZEHNDER, A. J. B. 1982 *Methanothrix soehngenii* gen. nov., sp. nov., a new acetotrophic non-hydrogen - oxidising methane bacterium. *Archives of Microbiology* 132, 1–9.

JERIS, J. S. & MCCARTY, P. L. 1965 The biochemistry of the methane fermentation using C_{14} tracers. *Journal Water Pollution Control Federation* **37**, 178–192.

JONES, W. J., WHITMAN, W. B., FIELDS, R. D. & WOLFE, R. S. 1983 Growth and plating efficiency of methanococci on agar media. *Applied and Environmental Microbiology* **46**, 220–226.

KELTJENS, J. T. & VOGELS, G. D. 1981 Novel coenzymes of methanogens. In *Microbial Growth on C_1 Compounds* ed. Dalton, H., pp. 152–158. London: Heyden.

KIRSOP, B. H. 1984 Methanogenesis. *CRC Critical Reviews in Biotechnology* **1**, 109–159.

MCCARTY, P. L. & MCKINNEY, R. E. 1961 Salt toxicity in anaerobic digestion. *Journal Water Pollution Control Federation* **33**, 399–415.

MAH, R. A. 1980 Isolation and characterisation of *Methanococcus mazei*. *Current Microbiology* **3**, 321–326.

MAH, R. A. & SMITH, M. R. 1981 In *The Prokaryotes* eds. Starr, M. P., Stolp, H., Trüper, H. G., Balows, A. J. & Schlegel, H. G., Vol. 1, pp. 948–977. Berlin: Springer-Verlag.

MINK, R. W. & DUGAN, P. R. 1977 Tentative identification of methanogenic bacteria by fluorescence microscopy. *Applied and Environmental Microbiology* **33**, 713–717.

POWELL, G. E. 1983 Interpreting the gas kinetics of batch cultures. *Biotechnology Letters* **5**, 437–440.

POWELL, G. E., HILTON, M. G., ARCHER, D. B. & KIRSOP, B. H. 1983 Kinetics of the methanogenic fermentation of acetate. *Journal of Chemical Technology and Biotechnology* **33B**, 209–215.

SMITH, P. H. & MAH, R. A. 1966 Kinetics of acetate metabolism during sludge digestion. *Applied Microbiology* **14**, 368–371.

SMITH, M. R. & MAH, R. A. 1978 Growth and methanogenesis by *Methanosarcina* strain 227 on acetate and methanol. *Applied and Environmental Microbiology* **36**, 870–879.

SMITH, M. R., ZINDER, S. H. & MAH, R. A. 1980 Microbial methanogenesis from acetate. *Process Biochemistry* **15**, 34–39.

STETTER, K. O., THOMM, M., WINTER, J., WILDGRUBER, G., HUBER, H., ZILLIG, W., JANE-COVIC, D., KONIG, H., PALM, P. & WUNDERL, S. 1981 *Methanothermus fervidus*, sp. nov. a novel extremely thermophilic methanogen isolated from an Icelandic hot spring. *Zentralblatt fuer Bakteriologie, Mikrobiologie und Hygiene, Abteilung 1. Originale* **C2**, 166–178.

WARD, D. M., MAH, R. A. & KAPLAN, I. R. 1978 Methanogenesis from acetate: a non-methanogenic bacterium from an anaerobic acetate enrichment. *Applied and Environmental Microbiology* **35**, 1185–1192.

WEIMER, P. J. & ZEIKUS, J. G. 1978 One-carbon metabolism in methanogenic bacteria. *Archives of Microbiology* **119**, 49–57.

WOESE, C. R. & FOX, G. E. 1977 Phylogenetic structure of the prokaryotic domain: the primary kingdoms. *Proceedings of the National Academy of Sciences of the United States of America* **74**, 5088–5090.

Woese, C. R., Magrum, L. J. & Fox, G. E. 1978 Archaebacteria. *Journal of Molecular Evolution* **11**, 245–252.

WOLFE, R. S. 1979 Methanogens: a surprising microbial group. *Antonie van Leeuwenhoek* **45**, 353–364.

WOLFE, R. S. 1982 Biochemistry of methanogenesis. *Experientia* **38**, 198–201.

ZEHNDER, A. J. B. & WUHRMANN, K. 1976 Titanium (III) citrate as a nontoxic oxidation-reduction buffering system for the culture of obligate anaerobes. *Science* **194**, 1165–1166.

ZEHNDER, A. J. B., HUSER, B. A., BROCK, T. D. & WUHRMANN, K. 1980 Characterisation of an acetate-decarboxylating, non-hydrogen-oxidising methane bacterium. *Archives of Microbiology* **124**, 1–11.

ZEIKUS, J. G. 1977 The biology of methanogenic bacteria. *Bacteriological Reviews* **41**, 514–541.

Mathematical Modelling of Methanogenesis in Sewage Sludge Digestion

F. E. MOSEY AND X. A. FERNANDES

Water Research Centre, Stevenage, Hertfordshire, UK

The methane fermentation that occurs naturally in the bottom sediments of rivers, lakes and marshes is widely used by the water industry for stabilising the sludges that arise from the purification of sewage (see also Kirsop *et al.* this volume). The fermentation is traditionally carried out in large concrete tanks fitted with floating steel gasholder roofs (Fig. 1). The fermentation gas, known in the trade variously as "digester gas", "sludge gas" or "biogas", typically comprises 65% CH_4 and 35% CO_2, contaminated with traces of H_2S (produced by sulphate-reducing bacteria) and H_2. The gas is normally burned to provide heat for the digestion tank and at some larger works is used for the generation of electricity.

The role of the traces of H_2 in the overall microbial ecology of this fermentation process is currently the subject of intensive research at the Water Research Centre (WRC), and mathematical modelling of the fermentation process is being used as a research tool to translate microbiological theory into experimentally testable predictions and to provide an operating aid for water industry personnel.

The Anaerobic Digestion Process

As shown in Fig. 2, the sludge digestion process is a strictly anaerobic, bacterial fermentation in which at least four groups of bacteria co-operate to provide a complete breakdown of complex organic biopolymers such as starch and cellulose into CO_2 and CH_4. The generation of energy for bacterial growth generally requires the biochemical oxidation of substrates in the

FIG. 1. Dorr–Oliver type BHG-S sludge digester at Rye Meads sewage treatment works (courtesy of Thames Water Authority).

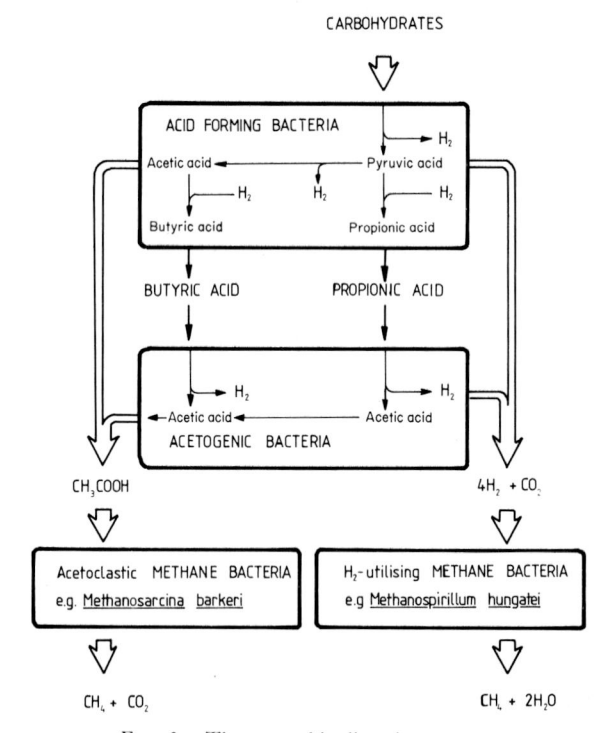

FIG. 2. The anaerobic digestion process.

complete absence of air. This is normally achieved by the extraction of oxygen from water followed by disposal of the surplus H_2. Anaerobic fermentations are always dominated by the need for a suitable H acceptor. In this particular fermentation one group of methane bacteria act as a terminal H acceptor for the other bacteria and in so doing provide a subtle regulatory system for the overall fermentation.

Hydrogen Transfer Reactions in Anaerobic Bacteriology

Bacteria that oxidise substrates to provide energy for growth accomplish oxidation by a series of dehydrogenase reactions which produce a stepwise transfer of H from the substrate to a carrier molecule, usually nicotinamide adenine dinucleotide (NAD).

$$NAD^+ + 2\,[H] \rightarrow NADH + H^+$$

Aerobic bacteria have a mechanism for reoxidising NADH to NAD^+ using the cytochrome chain and dissolved O_2 from the air. Anaerobic bacteria have no such mechanism and the products of anaerobic fermentations are determined by the availability of a suitable acceptor molecule to absorb the "surplus" H generated by the energy-yielding reactions.

The Role of the Acid-Forming Bacteria

The dominant acid-forming bacteria of the sludge digestion process use the Embden–Meyerhof pathway and the preferred product is acetic acid, e.g.

$$C_6H_{12}O_6 + 2\,H_2O \rightarrow 2\,CH_3COOH + 2\,CO_2 + 4\,H_2$$

This reaction provides the bacteria with their largest yield of energy (4 mol ATP/mol glucose) but requires the simultaneous production of large quantities of H_2 and is possible only in environments containing very small concentrations of H_2 (<1000 ppm in the gas phase).

One response of these bacteria to large concentrations of H_2 or low pH values is the production of butyric acid instead of acetic acid, i.e.

$$C_6H_{12}O_6 \rightarrow CH_3CH_2CH_2COOH + 2\,CO_2 + 2\,H_2$$

This reduces the production of H_2 and acid in the growth medium.

Another response is the production of propionic acid which consumes H_2, i.e.

$$C_6H_{12}O_6 + 2\,H_2 \rightarrow 2\,CH_3CH_2COOH + 2\,H_2O$$

The conversion of glucose to propionic acid is used by the bacteria as a method of removing surplus H_2 from the system.

The Role of the H_2-Utilising Bacteria

Hydrogen-utilising bacteria are the "H_2 scavengers" of the sludge digestion process. They normally control the redox potential of the fermentation by conversion of H_2 and CO_2 to CH_4:

$$4 H_2 + CO_2 \rightarrow CH_4 + 2 H_2O$$

Carbon dioxide is normally present in excess and some of the surplus unreacted CO_2 dissolves in the growth medium to form bicarbonate ions which provide the major constituents of the pH buffer. Hydrogen is the rate-limiting substrate for these bacteria and is almost completely consumed, but the traces that remain exert a strong regulatory effect on the metabolism of the acid-forming bacteria.

The Role of the Acetoclastic Methane Bacteria

Acetoclastic methane bacteria are the most important and least studied bacteria in the sludge digestion process. They are responsible for most of the CH_4 produced in the fermentation which they form exclusively from acetic acid:

$$CH_3COOH \rightarrow CH_4 + CO_2$$

They control the pH value of the growth medium by preventing an accumulation of acetic acid. Their very slow growth rate (minimum doubling time of 2–3 days) is often the rate-limiting step in the fermentation. In contrast they are of only minor importance in the rumen.

The Role of the Acetogenic Bacteria

Acetogenic bacteria ferment higher fatty acids, notably propionic and butyric acid, to produce acetic acid:

$$CH_3CH_2COOH + 2 H_2O \rightarrow CH_3COOH + CO_2 + 3 H_2$$
$$CH_3CH_2CH_2COOH + 2 H_2O \rightarrow 2 CH_3COOH + 2 H_2$$

They do not compete well with other bacteria as they are only able to grow in the presence of higher acids and the absence of H_2. The very small initial numbers of these bacteria was largely responsible for the long-term persistence of propionic acid in the computer simulation of a batch digester shown later in this chapter.

Mathematical Modelling of the System

The system has been modelled using classical Monod rate equations which have been described in detail by Mosey (1983). The only unusual feature of the model is the use of a regulator function which regulates a classical Monod rate equation according to the concentration of H_2 in the digester gas. In this model, the rate of fermentation is reduced by an abundance of H_2 to the same extent that respiration would be slowed down by a shortage of O_2. This is perhaps best illustrated for the conversion of glucose to pyruvic acid, i.e.

$$C_6H_{12}O_6 \rightarrow 2\ CH_3COCOOH + 2\ H_2$$

The classical Monod rate equation would be

$$R_G = 1.5X_G[\text{gluc}]/(0.128 + [\text{gluc}])$$

where R_G = unregulated rate of uptake of glucose (mmol/litre/day); X_G = concentration of glucose-fermenting bacteria (mg/litre); [gluc] = concentration of glucose (mmolar); numerical values of the constants are for fermentation at 35°C.

The proposed regulator function used in this model is $1/(1 + k'H)$, where H = concentration of H_2 in the digester gas (ppm); $k' = 1.5 \times 10^{-3}$ ppm.

The value for constant k' was derived theoretically from the known redox potential of the carrier molecule NAD.

The overall rate equation for the conversion of glucose to pyruvic acid then becomes

$$R = \frac{1.5X_G[\text{gluc}]}{(0.128 + [\text{gluc}])(1 + 1.5 \times 10^{-3}\ H)}$$

The effect of this regulator function on the rate of anaerobic carbohydrate metabolism by the Embden–Meyerhof pathway is shown in Fig. 3. At baseline concentrations of H_2 of ca. 30–80 ppm in the digester gas, the regulated rate of carbohydrate breakdown is 90–95% of the unregulated rate and the concentration of H_2 has little or no effect on the overall fermentation.

During surge loads or in the presence of inhibitors, the concentration of H_2 in the gas rises and the rate of carbohydrate metabolism slows down. For example, with 670 ppm H_2, the rate of carbohydrate breakdown would be only half its normal, unregulated value. This is important because the production of acidic intermediates such as acetic and propionic acids is slowed down further, which helps to prevent overloading the natural buffering capacity of the fermentation. To a large extent, it is this type of sophisticated self-regulating feature which makes the fermentation easy to operate and accounts for the widespread popularity of the anaerobic digestion process for sludge stabilisation in the water industry.

Computer Simulations

Two computer simulations of the methane fermentations of glucose are shown in Figs. 4 and 5. Steady-state concentrations of H_2 and of short-chain

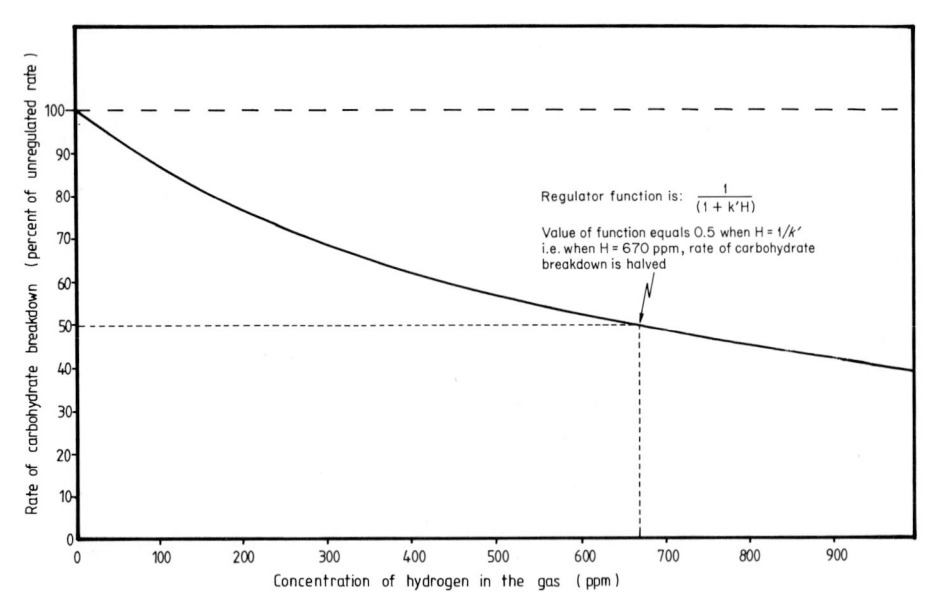

FIG. 3. Regulation of acid-forming bacteria by H_2 content of the digester gas.

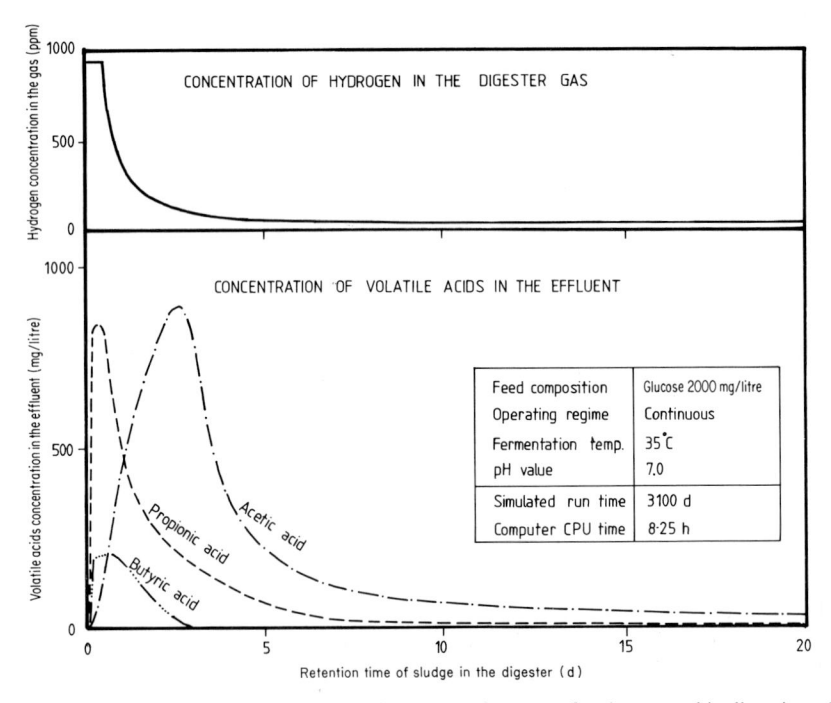

FIG. 4. Computer simulation of the steady-state performance for the anaerobic digestion of glucose.

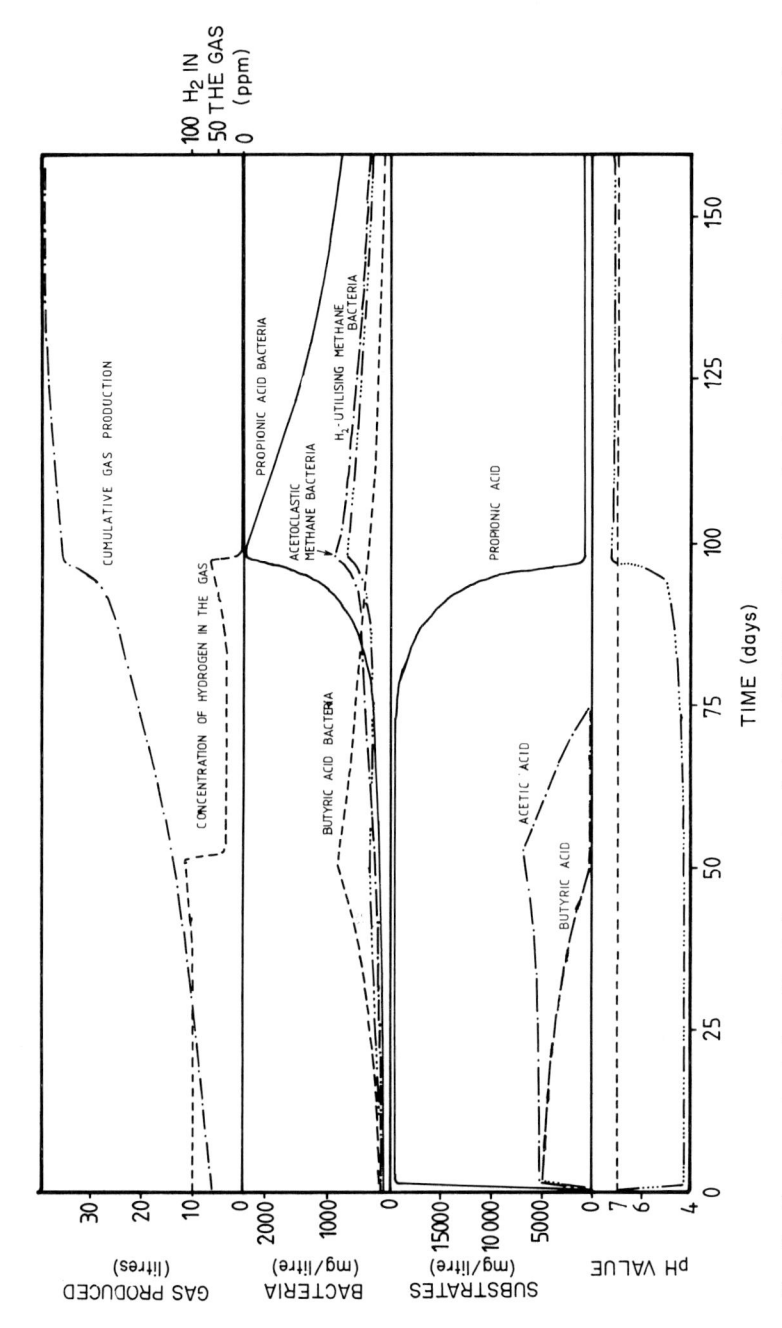

FIG. 5. Computer simulation of batch fermentation of glucose (50,000 mg/litre) by a mixed culture of methanogens and acid-forming bacteria.

fatty acids in a continuous chemostat reactor are simulated in Fig. 4. The predicted performance of a heavily overloaded batch fermenter is simulated in Fig. 5, in which the concentration of H_2 in the gas behaves almost as a monitor of the process, decreasing stepwise as the fatty acids are sequentially metabolised.

Field Measurements of H_2 in Digester Gases

The instrument used for the measurement of trace concentrations of H_2 is shown in Fig. 6. It was developed originally at the University of Newcastle for a medical application (Bartlett *et al.* 1980; Reid *et al.* 1980). A polarographic cell is used as the sensing element which also reacts very sensitively to H_2S. Removal of H_2S before analysis is most easily achieved by injecting the sample through a lead acetate gas-indicating tube connected to the inlet port of the instrument. Typical concentrations of H_2 found in digester gases during a recent WRC survey are shown in Fig. 7. The unusually large concentration recorded at Mansfield is believed to be caused by the presence of chloroform in the digesting sludge, but the other values provide a reasonable indication of the baseline concentrations of H_2 in digester gases.

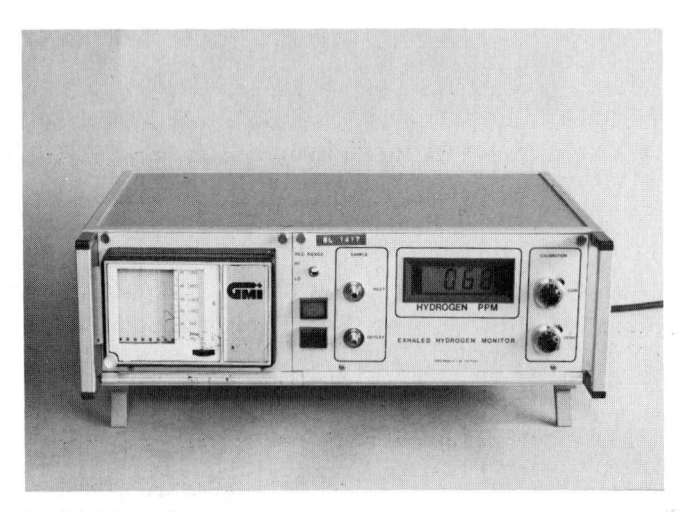

FIG. 6. "Exhaled hydrogen monitor" (GMI Medical Ltd, Renfrew) used to measure trace concentrations of H_2 in digester gas.

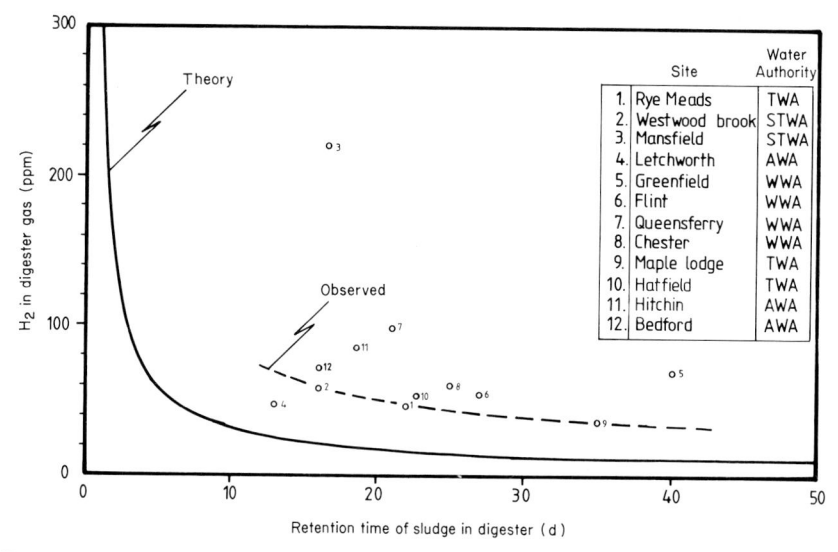

FIG. 7. Concentrations of H_2 in digester gases (WRC survey). (TWA) Thames; (STWA) Severn-Trent; (AWA) Anglia; (WWA) Welsh.

Conclusions

Methanogenesis in sewage digestion is brought about by a symbiotic and self-regulating community of anaerobic bacteria. Discussion of the microbial ecology of this fermentation and mathematical modelling techniques used to investigate H transfer between the individual groups of bacteria indicates a role for H_2 concentration of digester gas in process control. Recent developments in analytical instrumentation permit rapid measurement of trace concentrations of H_2 in the digester gas and offer the possibility of direct on-line monitoring and control of the anaerobic digestion process.

Acknowledgements

The authors would like to express their grateful thanks to the staff of Thames Water Authority, Severn-Trent Water Authority, Anglian Water Authority and Welsh Water Authority for their assistance in the sampling and analysis of digester gases and especially to Mr Collaby (TWA), Mr Noone (STWA), Mr Sale (AWA) and Mr Wheeler (WWA) without whose assistance the measurements would not have been possible. They wish to thank the Water Research Centre for permission to publish this paper.

References

BARTLETT, K., DOBSON, J. V. & EASTHAM, E. 1980 A new method for the detection of hydrogen in breath and its application to acquired and inborn sugar malabsorbtion. *Clinica Chimica Acta* **108,** 189–194.

MOSEY, F. E. 1983 Mathematical modelling of the anaerobic digestion process. Regulatory mechanisms for the formation of short-chain volatile acids from glucose. *Water Science and Technology* **15,** 209–232.

READ, N. W., MILES, C. A., FISHER, D., HOLGATE, A. M., KIME, D. N., MITCHELL, M. A., REEVE, A. M., ROCHE, T. B. & WALKER, M. 1980 Transit of a meal through the stomach, small intestine, and colon in normal subjects and its role in the pathogenesis of diarrhoea. *Gastroenterology* **79,** 1276–1282.

Computer-Controlled Fermenters for Simulation of the Activated-Sludge Process

A. R. Paskins and G. L. Jones

Water Research Centre, Stevenage, Hertfordshire, UK

To investigate microbial responses in natural environments it is often necessary to simplify the system by reducing and controlling the number of variables which are normally present. However, great care must be taken to ensure that the modifications carried out do not change the conditions too much from those occurring naturally.

In a full-scale open system such as the activated-sludge process for waste water treatment there are a number of variables over which no control can be exerted for experimental purposes, e.g. concentration of influent substrate, composition of influent sewage, temperature, pH, flow rate. In the laboratory these variables can be regulated by the use of synthetic sewage and controlled conditions. Other variables occurring in the full-scale process, such as the residence time of the biomass in the settlement tank, remain unrecognised or are ignored when the process is reduced to laboratory scale, although they may have a significant effect on the process.

The microorganisms in activated-sludge systems are subjected to periodic changes in prevailing conditions (Anonymous 1963). The changes occur in the concentrations of nutrients and O_2 available as the bacteria pass through the aeration and sedimentation tanks (Fig. 1). The changes occur with a relatively low frequency which cannot easily be reproduced when the process is scaled down. At some frequencies periodic changes in cultural conditions have been shown to have significant effects on the physiology of microorganisms (Mitchison 1971; Tempest *et al.* 1982). To study the effects of low frequencies of periodic changes experienced in the activated-sludge process, a pure culture of a bacterium in a computer-controlled fermenter was used to simulate on the laboratory scale the changes occurring in full-scale plant.

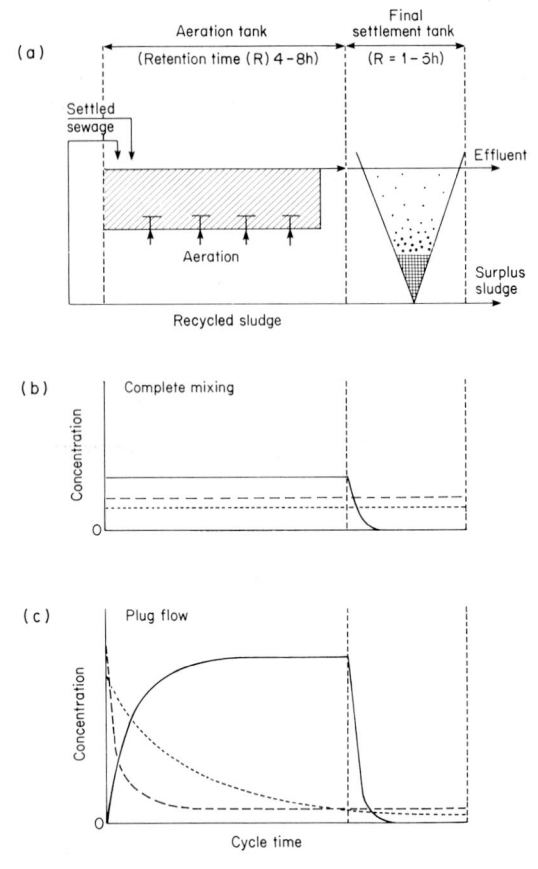

FIG. 1. (a) Diagram of activated-sludge plant; concentration profiles for (b) complete-mixing and (c) plug-flow systems. (——), Dissolved oxygen; (– –), BOD_5; (----), ammoniacal-N.

Experimental Procedure

Cultural Conditions

The fermenters used were 3-litre Porton-type fermenters (LH Engineering Co. Ltd., Stoke Poges, type LHE CC1500) with associated instruments for measuring pH, dissolved O_2, temperature, atmospheric pressure and concentrations of O_2 and CO_2 in the influent and effluent gases (Figs 2 and 3).

The culture used during the simulation experiments was a pure culture of a gram-negative, rod-shaped bacterium isolated from settled sewage, tentatively identified as a coliform member of the Enterobacteriaceae, able to use

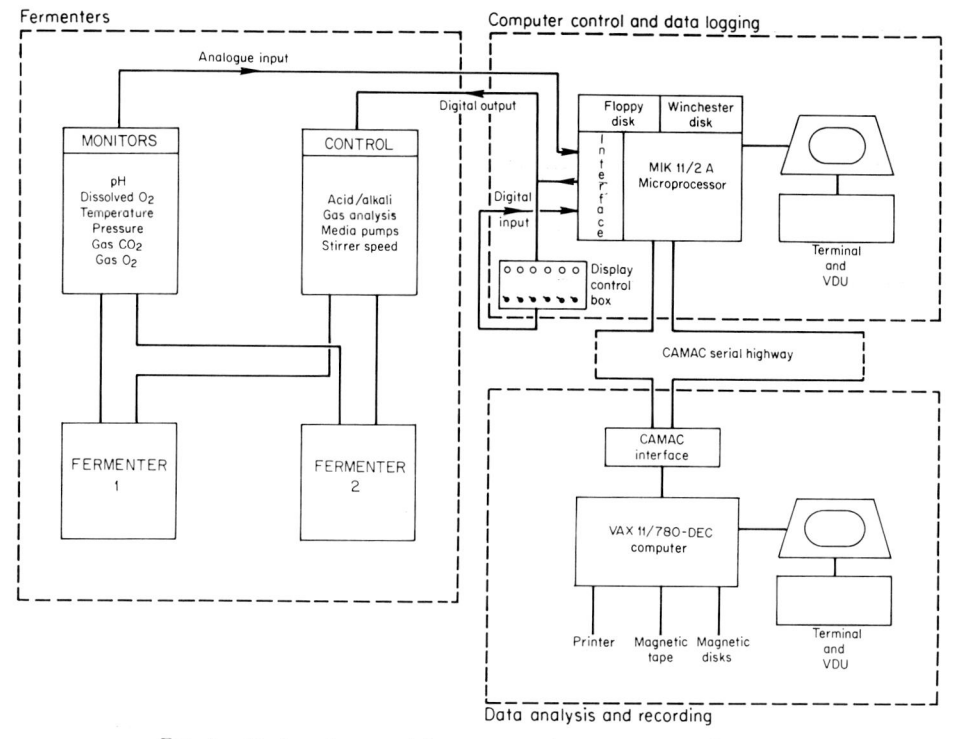

FIG. 2. Outline diagram of fermenters and computer control system.

glucose as sole source of carbon and energy. The culture was maintained on casitone–glycerol–yeast-extract agar medium (CGY) slopes (Pike *et al.* 1972) at 4°C and subcultured on plates at 2-month intervals.

The growth medium used was a glucose–mineral salts medium with sufficient glucose (10 g/litre) to support a dry weight of biomass similar to the dry weight of activated sludge achieved in a treatment plant with recycle. The mineral salts medium (Jones *et al.* 1975) was made up in 20-litre batches and the pH adjusted to 7.0 with 50% (w/v) NaOH before sterilisation for 15 min at 121°C. Glucose was autoclaved separately in concentrated aqueous solutions, to avoid caramelisation, and was aseptically added to the mineral salts solution.

Computer Control

Computer control of the fermenters is used to simulate the cyclic transitions between aerobiosis and anaerobiosis and the discontinuities in supply of substrate which are experienced by microorganisms in the activated-sludge

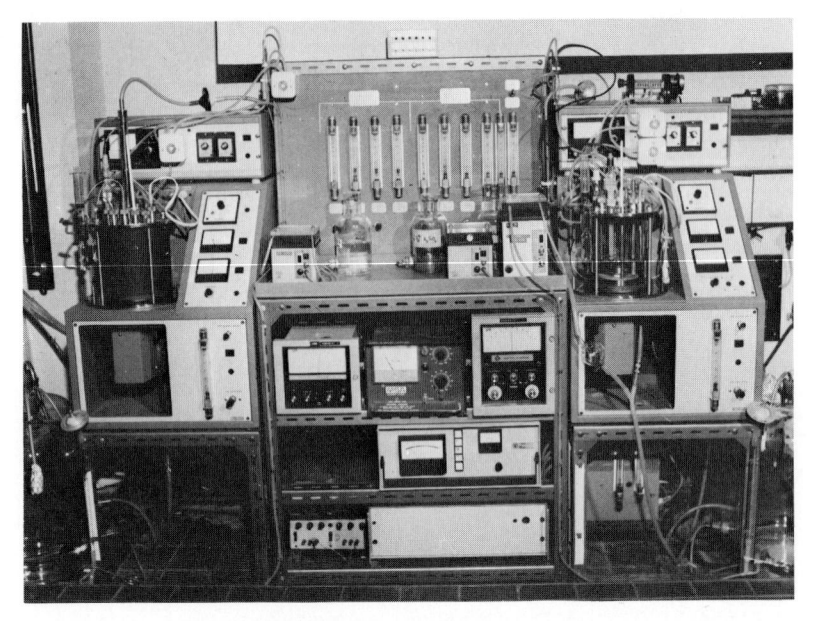

FIG. 3. Fermenters and associated equipment.

process treatment of sewage. Pre-determined cyclic changes in dilution rate and aeration are imposed upon cultures by the computer, together with control of temperature and pH. The computer also interrogates the various monitoring instruments, records the data and is able to recall and print historic data or to compute the values of derived parameters.

Control and data logging of the fermenters are effected with a DEC LSI-11/02-based MIK 11/2A microprocessor mounted in a Standard Engineering CAMAC crate (Fisher Controls; Figs 2 and 4). Modules to accept analogue and digital inputs and to provide digital outputs are also mounted in the crate and are under direct control of the MIK-11. The operating system used is RT-11; disk storage is provided by a 5-megabyte fixed Winchester disk and a 0.5-megabyte 8 in floppy disk (Data Systems DSD880).

Control variables, e.g. control pH, cycle times, are entered into the system via the VT100 terminal. Fixed digital input, such as whether data collection and control are required for a particular fermenter, is entered by operating a switch on the display/control box (Figs 2 and 5). With the microprocessor operational, the fermenters are brought on-line by the "standby/run" switches and this is indicated on the "chemostat" display. The "calib/run" switch disables the gas analysis system to allow calibration of the gas monitors whilst the rest of the control system functions normally; the "run" display is extinguished during calibration. When one of the "normal/experimental" switches is activated, data are collected from a fermenter

FIG. 4. VT100 terminal with visual display unit and CAMAC crate showing floppy disk and fixed Winchester disk above crate, MIK 11/2A microprocessor and interface unit to the serial highway (right side of crate) and analogue to digital converter, output driver and parallel register (left side of crate).

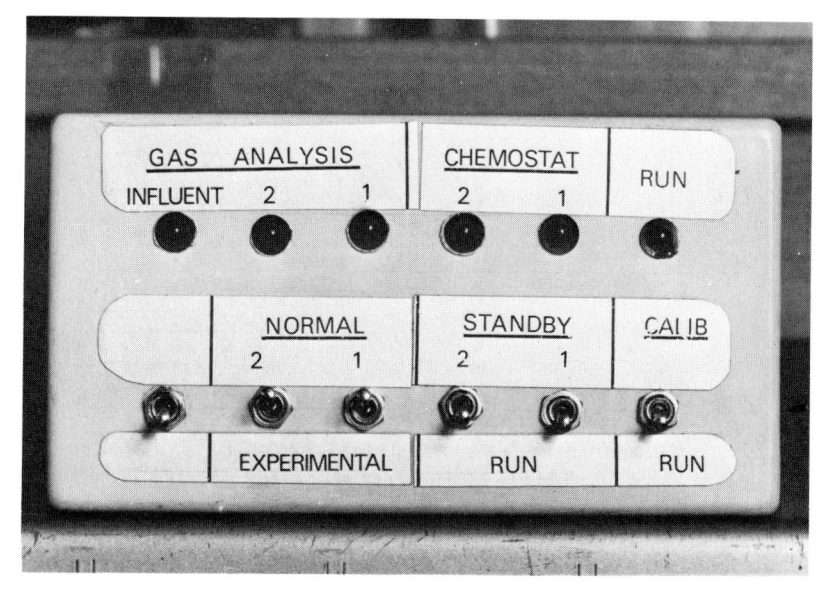

FIG. 5. Display control box. The six light-emitting diodes (top row) display the status of monitoring and control; the switches are used to enter fixed digital input.

at intervals of 20 s instead of the normal 20 min. The "gas analysis" display shows the fermenter being monitored and whether influent (aeration) or effluent (exhaust) gas is being analysed. Either effluent gas is analysed continuously and influent gas once a day or, when the left-hand switch (unlabelled) is activated, influent and effluent gas are analysed alternately.

The system as described can operate independently but is part of a much larger laboratory automation system (Fig. 2). The CAMAC crate is linked together with other crates via a serial highway and interface to the main computer (VAX 11/780-DEC). This permits more extensive analysis and recording of data. Direct access to the recorded data, which is usually stored on magnetic tape, is through interactive terminals to the main computer.

Control is by activation of TTL (Transistor Transistor Logic) logic controllers within the output driver (digital output) board of the CAMAC crate, which produces sufficient current or voltage output to activate a relay in the fermenter system which in turn switches power to the relevant pump, switch or other process. The microprocessor interrogates the various monitoring instruments and switches on the control box and control software at 5-s intervals to determine whether any control action is required, e.g. addition of acid or alkali to control pH.

Collection of Data

Data are normally recorded at 20-min intervals; more frequent data collection was found to be unnecessary for these experiments and consumed too much storage space. The 20-min data are transferred on a daily basis to the main computer where it is recorded on magnetic tape. There is also a daily output on a log sheet. Where frequent data collection is required, data can be collected at 20-s intervals by operating the experimental switch of either fermenter on the control box (Fig. 5). When activated, this causes data to be collected at 20-s intervals until the switch is turned off.

Recent data can be displayed on the VT100 terminal of the CAMAC crate when required, or recorded data can be examined on a VT100 interactive terminal to the main computer (Fig. 2). Most data manipulations are carried out via the main computer because of the greater computing power available to the operator.

Simulations

In a conventional plug-flow system, recycled sludge is mixed with the settled sewage entering the aeration tank (Fig. 1). Substrate removal during the passage of mixed liquor through the aeration vessel approximates to substrate removal in a batch culture system (Pirt 1975). The dissolved O_2 profile of the aeration tank approximates to the reciprocal of the substrate removal curve, i.e. with less substrate available the oxygen demand of the sludge is less and the concentration of dissolved O_2 increases until an equilibrium state is reached. The aeration period of mixed liquor in the aeration tank of a conventional plant is between 4 and 8 h.

After leaving the aeration tank the mixed liquor passes into the final settlement tank where it is allowed to settle. The sludge may take 1–5 h to pass through the settlement tank with an average residence period of 2–3 h (Bolton & Klein 1971). Because of the basal or "endogenous" respiration rate of the sludge, all the dissolved O_2 is consumed by the end of the first hour, leaving the sludge in the settlement tank under conditions of anoxia/anaerobiosis for up to 4 h.

With a complete-mixing system, the nutrient reaches a small, stable concentration and dissolved O_2 reaches a relatively large, stable concentration, with the exception of minor perturbations caused by variations in flow and strength of the sewage. Conditions in the settlement tank are similar to those for the plug-flow system.

To simulate these systems in the 3-litre fermenters, the microprocessor was programmed to switch the nutrient feed pump on and off cyclically and to switch the stirring rate between two pre-set levels. Conditions in the aeration tank were simulated by operating the stirrer at the higher of the two pre-set rates (650 rev/min), allowing vortex aeration. The complete-mixing system was simulated by supplying nutrient medium throughout the aeration period of 7 h. The plug-flow system was simulated by adding the same volume of nutrient medium to the culture in one cycle of operation, but during the first 15 min of the aeration period (Fig. 6).

In both cases, the settlement period was simulated by reducing the stirring speed to 100 rev/min. This was sufficiently fast to keep the culture in suspension but without aeration taking place. Wastage of surplus culture occurred during the periods of media addition by overflow over a fixed weir outlet. The experiments were done at pH 7.0 maintained by the addition of either 20% (w/v) NaOH or 6.7% (v/v) H_2SO_4, and at 20°C.

To calculate the yield coefficient (cell dry wt/unit wt glucose utilised in one cycle) the concentration of glucose in the fermenter and the cell dry wt were monitored during a cycle. Samples (50 ml) were taken aseptically from the fermenter vessel into a polycarbonate centrifuge tube and rapidly frozen by placing the tube in a freezing mixture of solid carbon dioxide and methylated spirits. The frozen samples were centrifuged in a refrigerated centrifuge (MSE High Speed 18, MSE Scientific Instruments, Manor Royal, Crawley, Sussex) for 15 min at 27,000 *g* with the temperature set at 15°C. This allowed the sample to melt and the biomass to form a pellet without the

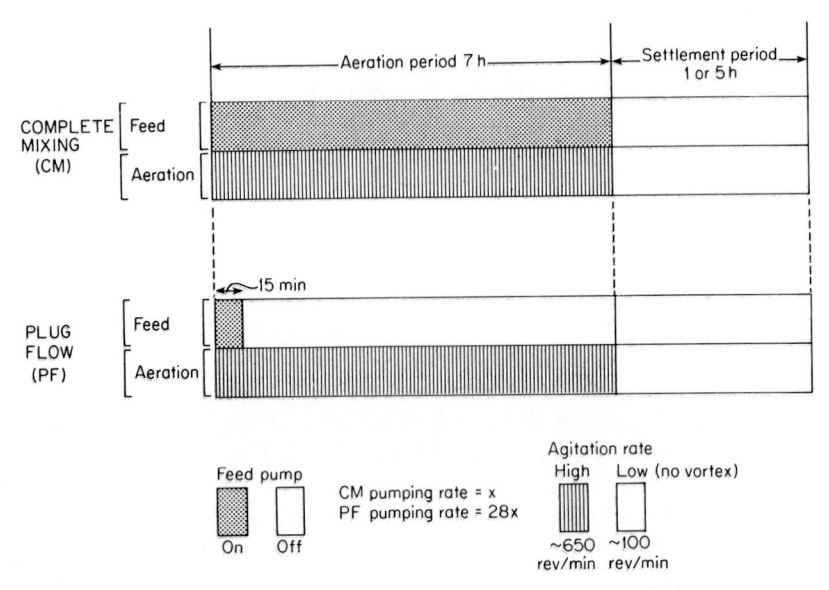

FIG. 6. Periodic changes used in simulation of complete-mixing and plug-flow activated-sludge processes.

temperature of the sample rising above 4°C. The supernatant liquid was removed and the pellet washed with distilled water, centrifuged, resuspended in a minimum of distilled water and dispensed to a pre-weighed Oxoid aluminium cap. The sample was dried overnight at 105°C and removed to a desiccator to cool before weighing. Glucose was determined by enzyme assay of the supernatant using the glucose oxidase–peroxidase method (Boehringer Glucose GOD-Period Test Combination, BCL Ltd., Bell Lane, Lewes, East Sussex). To measure the degree of fermentation occurring, the concentration of acetate in the supernatant was determined by direct gas–liquid chromatographic analysis of the acidified supernatant (Jones *et al.* 1975).

Performance of the System

Some typical results obtained by simulating activated sludge with a pure culture are shown in Figs 7 and 8. Residual dissolved O_2 carried over from the aeration period was rapidly consumed by the culture. No nutrient medium was added during this period, which lasted for either 1 or 5 h. Typical dissolved O_2 profiles obtained are shown in Fig. 7.

The fermenters correctly simulated the high rate of respiration experi-

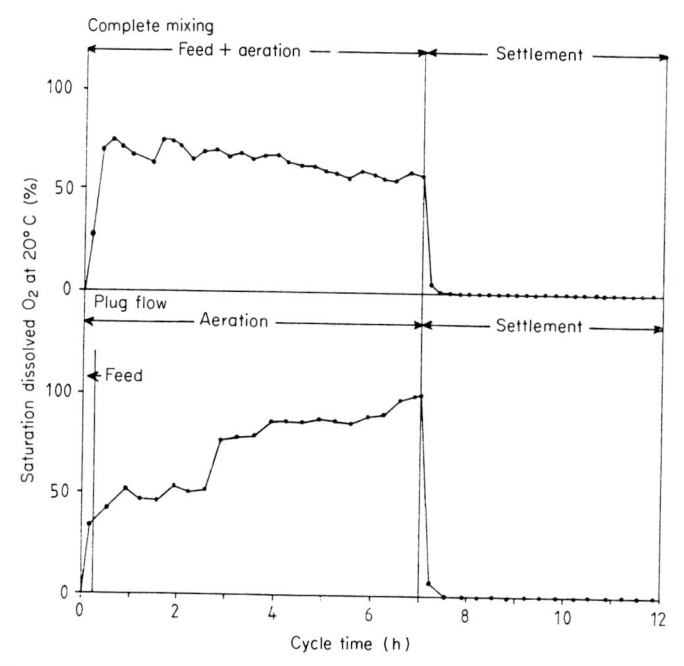

FIG. 7. Typical dissolved O$_2$ profiles of complete-mixing and plug-flow simulations with 5-h settlement phases.

enced when the concentrated returned sludge is mixed with incoming settled sewage and aerated, and the decline in respiration rate in plug-flow plants with increasing period of aeration (Fig. 8).

Potential respiration rates were determined at various times in the cycles using a Rank respirometer (Rank Bros., Bottisham, Cambridge) (Harrison & Loveless 1971). With initial respiration rates addition of glucose was omitted. *In situ* respiration rates were determined by measuring the O$_2$ consumed by the culture in the fermenter. This was measured as the difference in concentration of oxygen between the influent and exhaust gases.

During the aeration periods, the rate of change of specific respiration rate was proportional to the quantity of substrate available to the organisms, i.e. in the complete–mixing system, the rate changed little during the aerobic stage and was close to the endogenous rate (ca. 20 mg O$_2$/g dry wt biomass/h). In the plug-flow systems, the change of respiration rate was more akin to that observed in batch culture. In both plug-flow and complete-mixing systems, significant increases in the initial and potential respiration rates were observed in the settlement phase. These were assumed to be caused by an accumulation of reduced pyridine nucleotide, which remained

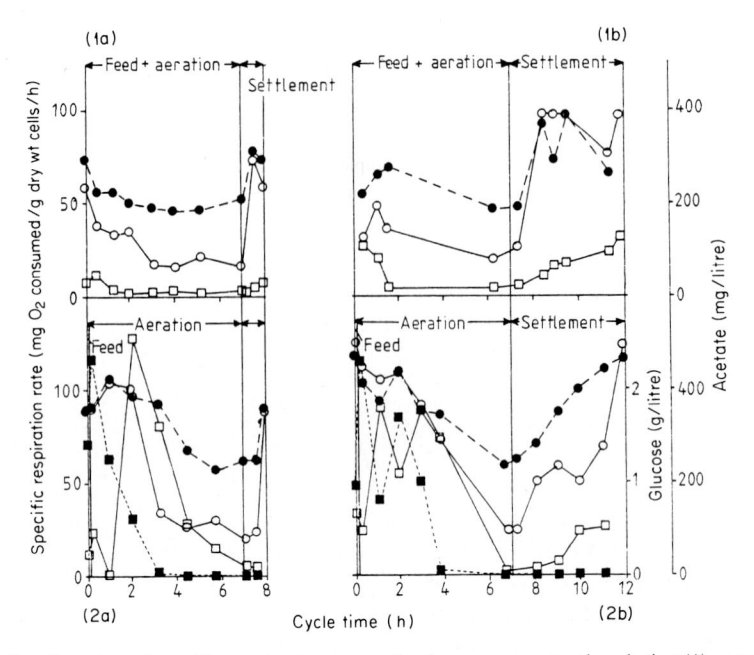

FIG. 8. Variation of specific respiration rate and substrate concentration during (1) complete-mixing and (2) plug-flow cycles with (a) 1-h and (b) 5-h settlement periods. ●, Potential respiration rate, excess glucose added as carbon source; ○, initial respiration rate, no additional carbon source added; ■, concentration of glucose; □, concentration of acetate.

unoxidised until aerobic conditions returned. The rapid oxidation of the pyridine nucleotides was assumed to be responsible for the extremely high *in situ* respiration rates observed during the first minutes of the aeration periods.

The levels of substrate detected in the medium reflect the ability of the aeration system to satisfy the oxygen demand of the culture. With complete mixing (Fig. 8) no glucose was detectable but acetate accumulated especially during the longer settlement phase. With plug-flow significant quantities of glucose and acetate accumulated during the aeration phase although the dissolved oxygen level was not limiting (Fig. 7).

Under simulated plug-flow conditions the yield of cells/unit mass of glucose oxidised was 16% less than with complete mixing (Table 1). With plug flow, flocculation of the culture occurred and was associated with production of an extracellular polymer, whereas with complete mixing the growth was dispersed. These two findings have great significance for the operation of the activated-sludge process since they imply that plug-flow systems will produce less sludge and will produce more rapidly settling solids. This is confirmed by the general acceptance in the waste water indus-

TABLE 1

Yield coefficients* of bacteria in simulated complete-mixing and plug-flow activated-sludge systems

Settlement period (h)	Complete mixing	Plug flow
5	0.30(1)†	0.25(3)
1	0.40(2)	0.34(2)
0	0.37(2)	n.d.‡

* Yield coefficient = mean cell dry wt/unit wt glucose utilised in one cycle.
† Number of times experiment repeated.
‡ Experiment not done.

try that plug-flow activated-sludge plants produce better settling sludges than do complete-mixing plants, although the explanation given is that the latter predispose to the growth of poorly settling filamentous bacteria, the condition termed "bulking" sludge (Garonszy & Barnes 1980; Houtmeyers *et al.* 1980; Verachtert *et al.* 1980, 1982; Rensink *et al.* 1982). The results obtained in the simulations suggest that better settlement is related to extracellular polymer production brought about by periodic feeding of substrate in addition to providing adverse conditions for the growth of filamentous bacteria. (See Lester *et al.*, this volume, for a discussion of the role of extracellular polymers in metal solubility.)

Discussion

The use of the computer-controlled fermenters has proved most successful in simulating the conditions occurring in conventional plug-flow or complete-mixing activated-sludge processes. By using this highly versatile system it should be possible to simulate a number of other processes for the treatment of waste water (or water), i.e. deep-shaft activated-sludge process, oxygen activated-sludge process, activated-sludge process with anoxic zones, incremental feeding (feeding with sewage at a number of points along the length of the aeration tank of the activated-sludge process), and anaerobic digestion. This system has already been used to investigate the effect of large concentrations of O_2 and CO_2 on nitrification (Paskins & Jones 1982) and is at present being adapted to investigate the metabolism of bacteria during anaerobic digestion.

The benefits of the computerised system are very versatile control, rapid transformation of data into usable form, manually acquired data entered and stored with ease, storage of large quantities of data in a small space, rapid and easy access to stored data for manipulation, frequency of data collection easily altered over a wide range, instrumentation rationalised by stream selection and automatic plotting of data (experimental and predicted from mathematical models) by graph plotter.

The use of computer-controlled fermenters can be applied to a much wider field of research than waste water treatment (e.g. Walach et al., this volume), but the versatility of the system is limited by the instrumentation available and our understanding of the process to be investigated. The use of such a system should prove invaluable in the validation or testing of mathematical models of the processes being simulated.

Acknowledgement

The authors wish to thank the Water Research Centre for giving permission to publish this paper.

References

ANONYMOUS 1963 *Water Pollution Research Annual Report 1962*, Department of Scientific and Industrial Research. London: HMSO.

BOLTON, R. L. & KLEIN, L. 1971 *Sewage Treatment - Basic Principles and Trends*, 2nd edn. London: Butterworths.

GARONSZY, M. C. & BARNES, D. 1980 Sequentially operated biological systems for bulking sludge control. *Process Biochemistry* 15, 42–48.

HARRISON, D. E. F. & LOVELESS, J. E. 1971 The effect of growth conditions on respiratory activity and growth efficiency in facultative anaerobes grown in chemostat culture. *Journal of General Microbiology* 68, 35–44.

HOUTMEYERS, J., VAN DEN EYNDE, E., POFFE, R. & VERACHTERT, H. 1980 Relations between substrate feeding pattern and development of filamentous bacteria in activated sludge process. Part 1: Influence of process parameters. *European Journal of Applied Microbiology and Biotechnology* 9, 63–77.

JONES, G. L., LOVELESS, J. E. & NOVAK, A. J. 1975 Kinetics of the utilisation of glucose by *Aeromonas* in the presence of the Bacterium NCIB 8250. *Water Research Centre, Technical Memorandum TM116*.

MITCHISON, J. M. 1971 *The Biology of the Cell Cycle*. Cambridge: University Press.

PASKINS, A. R. & JONES, G. L. 1982 Influence of high partial pressures of carbon dioxide and/or oxygen on nitrification. *Journal of Chemical Technology and Biotechnology* 32, 213–223.

PIKE, E. B., CARRINGTON, E. G. & ASHBURNER, P. A. 1972 An evaluation of procedures for enumerating bacteria in activated sludge. *Journal of Applied Bacteriology* 35, 309–321.

PIRT, S. J. 1975. *Principles of Microbe and Cell Cultivation*. Oxford: Blackwell.

RENSINK, J. H., DONKER, H. J. G. W. & IJWEMA, T. S. J. 1982 The influence of feed pattern on sludge bulking. In *Bulking of Activated Sludge—Prevention and Remedial Methods* eds. Chambers, B. & Tomlinson, E. J., pp. 147–165. Chichester: Harwood.

TEMPEST, D. W., NEIJSSEL, O. M. & LEEGWATER, M. P. M. 1982 Aspects of microbial physiology in relation to process control. *Journal of Chemical Technology and Biotechnology* **32**, 92–99.

VERACHTERT, H., VAN DEN EYNDE, E., POFFE, R. & HOUTMEYERS, J. 1980 Relations between substrate feeding pattern and development of filamentous bacteria in activated sludge processes. Part 2. Influence of substrate present in influent. *European Journal of Applied Microbiology and Biotechnology* **9**, 137–149.

VERACHTERT, H., HOUTMEYERS, J. & VAN DEN EYNDE, E. 1982 Relations between substrate feeding pattern and development of filamentous bacteria in activated sludge. In *Bulking of Activated Sludge—Prevention and Remedial Methods* eds. Chambers, B. & Tomlinson, E. J., pp. 128–146. Chichester: Harwood.

Control of the Activated-Sludge Process Using Adenosine Triphosphate Measurements

I. R. JOHNSON*

Welsh Water Authority, Bridgend, UK

D. A. STAFFORD

Department of Microbiology, University College, Cardiff, UK

The activated sludge process is designed to increase the purification of waste water by contact with a large concentration of microorganisms in the presence of sufficient dissolved O_2.

Efficient operation and control of the activated sludge system has three major requirements: (1) recognition of the fundamental basis for control, (2) knowledge of the system constraints and limitations and (3) utilisation of a control strategy consistent with requirements (1) and (2). While there have been great advances in our ability to meet the first two requirements, there has been a large technical gap in achieving the third. This is due in part to the complexity of the system and difficulty in accurate modelling.

Factors Affecting Performance of the Activated-Sludge Process

Mathematical models of the activated-sludge process are available which use our knowledge of bacterial growth and utilisation of substrate at the very low specific growth rates encountered in the process.

The two parameters which can be used most readily for the control of an activated-sludge system are (1) the rate of wastage of activated sludge and (2) the rate of recycle of activated sludge. Within the limits set by aeration

---

*Present address: Cardiff Laboratories for Energy and Resources Ltd., Lewis Road, East Moors, Cardiff, Glamorganshire CF1 5EG, UK.

capacity and the capacity of the final settlement tank, these two parameters provide an immediate method for the control of the system.

Rate of Wastage of Activated Sludge

For equilibrium to be maintained to a dynamic system, input must be matched by output. In the case of the activated-sludge system, therefore, growth of new biomass is balanced by the rate at which solids are wasted from the system, i.e. specific growth rate = specific wastage rate = 1/sludge age. Downing & Knowles (1966) developed a set of empirical equations which allow the calculation of the yield of sludge from biochemical oxygen demand (BOD) for a given specific growth rate.

Simulations with a mathematical model showed that variations in effluent quality were reduced if the volumetric flow rate of wasted sludge was kept constant, instead of continually adjusting the wastage rate in an attempt to keep the concentration of active solids constant (Curds 1973).

Rate of Recycle of Activated Sludge

In addition to returning solids from the final settlement tank to the aeration tank, the rate at which sludge is recycled also affects the contact time between the incoming waste water and the microorganisms in the activated sludge. Simulations with the mathematical model show that for a given retention period and specific sludge wastage, variability in effluent quality and the mass of BOD discharged in 24 h is considerably reduced by reducing the rate of recycle (Jones 1976). From this it may be concluded that provided solids do not build up excessively in the settlement tank and impair effluent quality, the lower the rate at which sludge is recycled the better will be the overall performance of the plant.

The discharge of waste water to watercourses requires improved operational control, and a number of relationships have been used to define organic removal characteristics and excess sludge production from municipal and industrial waste waters using the activated-sludge process. The fraction of volatile, suspended, active biomass is related to both the volatile suspended solids present in the raw waste water and to the sludge age. Increasing the sludge age increases the non-active, volatile solids accumulation in the system.

Measurement of Microbial Activity in the Activated-Sludge Process

Since it is important to understand the fundamental (biological) nature of the secondary waste treatment processes in order to control the overall process, biochemical parameters have been proposed as measures of both biomass and bioactivity in waste treatment systems.

The most widely used methods for estimating activated-sludge cell concentration are total suspended solids (TSS) and volatile suspended solids (VSS). Such measures would be a satisfactory estimate of the active microbial mass if the activity and viability per unit mass of solids were constant. A more reliable estimate is required that is related to the biological activity of the activated sludge.

If the measurement of a specific cell constituent such as adenosine triphosphate (ATP) is to be used to calculate such total microbial biomass, the following requirements must be satisfied:

(1) The measured compound must be present in all living cells and must be absent from all dead cells.
(2) It must not be found associated with non-living, detrital material.
(3) It should exist in fairly uniform concentrations in microbial populations, but may be influenced by environmental stresses.
(4) The analytical technique for the measurement of the biomass indicator must be capable of measuring milligram quantities and must be relatively quick and easy to use for large-scale programmes involving numerous samples.

Possible Uses for ATP Measurements in the Activated-Sludge Process

The role of ATP as a mediator between energy-producing and energy-consuming reactions is well established (Atkinson 1969). The pool size of ATP in living cells is influenced by the metabolic activity of the cell and is therefore dependent on the availability of substrate and O_2. Such relationships between environmental conditions and the size of the ATP pool have been investigated with pure cell suspensions of many different organisms (Hamilton & Holm-Hansen 1967; Hobson & Summers 1972; Lazdunski & Belaich 1972). It appears that the endogenous ATP pool is reasonably constant intergenerically and may yield a good estimate of viable biomass, with an endogenous ATP of ca. 2 μg/mg cell material. Patterson *et al.* (1970) indicated that the ATP pool is affected by the metabolic activity of an activated-sludge culture and may be expected to respond rapidly and decisively to an increase in substrate loading, while being only gradually reduced as the organisms enter an endogenous phase. The ATP pool responds rapidly to changes in the metabolic activity of activated sludge, as demonstrated by both laboratory and field studies. This may be the most significant aspect of this approach since a metabolic activity parameter is more appropriate than a biomass parameter in the potential control of a biological waste treatment process.

There are five important areas of where this technique may be applied.

(1) Measurements of ATP may be used to establish whether the mixed-liquor suspended-solids (MLSS) concentration (presently used for process control) provides an accurate representation of the active biomass present in activated-sludge plants.
(2) The influence of such factors as dissolved O_2, suspended solids, sludge age and diurnal variations in flow of sewage may be investigated.
(3) The present method of controlling the biomass by maintaining a constant loading factor (food/mass ratio) of ca. 0.3 kg applied BOD/kg sludge may be studied. This is achieved by adjusting the wastage rate of the activated sludge to keep the MLSS within a predetermined range. Up to 40% of the sludge in the system may have been wasted during this period. This stresses the remaining biomass and invariably leads to a period of poor BOD removal. Therefore, there exists the need for a more accurate form of controlling the plant, in particular the wastage rate. Using ATP as a measure of the active biomass content of the sludge, closer control of plant performance may result in optimising the wastage rate of sludge.
(4) By measuring the changes in microbial activity in the mixed liquor and returned activated sludge, the effect of toxic discharges may be monitored and then be diverted and modified before biological treatment.
(5) At present there is little or no process control in aerobic digestion units. Measurement of O_2 uptake rates is useful but there are problems due to temperature dependence. A measurement of the active biomass concentration may permit some control over the process.

Procedure for ATP Assays

Extraction of ATP

A modification of the method of Dhople & Hanks (1973) was used in experiments at an operational sewage works (see p. 188). Duplicate samples of mixed liquor and returned activated sludge (1 ml) were whirl-mixed (Hook & Tucker Ltd., Croydon, Surrey) with chloroform (0.3 ml) for 10 s in a tall, thin, glass test tube. Denaturation of enzymes and removal of chloroform was achieved by immersing the base of the test tube in a boiling waterbath for 2.5 min. The test tube was well shaken during the last minute of this step to expel the last traces of chloroform. Under these conditions no significant evaporation of water occurred and adenylates were extracted quantitatively. Extracts prepared in this way are stable at 0°C for several hours and at −20°C for several weeks (Simpson *et al.* 1976). The extracts contain no organic solvents or added chemical reagents which might interfere with subsequent adenylate assay procedures. Before assay for ATP, extracts were centrifuged for 3 min in a bench centrifuge (MSE Ltd., Crawley, Sussex) to sediment particulate material.

Measurement of ATP

The equipment and method of ATP estimation described by Statham & Langton (1975) was used. The peak light intensity was measured in a specially designed photometer utilising a low-background photo-muliplier tube (EMI 9524S). However, an operational amplifier between the photometer and the chart record (Phillips PM 8220) was not required.

Assay Procedure

Arsenate-buffered, freeze-dried luciferase solution in distilled water was prepared on the day before it was required and stored until use at 4°C in order to reduce background light emission. Before use this preparation was centrifuged to give a clear solution. All assays were done at room temperature (18° ± 2°C). Luciferase (0.2 ml) was pipetted into a cuvette which was placed in position above the photomultiplier in the photometer. A sample (10 µl) was rapidly injected into the cuvette through the light-tight rubber seal, using a Hamilton CR700 constant-rate syringe (The Hamilton Co., Reno, Nevada, USA). The peak light intensity was noted on the chart recorder. Results are expressed as ng ATP/10 µl sample.

ATP Standards

Since luciferase activity varies from batch to batch with age and conditions of storage, it is necessary to prepare a calibration curve with standard ATP each day.

Crystalline ATP (disodium salt, mol. wt 551.2) was weighed in 55.12-mg portions and dissolved in 100 ml distilled water to give a 1-mM solution. This stock solution was poured into individual test tubes, which were capped and stored at −20°C until required. Storage for 2 months showed no loss of strength in the ATP stock solution. When ATP standards were required, a test tube of stock solution was thawed and diluted with distilled water to the required concentrations.

The mixed-liquor and returned activated-sludge ATP concentrations were obtained using the calibration curve (Fig. 1).

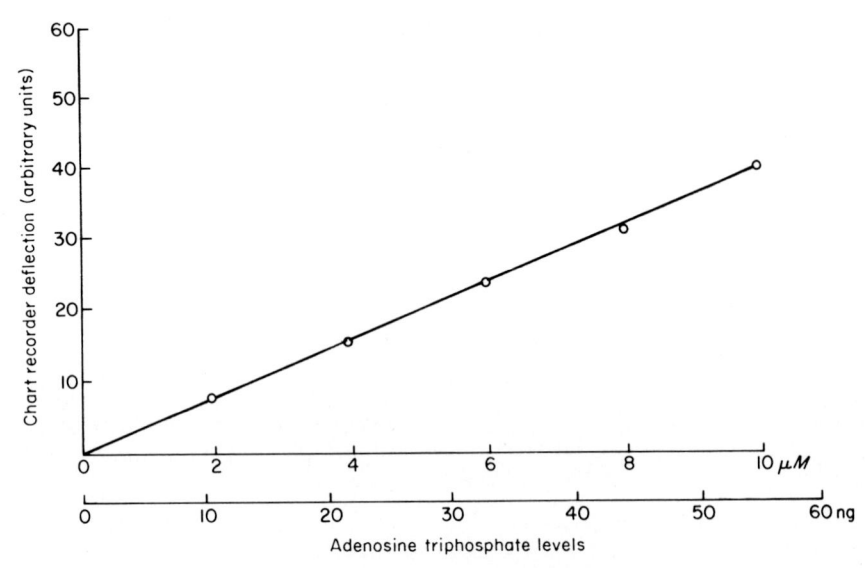

FIG. 1. A typical adenosine triphosphate calibrated curve.

Application of ATP Measurements to an Operational Activated-Sludge Plant

Source of Activated Sludge

Gowerton sewage treatment works within the Welsh Water Authority receives a dry weather flow (DWF) of ca. 7100 m³/day from a population of 23,400 and various industries (predominantly metal finishing) from an area of 130 ha. Approximately 26% of the flow is of trade origin.

Crude sewage samples were taken after screening and degritting. Samples of mixed liquor were taken at the outlet weir of the aeration tank. Samples of returned activated sludge were taken at the point of entry into the aeration lane. The final effluent was sampled at the final effluent outlet chamber of the works.

Analysis of Data

Sludge loading is calculated using the crude sewage BOD load (kg/day) and the mass of activated sludge (kg) held in the aeration tanks:

Sludge loading factor = BOD load/Mass of sludge

Wastage rate is calculated from the following formula which applies only to a system where sludge is wasted within one sludge passage.

$$\text{Wastage rate (kg/day)} = \frac{\text{Mass of sludge wasted (kg/day)}}{\text{Mass of sludge in system (kg)}}$$

Sludge age is the reciprocal of wastage rate. *Active biomass content (ATP pool)* of mixed liquor and returned activated sludge (RAS) is expressed as µg ATP/mg VSS.

Determination of Optimum Sludge Age

In attempting to keep MLSS constant, the quantity of surplus activated sludge removed from the system periodically may vary considerably, resulting in sludge ages of 1–15 days. During a period from January to April changes in sludge age, active biomass content of the mixed liquor, sludge settleability and effluent quality were monitored (Figs 2 and 3). There was a distinct peak in the active biomass concentration, sludge volume index (SVI) and stirred specific volume index (SSVI) for sludge ages of 3–7 days. Final effluent quality as measured by BOD, optical density and total suspended-solids (TSS) values deteriorated coincidentally with the peak in active biomass (ATP pool). Thus it appears that sludge ages of 3–6 days should be avoided. Therefore it was decided to operate the sewage treatment works at a sludge age of 7.5 ± 2.5 days. A sludge age of less than 3 days was considered to be unsuitable for effective and stable treatment, as indicated by the second peak in SVI and SSVI (Fig. 2).

Operation Control of Optimum Sludge Age

As operational control is now based on sludge age, it was necessary to devise a simple means whereby operators could determine the volume of sludge wastage from other plant data to achieve a constant sludge age. The choice of a particular sludge age determines that particular constant proportion of the mass of sludge that is wasted from the system each day.

Sludge age and the mass of sludge to be wasted daily may be calculated as follows:

$$\text{Sludge age (days)} = \frac{\text{Mass of sludge in system (kg)}}{\text{Mass of sludge wasted (kg/days)}} \quad (1)$$

i.e.

$$\text{Mass of sludge wasted (kg/day)} = \frac{\text{Mass of sludge in system (kg)}}{\text{Sludge age (days)}} \quad (2)$$

The volume of activated sludge wasted (or returned activated sludge, RAS) may be calculated from the value of its mass given by Eq. (2). Hence, as mass = volume × density, Eq. (2) becomes

$$\frac{\text{Volume sludge}}{\text{wasted daily (m}^3)} \times \frac{\text{Wasted sludge (RAS)}}{\text{suspended solids (mg/litre)}} = \frac{\text{MLSS (mg/litre)} \times \text{aeration tank volume (m}^3)}{\text{Sludge age (day)}} \quad (3)$$

i.e.

$$\frac{\text{Volume sludge}}{\text{wasted daily}} = \frac{\text{MLSS}}{\text{RAS suspended solids}} \times \frac{\text{Aeration tank volume}}{\text{Sludge age}} \quad (4)$$

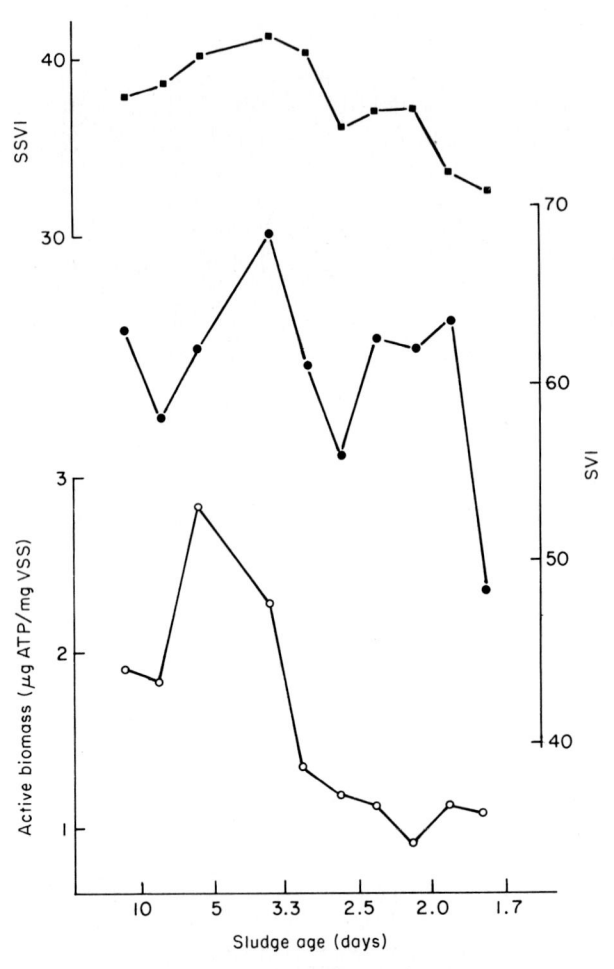

Fig. 2. Changes in active biomass, sludge volume index (SVI) and stirred specific volume index (SSVI) of aeration tank mixed-liquor output with sludge age.

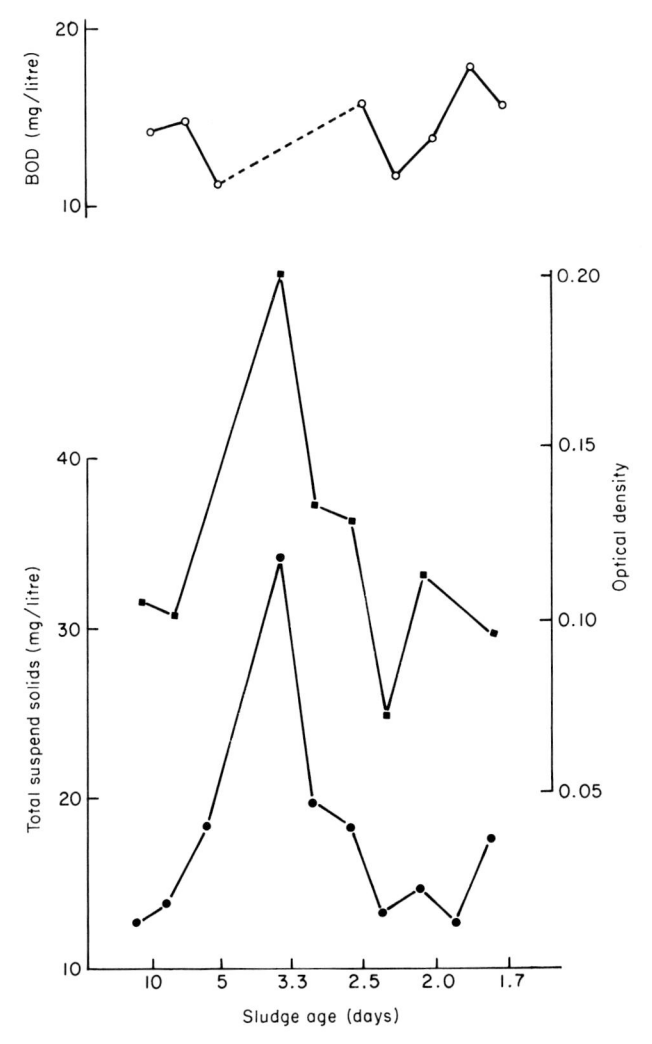

FIG. 3. Changes in final effluent quality with sludge age.

Preliminary indications that ML:RAS (waste sludge) suspended-solids values may yield useful information is shown by the relationship between this ratio and sewage flow (Fig. 4). Equation (4) shows that the volume of activated sludge wasted is proportional to (ML:RAS suspended solids values) × f, where f = Aeration tank volume/Sludge age.

A value for f of 352 m³/day was derived from the following plant conditions: mass of sludge in system, 7821 kg; sludge age, 7.5 days; RAS suspended solids, 7693 mg/litre; MLSS 2950 mg/litre. A mean value of 350 m³/

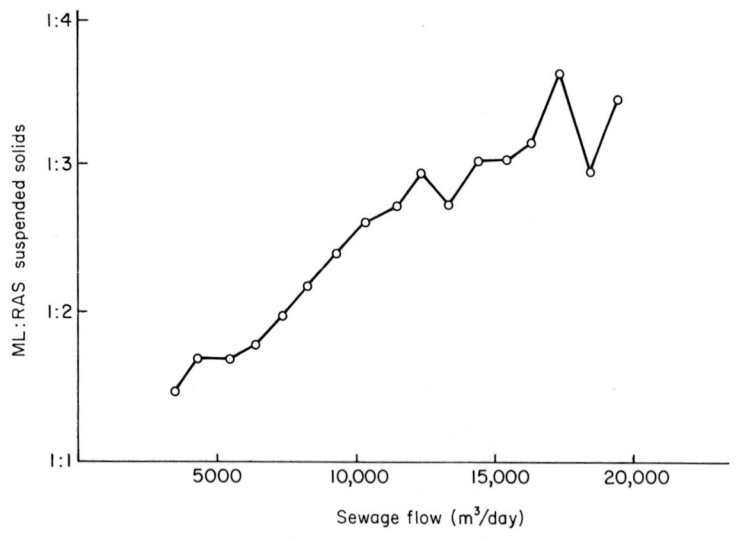

FIG. 4. Relationship between mixed-liquor (ML): returned activated-sludge (RAS) suspended solids and sewage flow.

day was obtained during operational variation in plant conditions. By using these data a graph may be constructed to give a direct reading of the volume of surplus activated sludge (m^3) for various ML:RAS suspended solids values (Fig. 5). Values for surplus sludge estimates from Fig. 5 compare well with values calculated from experimental results over a period of 1 month, being on average 9% greater than the calculated value for a 7.5-day sludge age. Effluent quality during this trial period was good, yielding an average monthly BOD of 16 mg/litre and TSS of 15 mg/litre.

Use of ML:RAS ATP Levels to Estimate Sludge Wastage Rate

Since ATP measurements of active biomass content are more accurate than suspended-solids determinations, the use of ML:RAS ATP will permit more accurate control of the activated sludge plant. Thus any alteration in ATP levels in response to changes in plant environmental conditions may be used to control the wastage of surplus activated sludge. From an average ATP level in the mixed liquor for the period January–April of 34 ng/10 μl (3.4 mg/litre) and an aeration tank volume of 2464 m^3 the mass of ATP in system was 8.65 kg. Therefore, the mass of ATP to be wasted to give a 7.5-day sludge age is 8.65/7.5 = 1.154 kg/day. From an average ATP level in the returned activated sludge of 77 mg/10 μl (7.7 mg/litre), the volume of sludge to be wasted is (1.154 × 1000)/7.7 = 150 m^3/day. Thus factor f, calculated

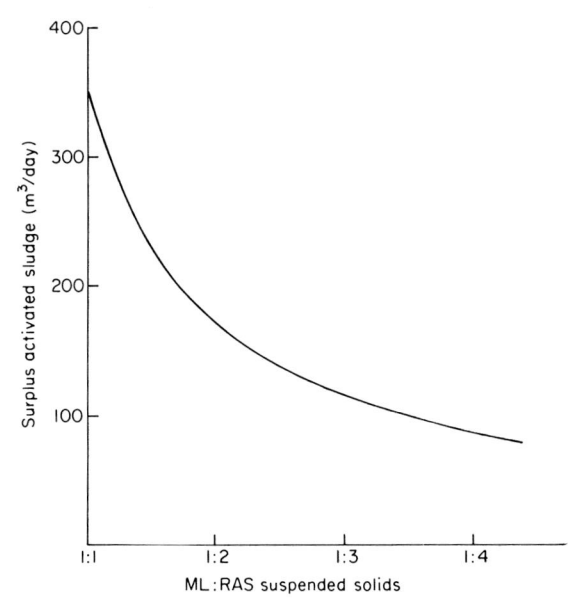

FIG. 5. Relationship between mixed-liquor (ML): returned activated-sludge (RAS) suspended solids and volume of surplus activated sludge for a sludge age of 7.5 days.

by substituting ATP levels for suspended solids values in Eq. (4), is (150 × 7.7)/3.4 = 340 m³/day. This value of 340 compares well with the value of 350 calculated on the basis of suspended-solids levels.

Use of ATP Level as an "Early Warning" System

The use of ML:RAS ATP levels in controlling the wastage of activated sludge provides a simple and effective early warning of perturbations in the incoming sewage. Thus from Eq. (4), reduction in the ML ATP level results in a reduction in sludge wastage rate.

Therefore, if a toxic discharge enters the aeration tank treatment lane and reduces ML and, eventually, RAS ATP levels, the plant operator can reduce the wastage rate thereby retaining more biomass in the system to combat the discharge. If ATP remains at a reduced level, the flow of sewage containing the toxic discharge can be diverted either for preliminary treatment or to storm tanks. When the discharge ends, the ATP levels increase again and the wastage rate may be steadily increased to its original value. This sequence is represented in Fig. 6. The sludge wastage rate may also be adjusted to accommodate diurnal variations in sewage volume and composition.

The ML:RAS ATP levels can therefore provide means for a simple and

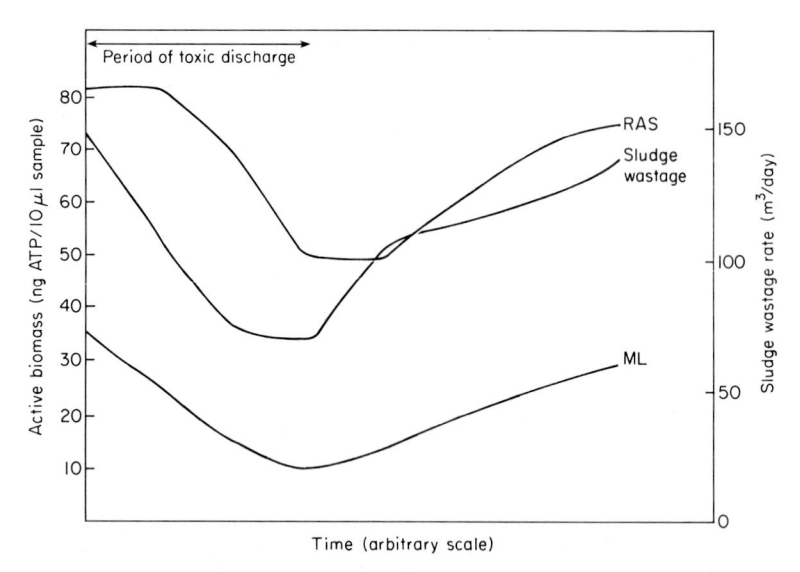

FIG. 6. Representation of effect of toxic discharge to sewage input on mixed-liquor (ML) and returned activated-sludge (RAS) active biomass and their influence on control of sludge wastage rate.

effective control of the activated sludge process, in that the primary control parameter, the wastage rate of sludge, may be continuously adjusted in response to changes in environmental conditions without the need for lengthy laboratory analyses for BOD and SS levels.

Conclusions

The use of ATP as an aid to process control of an activated-sludge plant has been advocated by many workers, e.g. Patterson *et al.* (1970), Chiu *et al.* (1973), but a lack of published data relating to its practical application in full-scale operational plants has made it difficult to draw any conclusions as to its practical usefulness. However, ATP measurement has become established as a measure of the active biomass content of activated sludge in laboratory-scale and pilot plant units, when it has been shown to reflect changes in environmental conditions.

Previous studies of operational activated-sludge plants involving measurements of ATP have concluded that while ATP can yield useful information as to the condition of the sludge, its interpretation for practical control is difficult and therefore there would be no immediate advantage in using this measurement over conventional control parameters (Ortman *et al.* 1977).

Indeed, it has been observed that no studies have been made to show the use of ATP levels to control residence time (Roe & Bhagat 1982). However, the present study at a sewage works has shown a useful correlation between ATP levels and cell residence time. Levels of ATP in sludge respond rapidly to changes in environmental conditions, and therefore the method provides a sensitive control of the treatment process. With elaborate instrumentation and process control, this system has been shown to have practical significance for full-scale treatment plants. To fully utilise the technique, automatic monitoring and control with a microprocessor will be necessary. This should be the subject of a large-scale development programme for the control of biological treatment systems, both aerobic and anaerobic.

References

ATKINSON, D. E. 1969 Regulation of enzyme function. *Annual Reviews of Microbiology* **23**, 47–68.

CHIU, S. Y., KAO, I. C., ERICKSON, L. E. & FAN, L. T. 1973 ATP pools in activated sludge. *Journal of Water Polution Control Federation* **45**, 1746–1758.

CURDS, C. R. 1973 A theoretical study of factors influencing the microbial population dynamics of the activated sludge process. II. A computer simulation study to compare two methods of plant operation. *Water Research* **7**, 1439–1452.

DHOPLE, A. M. & HANKS, J. H. 1973 Quantitative extraction of adenosine triphosphate from culturable and host-grown microbes: calculation of ATP pools. *Applied Microbiology* **26**, 399–403.

DOWNING, A. L. & KNOWLES, G. 1966 Population dynamics in biological treatment plants. *Proceedings of Third International Conference on Pollution Research, Munich* **2**, 117–142.

HAMILTON, R. D. & HOLM-HANSEN, O. 1967. *Limnology and Oceanography* **12**, 319–324.

HOBSON, P. N. & SUMMERS, R. 1972 ATP pool and growth yield in *Selenomonas ruminantium*. *Journal of General Microbiology* **70**, 351–360.

JONES, G. L. 1976 Microbiology and activated sludge. *Process Biochemistry* Jan.Feb. pp. 3–5.

LAZDUNSKI, A. & BELAICH, J. P. 1972 Upcoupling in bacterial growth: ATP pool variation in *Zymomonas mobilis* cells in different uncoupling conditions of growth. *Journal of General Microbiology* **70**, 187–197.

ORTMAN, C. LAIB, T. & ZICKEFOOSE, C. S. 1977 TOC, ATP and Respiration rate as control parameters for the activated sludge process. *U.S. Environmental Protection Agency, Office of Research and Development, [Report]* EPA-600/2-77-142.

PATTERSON, J. W. BREZONIK, P. L. & PUTMAN, H. D. 1970 Measurement and significance of adenosine triphosphate in activated sludge. *Environmental Science and Technology* **4**, 569–575.

ROE, P. C. & BHAGAT, S. K. 1982 Adenosine triphosphate as a control parameter for activated sludge processes. *Journal of Water Pollution Control Federation* **54**, 244–254.

SIMPSON, B., STATHAM, M., HUDSON, P. & HUGHES D. E. 1976 A study of the biological treatability of the trade effluent produced from the manufacture of ion exchange resins: the use of adenosine triphosphate and adenylate charge measurements. *Process Biochemistry* Jan./Feb. pp. 21–24.

STATHAM, M. & LANGTON, D. 1975 The use of adenosine triphosphate measurements in the control of the activated sludge process by the solids retention time method. *Process Biochemistry* **10**, 25–28.

Assessment of the Role of Bacterial Extracellular Polymers in Controlling Metal Removal in Biological Waste Water Treatment

J. N. LESTER, R. M. STERRITT, THOMASINE RUDD AND MELANIE J. BROWN*

Public Health Engineering Laboratory, Imperial College of Science and Technology, London, UK

The removal of toxic metals in biological waste water treatment processes is important in maintaining the quality of receiving waters so that they comply with standards governing their subsequent abstraction for drinking water (European Economic Community 1975). In particular, it is the capacity of the activated-sludge biomass which contributes significantly to the efficiency of toxic metal removal (Oliver & Cosgrove 1974). Bacterial extracellular polymers present in flocculating activated-sludge biomass have been implicated in the removal of soluble metal ions (Dugan & Pickrum 1972; Brown & Lester 1979). Strains of *Zoogloea ramigera*, which are believed to be common inhabitants of activated sludge, have been shown to have a high affinity for soluble metal ions (Friedman & Dugan 1968), and polymeric materials extracted from activated sludge have been shown to bind with metal ions (Steiner *et al.* 1976).

Two major approaches to the assessment of the importance of metal binding by bacterial extracellular polymers have been adopted. Some studies have been based on comparisons of metal uptake by capsulated and non-capsulated strains of the same species (Friedman & Dugan 1968; Bitton & Friehofer 1978), while others have employed extraction methods for the isolation of polymers prior to determining metal complexation in the absence of bacterial cells (Brown & Lester 1982a,b).

Several methods for the extraction of extracellular polymers have been described, and these have been summarised by Novak & Haugan (1981). In

*Present address: QMC Industrial Research Ltd., 229 Mile End Road, London E1, UK

most cases it has been impossible to assess if any of these methods effected a complete extraction since there is currently no universal method for providing quantitative recovery of polymers from activated sludge. Thus, the precise importance of metal binding by extracellular polymers as a mechanism of metal removal in activated sludge is still uncertain.

Cultivation of Polymer-Forming Organisms

Laboratory Scale Activated-Sludge Simulation

A flocculating activated-sludge biomass may be grown in the laboratory ("laboratory activated sludge") in an activated-sludge simulation similar to that described by Stoveland & Lester (1980) using synthetic sewage. The apparatus consists of a borosilicate glass aerator with a capacity of 3.1 litres which is connected by flexible tubing to a 2-litre borosilicate settler with a conical base. The activated-sludge biomass is mixed and aerated by the introduction of compressed air through a glass sinter mounted in the base of the aerator. The sludge settles under quiescent conditions in the settler and is continuously withdrawn from the base and recycled back to the aerator via a peristaltic pump. Medium (synthetic sewage) is introduced at a constant flow rate of 1 litre/h to the aerator and the effluent leaves via an overflow weir at the top of the settler.

The composition of the synthetic sewage used is (mg/litre demineralised water): peptone (Oxoid, L34), 234; Lab Lemco (Oxoid, L29), 156; NaCl, 9.5; $CaCl_2$, 4.8; $MgSO_4 \cdot 7H_2O$, 1.5; NH_4Cl, 72; KH_2PO_4, 6.5. The synthetic sewage may be prepared as a concentrated solution and sterilised in 10-litre batches at 121°C for 1 h. The concentrate is diluted with tap water just before addition to the aerator to give a chemical oxygen demand (COD) of ca. 400 mg/litre.

The biomass initially develops from air-borne microorganisms which are allowed to inoculate the synthetic sewage. After 1 or 2 weeks the sludge recycle can be started and the desired growth rate (sludge age) of the biomass can be maintained by removing an appropriate quantity of mixed liquor from the aerator daily.

The flocculating biomass produced in such a system is very similar qualitatively to that produced in a full-scale activated-sludge plant. It also has several advantages over activated sludge taken from a full-scale plant for the study of the role of extracellular polymers in the removal of metals. The suspended solids are almost entirely of biological origin, whereas in a full-scale plant some inert non-flocculating particles may enter the activated sludge process in the effluent produced from the sedimentation of raw sewage. The laboratory system also affords a greater degree of operational control, permitting the establishment of a range of sludge ages and the maintenance of a consistent influent substrate concentration and flow rate.

Pure Cultures

Pure cultures of a polymer-forming strain of *Klebsiella aerogenes* NCTC 8172 were grown in a liquid culture medium of the following composition (g/litre demineralised water): sucrose, 10; $(NH_4)_2SO_4$, 0.3; $NaH_2PO_4 \cdot 2H_2O$, 2; $K_2HPO_4 \cdot 3H_2O$, 2; K_2SO_4, 1; NaCl, 1; $MgSO_4 \cdot 7H_2O$, 0.2; $CaCl_2 \cdot 6H_2O$, 0.02; $FeSO_4 \cdot 7H_2O$, 0.001. Some of the cultures of *K. aerogenes* (studies reported here) were also grown in a medium containing (g/litre) trisodium citrate, 2; NH_4Cl, 1; K_2HPO_4, 0.2; KH_2PO_4, 0.15; $MgSO_4 \cdot 7H_2O$, 0.1; $CaCl_2 \cdot 6H_2O$, 0.01; $MnSO_4 \cdot 4H_2O$, 0.005; $FeSO_4 \cdot 7H_2O$, 0.005.

For the comparison of polymer extraction methods and for studies of metal uptake by untreated cells, by cells from which polymer was removed and by extracted extracellular polymers, batch cultures were grown in citrate medium at 25 ± 2°C and harvested after 5 days. The medium containing sucrose as the sole carbon and energy source yielded greater quantities of polymer from *K. aerogenes* than the medium containing citrate. The sucrose medium was used to cultivate *K. aerogenes* in continuous culture in a chemostat at a dilution rate of 0.05/h with a pH of 6.8, a dissolved oxygen concentration of ca. 6.3 mg/litre and at a temperature of 25 ± 2°C from which polymer was obtained which could be fractionated into soluble and colloidal forms and which was used to determine the stability constants of metal–polymer complexes.

Extraction and Assay of Polymer

Polymer Extraction

Physical Methods

It has been shown that high-speed centrifugation at up to 33,000 g can remove extracellular material from a capsulated strain of *K. aerogenes* (NCTC 8172) grown in liquid culture (Brown & Lester 1980; Rudd *et al.* 1982). Samples of the culture are centrifuged for 10–15 min, and the soluble polymer present in the supernatant may be separated from the cell debris by decantation or filtration through a 0.2-μm pore size membrane filter. In order to maximise polymer recovery, the pellet may be resuspended and centrifuged again for a further 10 min in order to increase the shear force exerted on the cells.

Ultrasonication has been used to separate polymers from cells both as a complete extraction method and as a pretreatment before high-speed centrifugation. Extraction may be achieved using a commercial ultrasonic bath or probe. A power output of up to 80 W for 10–15 min may be used, although Rudd *et al.* (1982) used ultrasonication at 18 W followed by centrifugation at 33000 g for 60 min as a preferred method for the extraction of polymer from cultures of *K. aerogenes*.

Polymers may also be extracted by heat treatment, either boiling or autoclaving. A relatively simple method, modified from that used by the Water Pollution Research Laboratory (1971), involves placing the sample in free-flowing steam for 10 min. The polymer is obtained in the supernatant fluid after centrifuging the samples while still hot in a cooled centrifuge at 8000 g for 10 min.

Chemical Methods

Polymers may be extracted from cultures of capsulated bacteria and activated sludge with NaOH (Tezuka 1973; Brown & Lester 1980). The cells are collected by centrifuging the sample at 2000 g for 20 min and the supernatant fluid is discarded. Extraction is effected by the addition of 2 volumes of 2-M NaOH followed by gentle agitation for 5 h at room temperature. The polymer remains in the supernatant fluid after centrifuging the sample at 2000 g for 20 min.

Nishikawa & Kuriyama (1968) have described the use of EDTA as an extractant. A modified form of this method (Brown & Lester 1980) employs the addition of an equal volume of 2% (w/v) EDTA to the sample which is left to stand for 3 h at 4°C. The polymer is recovered by centrifugation or filtration.

Purification and Recovery of Extracellular Polymers

Extracted soluble polymers obtained in the supernatant fluid or filtrate will be contaminated to some degree by residual components of the culture medium and by intracellular material from ruptured cells. Low-molecular-weight contaminants may be removed by dialysis in visking tubing against several changes of distilled water for 24 h. The removal of high-molecular-weight contaminants is more difficult and if relatively pure preparations are required, the choice of an extraction method that causes minimal disruption of the cells with maximum recovery of the polymer is probably the best approach.

Polymers can be obtained in a dry form for storage or further analysis by precipitation in ethanol (Pavoni *et al.* 1972). Two volumes of 96% (v/v) ethanol are added to one volume of polymer extract (i.e. the supernatant fluid or filtrate obtained after extraction), and the polymer is left to precipitate for 48 h at 4°C. Alternatively, acetone may be used to precipitate the polymer.

Assay of Extracellular Polymers

Since most extracellular material from capsulated bacteria and activated sludge is polysaccharide, it may conveniently be assayed, for comparative purposes, on material from the same source by the phenol–sulphuric acid method for hexoses (Dubois *et al.* 1956). Other polymeric materials found in extracts, i.e. DNA and protein, may be determined using the diphenylamine assay of Burton (1956) and the method of Lowry *et al.* (1951), respectively. Alternatively, if the polymer is collected by precipitation in ethanol, the yield can be assessed gravimetrically.

Comparison of Extraction Methods

Since no empirical method exists for the quantification of extracellular polymers, the efficiencies of the various extraction methods could be compared only on a relative basis. Furthermore, since it may be necessary to obtain relatively pure preparations of extracellular polymers, some measure of the degree of contamination by intracellular components of the matrix is necessary. Relatively large concentrations of protein and, especially, DNA in the extracts may be considered to be indicative of cell lysis.

The quantities of polysaccharide (determined as hexose sugar), protein and DNA extracted by a range of methods applied to pure cultures of *K. aerogenes* (NCTC 8172), activated sludge and laboratory activated sludge developed on a synthetic sewage are shown in Table 1. The best extraction methods are those which extract the greatest quantity of hexoses with relatively small quantities of DNA and protein.

Ultrasonication alone was not effective as an extraction method for any of the materials examined, although it may be useful as a pretreatment in conjunction with other methods. High-speed centrifugation alone was also ineffective for the two forms of activated sludge, although it effected a considerable degree of extraction from the pure culture of *K. aerogenes*. A combination of these two methods also had very little effect on the activated sludges, but was relatively effective for the pure culture. Steam, NaOH and EDTA treatment each released large quantities of hexoses from the activated sludges. However, NaOH treatment released three times as much DNA from the activated sludge, as did the other two methods, and NaOH and EDTA treatment each released the largest quantities of protein from the sludges, indicating a considerable degree of cellular disruption.

TABLE 1

Extraction of hexoses, protein and DNA from cultures of Klebsiella aerogenes *NCTC 8172 and activated sludge by various methods*

		Quantity extracted (mg/g/litre suspended solids)		
Material	Extraction method	Hexoses	Protein	DNA
K. aerogenes	Centrifugation (33,000 g, 10 min)	93.2	11.5	3.2
	Ultrasonication (18 W, 10 min)	39.5	13.0	0.6
	Centrifugation and ultrasonication	88.8	20.3	1.8
	Steaming	114	49.5	6.2
	Sodium hydroxide	12.4	252	8.0
	EDTA	139	n.d.*	9.4
Activated sludge	Centrifugation (33,000 g, 10 min)	0.4	2.9	0.3
	Ultrasonication (18 W, 10 min)	0.2	0.1	0.1
	Centrifugation and ultrasonication	0.4	1.2	0.2
	Steaming	19.9	75.1	3.7
	Sodium hydroxide	63.2	528	13.1
	EDTA	23.9	118	4.3
Laboratory activated sludge	Centrifugation (33,000 g, 10 min)	0	0.1	1.5
	Ultrasonication (18 W, 10 min)	0.5	3.6	1.2
	Centrifugation and ultrasonication	0.6	4.4	0.5
	Steaming	18.3	47.6	4.3
	Sodium hydroxide	20.6	222.6	0.5
	EDTA	21.0	68.0	3.0

* n.d., not determined.

The degree of disruption of floc structure in the laboratory activated sludge is demonstrated in Fig. 1. It appears that treatment with EDTA had little effect on the floc structure. Ultrasonication caused some fragmentation, although a separation of the cells from the matrix was not evident; high-speed centrifugation alone and with ultrasonication showed a greater degree of disruption, but steaming had the greatest effect.

Integrity of Capsular Polymers

Some extraction methods may recover extracellular polymers in a soluble form; this is especially true if filtration is used to separate the polymer from cell debris. If the extracted material is representative of the extracellular matrix of activated sludge, it must form an integral part of the settleable solids. However, there is no certainty that extracted soluble polymer is representative of material forming part of the sludge floc, since the soluble material and the matrix clearly have different properties (Novak & Haugan

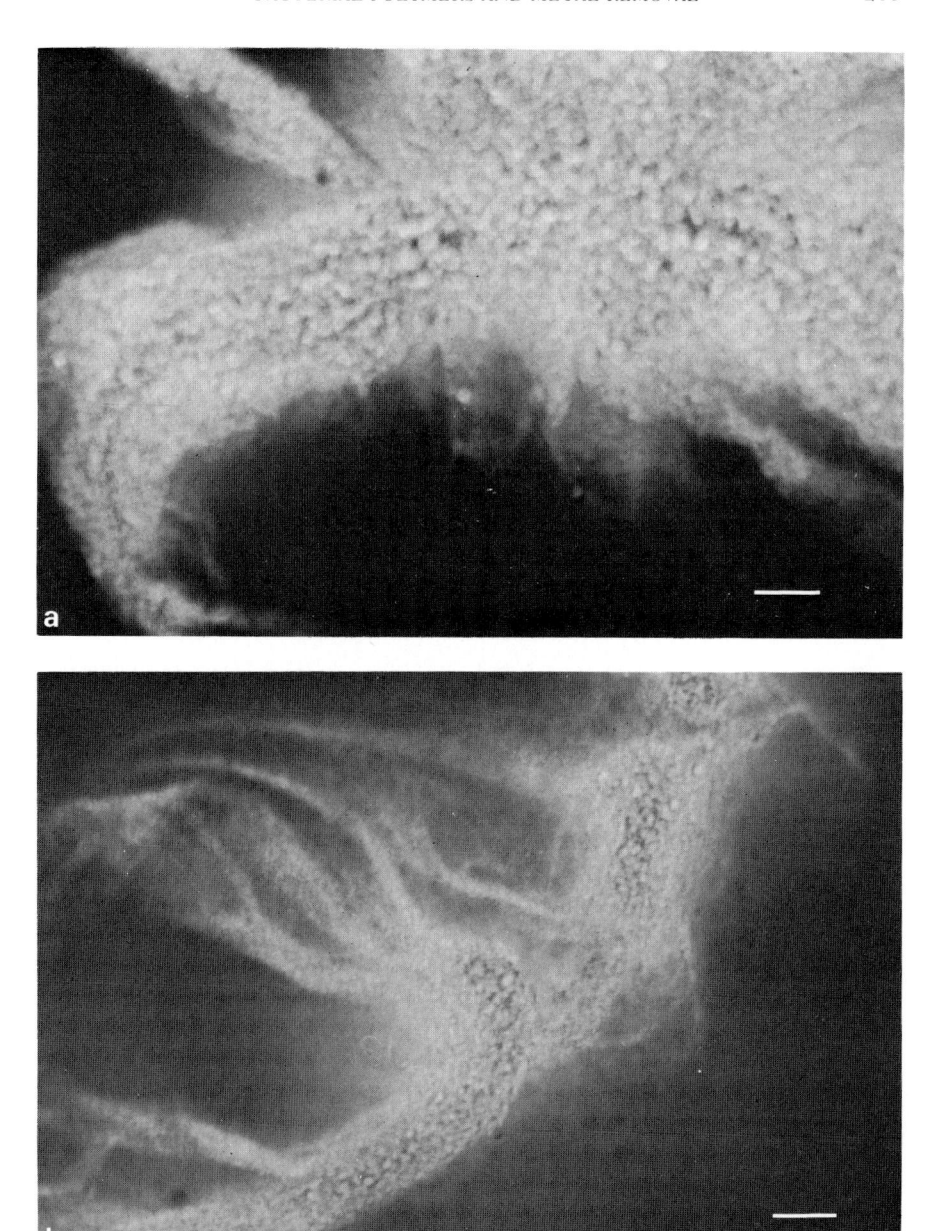

FIG. 1. Effect of various methods of disruption on floc structure of laboratory activated sludge grown in synthetic sewage. (a) Untreated; (b) after treatment by EDTA; (c) ultrasonication; (d) high-speed centrifugation; (e) ultrasonication and high-speed centrifugation; (f) steaming. Negative stains with India ink. (Bar = 50 μm).

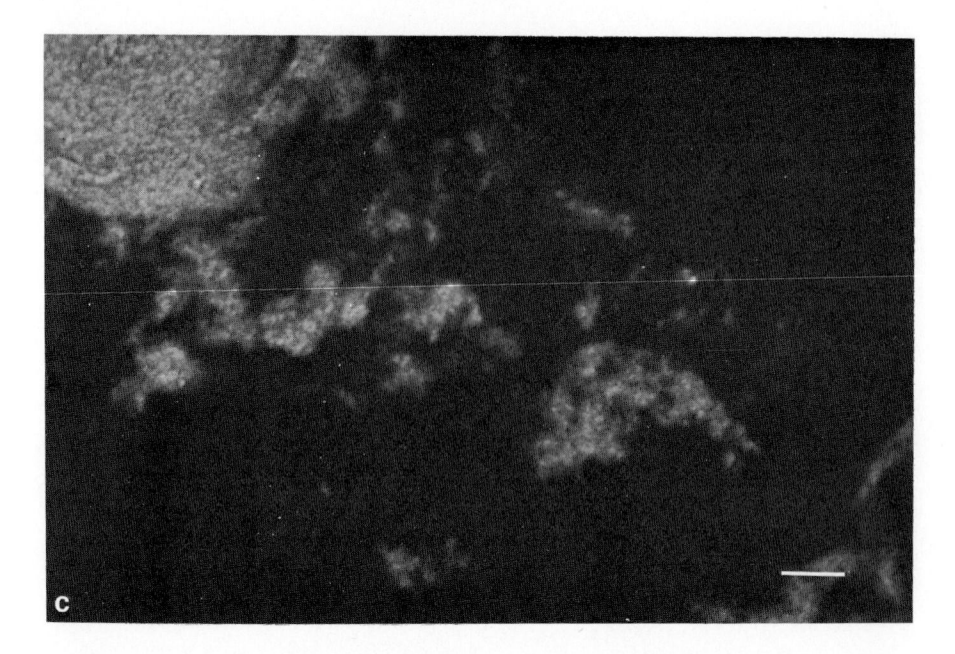

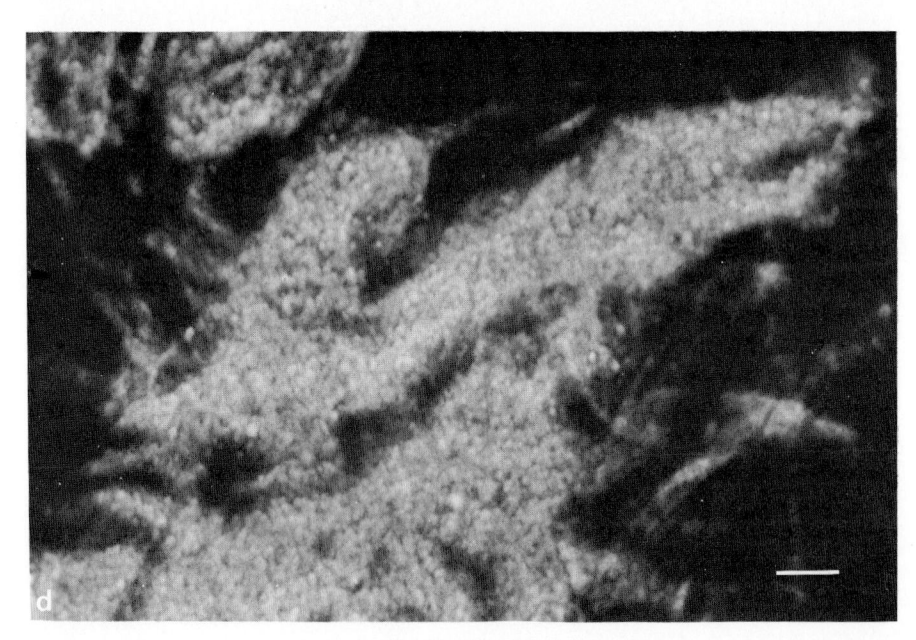

FIG. 1 (*Continued*)

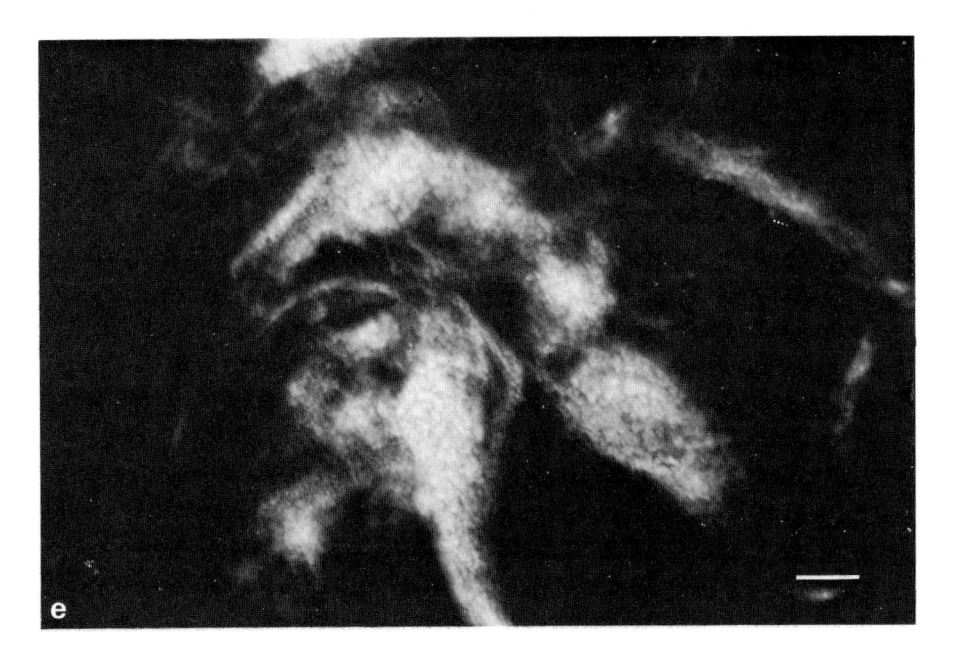

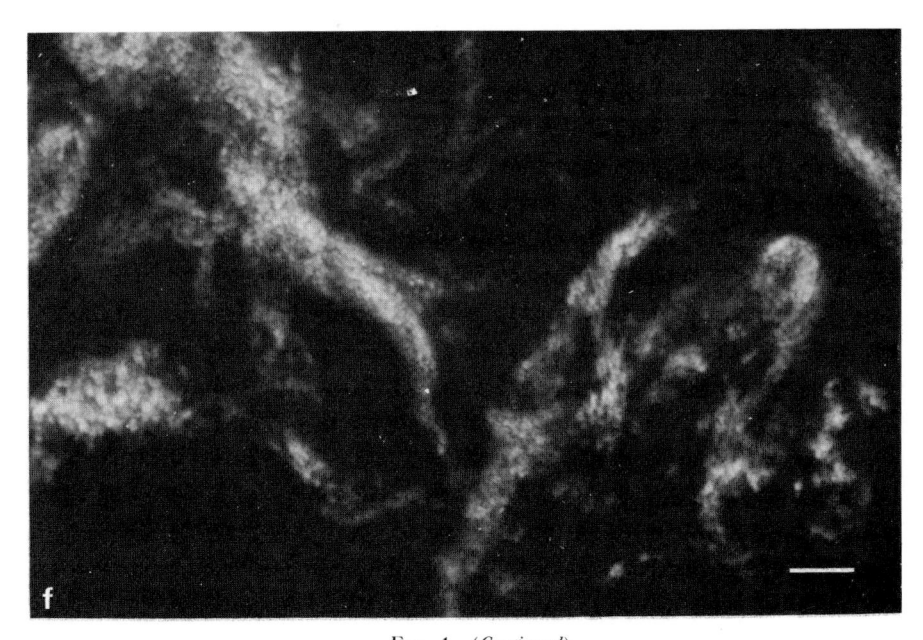

FIG. 1 (*Continued*)

1981), although the possibility of the two materials being the same, with small structural modifications during extraction accounting for the transformation from an insoluble to a soluble form, cannot be discounted.

Two distinct types of polymer have been extracted from cultures of *K. aerogenes*, a colloidal form which does not pass a 0.2-μm filter, and a soluble form. The two forms may be obtained by careful decantation of the supernatant obtained by centrifuging the sample, which has been pretreated by ultrasonication, at 33,000 g for 60 min. Filtration of the extracted polymer separates the soluble form, which is recovered in the filtrate, from the colloidal form, which is trapped on the filter. If this material is included in the quantity of polymer components obtained by various extraction methods presented in Table 1, the yield of total polymer would be ca. five times greater than that of the soluble material alone.

The colloidal form of the polymer is not readily solubilised since it can be resuspended in distilled water and recovered by further filtration with an efficiency of ≥ 80%. It is probable that this colloidal material is more closely related than the soluble polymer to the insoluble matrix of activated sludge. Moreover, the relative quantities of colloidal and soluble polymer produced by *K. aerogenes* in continuous culture change greatly with dilution rate. The effect of dilution rate on the quantities of colloidal and soluble polymer produced by *K. aerogenes* is shown in Table 2. At the lower dilution rates the colloidal form predominated, whereas at the higher rates the extract consisted largely of soluble polymer. This may have important implications for metal removal in biological sewage treatment processes running at different cell residence times, depending on the relative affinities of metal ions for the two types of polymer.

TABLE 2

The effect of dilution rate on the production of soluble and colloidal forms of extracellular polymer by Klebsiella aerogenes *NCTC 8172 in continuous culture*

Dilution rate (/h)	Polymer extracted (mg/litre)	Percentage composition of polymer	
		Soluble	Colloidal
0.01	2580	7	75
0.05	648	9	77
0.10	219	17	62
0.25	128	91	23*
0.50	57	95	14*

* Subject to experimental error.

Metal Complexation by Soluble and Matrix Extracellular Polymers

Determination of Metal Binding

The complexation of metal ions by extracellular polymers may be determined by adding a suitable concentration of a soluble metal salt to a solution of extracted polymer followed by shaking for 3 h to achieve equilibrium. The polymer–metal complex may be separated from the free metal ion by using gel-filtration chromatography.

A suitable gel for the separation of the free and complexed metal is one on which the polymer is eluted at the void volume (V_0); examples of such gels are sephadex G-10 and G-15 (Pharmacia Ltd.) and Biogel P-type medium (Biorad Laboratories). The composition of the eluent may be adjusted to obtain the desired conditions of pH and ionic strength, but in any case should have an ionic strength of at least 0.02 M (obtained by the addition of NaCl or another alkali metal salt) to prevent excessive retardation or irreversible adsorption of the metal ion on passage through the gel bed. The pH of the eluent may be adjusted with, e.g., 0.02-M NaOH or HCl, or a dilute buffer which does not form complexes or precipitate with the metal ion.

The gel is swollen in the eluent, poured into a suitable column (e.g. 1 × 100 cm) and equilibrated by the passage of several column volumes of eluent. Flow rates used for each gel are those recommended by the manufacturer. An aliquot of the sample of approximately 3% of the total volume of the gel bed (V_t) is applied to the column and fractions eluted are collected in a fraction collector. Concentration of metals were determined by flameless atomic absorption spectroscopy. A typical elution profile of the separation between metals complexed with *K. aerogenes* polymer and free metal ions is shown in Fig. 2.

Some metal ions behave in an abnormal manner in certain eluents; the profiles shown in Fig. 2 were obtained using the most appropriate eluent for each metal. The elution of the free metal ion should be checked in each case and if it elutes too close to V_0 or is significantly retarded, another eluent should be sought. The use of Sephadex G-25 may be advantageous if abnormal elution occurs since Posner (1963) has shown that it adsorbs metals to a lesser degree than other Sephadex G-types.

Determination of Stability Constants

The fate of a particular metal ion in the activated-sludge process will depend on the relative stability of the complexes it may form with various components of the system. Stability constants are, by definition, precise ex-

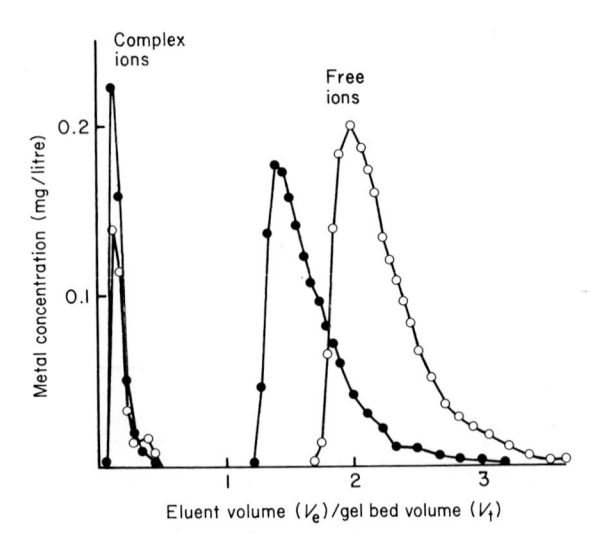

FIG. 2. Separation of free metal ions and metal ions complexed with *Klebsiella aerogenes* extracellular polymer by gel filtration using Sephadex G-10. (○) Nickel; (●) cobalt.

pressions of the stability of such complexes and can therefore be used to quantify the extent of metal–polymer interactions. However, stability constants are only valid for the particular experimental conditions under which they are determined. Commonly such conditional constants are recorded for the pH value and ionic strength of the medium in which they are determined.

The stability constants of metal ion–polymer complexes can be determined by a gel-filtration method (Mantoura & Riley 1975). The method involves eluting the polymer–metal complex from a chromatographic column with an eluent containing the metal at the same concentration as that of the original metal–polymer mixture. The eluent may be buffered to a desired pH value and ionic strength, using, e.g. 0.01-M Tris–HCl pH 6.8 (Perrin & Dempsey 1974). If a component of the buffer complexes the metal ion, the concentration of the free metal ion which is available for complexation with the polymer can be calculated from the pH value of the buffer, the pK_a of the ligand and the stability constant for the metal–ligand interaction (Mantoura & Riley 1975).

A gel column is equilibrated with the eluent containing the required metal concentration. The dried polymer is dissolved in a small volume of the eluent to give a concentration of 1–3 μg/ml, allowed to equilibrate, applied to the column, and eluted in the normal way. An example of the type of elution profile obtained for a complex of cobalt with *K. aerogenes* polymer is shown in Fig. 3. The area of the peak in the void volume is equivalent to the amount of

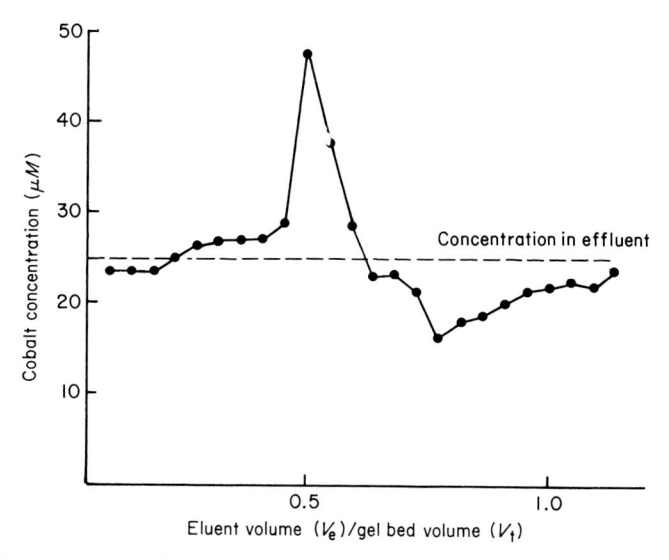

FIG. 3. Elution profile of a cobalt–*Klebsiella aerogenes* polymer complex using Tris buffer, pH 6.8, containing 20-μM (2×10^{-5} M) cobalt on Sephadex G-25.

metal bound by the polymer. Given the molecular weight of the polymer, which can be determined by gel-filtration chromatography (Churms & Stephen 1974), the conditional stability constant, K, is given by

$$M_b/(M_p[M_f]) = K[n - (M_b/M_p)] \qquad (1)$$

where M_b = moles metal bound, M_p = moles polymer added, $[M_f]$ = concentration of free metal ion, and n = number of binding sites per molecule of polymer. If the experiment is repeated using a range of values for $[M_f]$, Eq. (1) can be solved by plotting $M_b/M_p[M_f]$ (y axis) against M_b/M_p (x axis). The gradient of the line gives $-K$ and the intercept on the x axis gives n. An example of the graphical determination of the stability constant for the formation of a complex between cobalt and *K. aerogenes* polymer is shown in Fig. 4.

The stability constants of complexes formed between cadium, copper, nickel and cobalt and the *K. aerogenes* polymer were determined in 0.05-M Tris buffer at pH 6.8 and an ionic strength of 0.05. The constants which were determined on a sample consisting predominantly of colloidal material recovered by precipitation of the extract with ethanol are shown in Table 3 together with the number of binding sites per molecule. The molecular weight of the polymer was taken to be 1.7×10^6 obtained from the data of Churms & Stephen (1974). From the graphical solutions of Eq. (1) only a single value for K was obtained in each case, since no biphasic curves with

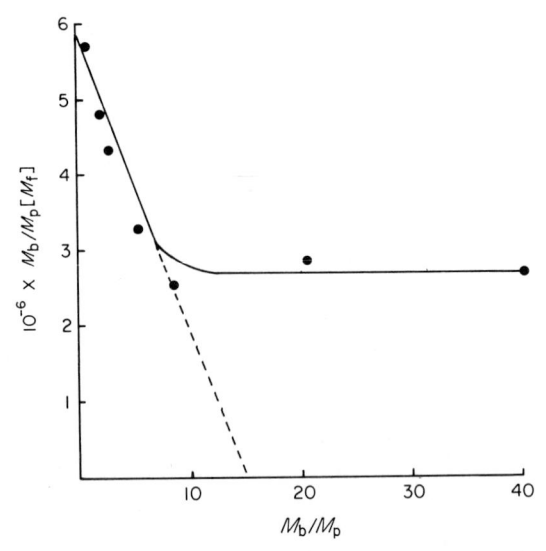

FIG. 4. Determination of the stability constant (*K*) of a cobalt–*Klebsiella aerogenes* polymer complex. *K* is given by the slope (broken line). (M_b), Moles metal bound; (M_p), moles polymer extracted; [M_f], concentration of free metal ion.

negative gradients were obtained, indicating that the polymer in this sample probably carried only one type of metal binding site.

Metal Removal by Biomass and Extracted Polymers

Some species of non-capsulated bacteria can immobilise large quantities of metals from solution (Sterritt & Lester, 1980). Therefore attempts were made to assess the extent of complexation of metals by extracellular poly-

TABLE 3

Conditional stability constants (K) *of complexes between* Klebsiella aerogenes *NCTC 8712 extracted extracellular polymer and metal ions and numbers of binding sites* (n) *per molecule of polymer*

Metal	$\log_{10}K$	n
Copper	7.7	7
Nickel	5.5	27
Cobalt	5.5	17
Cadmium	5.2	51

TABLE 4

The removal of metals from solution by untreated cells, stripped cells and extracted extracellular polymer of Klebsiella aerogenes *NCTC 8172*

Metal	Metal concentration added (mg/litre)	Metal removed (%) Untreated cells	Stripped cells	Extracted polymer
Cadmium	0.01	95	89	0
	0.1	84	84	1.0
	1	98	78	0.50
	10	81	58	n.d.*
Nickel	0.01	100	60	50
	0.1	64	52	27
	1	74	37	2.10
	10	71	30	0.40
Cobalt	0.01	100	70	7.0
	0.1	88	36	1.3
	1	83	21	0.10
	10	89	29	0.05
Manganese	0.01	95	71	21
	0.1	94	67	2.50
	1	89	52	0.80
	10	94	55	0.10

* n.d., not determined.

mers of *K. aerogenes* and laboratory activated sludge in comparison with mechanisms of uptake not mediated by the polymers. This was done by comparing the removal of metals from solution by untreated cells, cells from which polymer had been removed ("stripped" cells) and extracted polymer. The percentage removal of cadmium, nickel, cobalt (added as nitrate salts) and manganese (added as the sulphate) at concentrations of 0.01–10 mg/litre in cultures of *K. aerogenes* are shown in Table 4. Untreated cells removed 80% of cadium cobalt and manganese at all concentrations added, and slightly less nickel (64–74%) at concentrations of 0.1–10 mg/litre. Removal of extracellular polymer reduced the capacity of the cells to remove all four metals, the effect being most marked in the case of nickel and cobalt. However, the concentrations of metal bound by the extracted polymers were generally less than the differences in metal uptake between untreated and stripped cells at most metal concentrations.

The results of a similar experiment with laboratory activated-sludge biomass are shown in Table 5. Except at the smallest concentration of cadmium and nickel, the removal of extracellular polymer had a marked effect on the ability of stripped biomass to remove the metals and this was reflected by the

TABLE 5

The removal of metals from solution by untreated biomass, stripped biomass and extracted extracellular polymer from laboratory activated sludge

Metal	Metal concentration added (mg/litre)	Metal removed (%)		
		Untreated biomass	Stripped biomass	Extracted polymer
Cadmium	0.001	100	100	90
	0.01	100	30	87
	0.01	93	12	81
	1	39	2	20
Nickel	0.01	100	94	40
	0.1	100	70	41
	1	72	52	33
	10	34	7	3
Manganese	0.01			10
	0.1	n.d.*	n.d.	5
	1			1
	10			0.2
Cobalt	0.01			90
	0.1	n.d.	n.d.	75
	1			82
	10			22

* n.d., not determined.

comparatively large amount of complexation for the metals by the extracted polymer. The greater quantities of cadmium complexed by untreated cells and extracted extracellular polymer in comparison with the uptake by stripped biomass suggests that the role of polymer in the removal of this metal in the activated-sludge process may be more important than any other mechanism.

Further experiments with chemostat cultures of *K. aerogenes* at dilution rates of 0.06–0.30/h and laboratory activated sludge at sludge ages of 3–18 days did not show any marked effect of these parameters on the pattern of metal complexation by untreated biomass, stripped biomass and extracted extracellular polymer.

Relative Affinities of Soluble and Colloidal K. aerogenes *Polymer for Metal Ions*

Since the soluble and colloidal types of polymer obtained from *K. aerogenes* may be only partially representative of the capsular material surrounding the cells, and since their relative proportions could potentially influence the ratio of soluble to insoluble forms of metal ions, and thus the degree of removal, the relative affinities of the soluble and colloidal polymers for six metal ions were determined. Metals were added to a polymer extract from a culture of *K. aerogenes* and allowed to equilibrate for 3 h. The soluble polymer was separated by filtration and the total and soluble polymers were fractionated by gel filtration to determine M_b. The metal bound by the colloidal fraction was determined by subtraction. The metals (as nitrate salts except for manganese sulphate) were added at a concentration of $10^{-5} M$, both individually and all together. The degree of complexation, expressed for comparative purposes as moles metal bound per gram polymer extracted per mole metal added, is shown in Table 6.

Nickel, when added separately, was complexed to the greatest extent, with five times as much being bound by the soluble polymer than by the colloidal polymer. Copper and cadmium were also more strongly bound by the soluble form than the colloidal form of the polymer; cobalt, manganese and thallium were only weakly bound.

When all the metals were added together to an aliquot of the extracted polymer, the degree of binding of all the metals (except thallium) to the soluble polymer was decreased, while the complexation of nickel by the colloidal polymer was unchanged and the complexation of copper and cadmium increased. This may indicate that in the case of the soluble polymer there is some degree of competition of the metals for available binding sites, while the complexation of copper and cadmium by the colloidal polymer may be enhanced by the presence of other metal ions.

TABLE 6

Affinities of soluble and colloidal forms of Klebsiella aerogenes *NCTC 8172 extracellular polymer for metal ions added individually and all together*

Metal	Complexation by polymer*			
	Metals added individually		Metals added together	
	Soluble	Colloidal	Soluble	Colloidal
Copper	793	49	82	310
Cadmium	250	41	17	448
Cobalt	48	18	1	0
Nickel	1200	230	420	189
Manganese	29	20	10	0
Thallium	0	0.2	0	0

* Micromoles metal bound per gram polymer per mole metal added.

Discussion

In a multicomponent system, such as the activated-sludge process, the behaviour of metal ions, particularly with respect to their adsorption or complexation by the biomass, is difficult to assess accurately *in situ*. Thus, the best approach towards gaining an insight into the role of the extracellular polymeric matrix in removing metal ions would appear to be to separate the bacterial floc into its component parts. Many workers have attempted to do this, but it is apparent that the various methods used all achieve different extraction efficiencies and different degrees of contamination of the extract by intracellular materials (Brown & Lester 1980). Moreover, it would appear that some methods described for the extraction of polymer from activated sludge remove a negligible quantity of the matrix surrounding the cells (Novak & Haugan 1981). From determinations of total polysaccharides of *K. aerogenes*, Rudd *et al.* (1982) found that the optimum extraction efficiency of physical methods was ca. 70%, but under certain cultural conditions it was only 20%.

Despite the probably low extraction efficiency with activated sludge here with conventional methods, the quantities of metal ions bound by polymer which have been separated from the floc matrix indicated that, particularly in the case of cadmium, the polymer may have been responsible for a large proportion of the removal achieved by the intact floc. The stability constants for complexes between metal ions and the *K. aerogenes* polymer were in the range 10^5–10^7. These values are comparable with stability constants for metal ions with humic and fulvic acids (Mantoura & Riley 1975; Sohn & Hughes 1981), the importance of which in the environmental transport of metal ions is well recognised (Sposito 1981). Since the activated-sludge polymer appeared to complex metals to a greater extent than the *K. aerogenes* polymer, possibly due to the greater density of charged groups attached to the polysaccharide chain in the former (Brown & Lester 1979), it is possible that the capacity of activated sludge for metal ions is greater than that of many other naturally occurring organic ligands.

The influence of cell residence time (sludge age) on the removal of metals in the activated-sludge process and in other bacterial systems has been the subject of several studies (Sterritt & Lester 1981; Sterritt *et al.* 1981; Brown & Lester 1982b) which have shown that an increase in metal removal with increasing sludge age may be due in part to changes in the quantity and composition of polymer produced by the biomass. Continuous cultures of *K. aerogenes* produce the greatest quantities of extractable polymer at lower dilution rates, although this does not greatly enhance metal removal. In the activated-sludge process, polymer concentrations increased in direct proportion to the increase in biomass concentrations with increasing sludge age (Brown & Lester 1982b), although relative proportions of slime polymers (loose, partially soluble materials) and capsular polymers may change (Saunders & Dick 1981). If slime or soluble polymers complex with metals to any important extent, their loss in sewage works effluent will have a detrimental effect on metal removal. However, capsular material will be retained in the system in the flocs and continue to be active in binding metals that will be removed by sedimentation. Thus, in studies of the role of polymers in metal binding, attention should be directed towards maintaining the integrity of the various physical forms in which they may exist and developing non-disruptive methods for their differentiation and separation.

An increased awareness of the need to control the quality of effluents discharged to receiving waters, particularly with respect to toxic metal ions, is reflected by a proposal for a directive relating specifically to cadmium (Commission of the European Communities 1981) in addition to an existing directive on the quality of surface waters (European Economic Community 1975). As concern for the quality of the environment increases there will be greater interest in processes which can be employed to limit concentrations of toxic pollutants in the aquatic environment. Dugan & Pickrum (1972) have suggested that it may be possible to develop a microbiological process, involving extracellular polymers, to remove metal ions from solution. Thus, an understanding of the role of extracellular polymers in removing metal ions in waste water treatment processes and their potential for application in other processes for the control of pollution by metals may be of value in overall pollution control strategy.

References

BITTON, G. & FREIHOFER, V. 1978 Influence of extracellular polysaccharides on the toxicity of copper and cadmium towards *Klebsiella aerogenes*. *Microbial Ecology* **4**, 119–125.

BROWN, M. J. & LESTER, J. N. 1979 Metal removal in activated sludge: the role of bacterial extracellular polymers. *Water Research* **13**, 817–837.

BROWN, M. J. & LESTER, J. N. 1980 Comparison of bacterial extracellular polymer extraction methods. *Applied and Environmental Microbiology* **40**, 179–185.

BROWN, M. J. & LESTER, J. N. 1982a Role of bacterial extracellular polymers in metal uptake in pure bacterial culture and activated sludge—I. Effects of metal concentration. *Water Research* **16**, 1539–1548.

BROWN, M. J. & LESTER, J. N. 1982b Role of bacterial extracellular polymers in metal uptake in pure bacterial culture and activated sludge—II. Effects of mean cell retention time. *Water Research* **16**, 1549–1560.

BURTON, K. 1956 A study of the condition and mechanisms of the diphenylamine reaction for the colorimetric estimation of deoxyribonucleic acid. *Biochemical Journal* **62**, 315–323.

CHURMS, S. C. & STEPHEN, A. M. 1974 Studies of molecular-weight distribution for hydrolysis products from some *Klebsiella* capsular polysaccharides. *Carbohydrate Research* **35**, 73–86.

COMMISSION OF THE EUROPEAN COMMUNITIES 1981 Proposal for council directive concerning the limit values for discharges of cadmium into the aquatic environment. *Com (81)* 56, Final.

DUBOIS, M., GILLES, K. A., HAMILTON, J. K., REBERS, P. A. & SMITH, F. 1956 Colorimetric method for determination of sugars and related substances. *Analytical Chemistry* **28**, 350–356.

DUGAN, P. R. & PICKRUM, H. M. 1972 Removal of mineral ions from water by microbially produced polymers. *Proceedings of the 27th Industrial Waste Conference, Purdue University, Engineering Extension Series* No. 141, pp. 1019–1038.

EUROPEAN ECONOMIC COMMUNITY 1975 Council Directive concerning the quality of water intended for the abstraction of drinking water in the member states (75/440/EEC), Official Journal of the European Communities L194/26-L194/31.

FRIEDMAN, B. A. & DUGAN, P. R. 1968 Concentration and accumulation of metallic ions by the bacterium *Zoogloea*. *Developments in Industrial Microbiology* **9**, 381–388.

LOWRY, O. H., ROSEBROUGH, N. J., FARR, A. L. & RANDALL, R. J. 1951 Protein measurement with Folin phenol reagent. *Journal of Biological Chemistry* **193**, 265–275.

MANTOURA, R. F. C. & RILEY, J. P. 1975 The use of gel filtration in the study of metal binding by humic acids and related compounds. *Analytica Chimica Acta* **78**, 193–200.

NISHIKAWA, S. & KURIYAMA, M. 1968 Nucleic acid as a component of mucilage in activated sludge. *Water Research* **2**, 811–812.

NOVAK, J. T. & HAUGAN, B. E. 1981 Polymer extraction from activated sludge. *Journal of the Water Pollution Control Federation* **53**, 1420–1424.

OLIVER, B. G. & COSGROVE, E. G. 1974 The efficiency of heavy metal removal by a conventional activated sludge treatment plant. *Water Research* **8**, 869–874.

PAVONI, J. L., TENNEY, M. W. & ECHELBERGER, W. F. 1972 Bacterial exocellular polymers and biological flocculation. *Journal of the Water Pollution Control Federation* **44**, 414–431.

PERRIN, D. D. & DEMPSEY, B. 1974 *Buffers for pH and Metal Ion Control*. London: Chapman & Hall.

POSNER, M. M. 1963 Importance of electrolyte in the determination of molecular weights by "Sephadex" gel filtration with especial reference to humic acid. *Nature, London* **198**, 1161–1163.

RUDD, T., STERRITT, R. M. & LESTER, J. N. 1982 The use of extraction methods for the quantification of extracellular polymer production by *Klebsiella aerogenes* under varying cultural conditions. *European Journal of Applied Microbiology and Biotechnology* **16**, 23–27.

SAUNDERS, F. M. & DICK, R. I. 1981 Effect of mean cell residence time on organic composition of activated sludge effluents. *Journal of the Water Pollution Control Federation* **53**, 201–215.

SOHN, M. L. & HUGHES, M. C. 1981 Metal ion complex formation constants of some sedimentary humic acids with Zn(II), Cu(II) and Cd(II). *Geochimica et Cosmoschimica Acta* **45**, 2393–2399.

SPOSITO, G. 1981 Trace metals in contaminated waters. *Environmental Science and Technology* **15**, 396–403.

STEINER, A. E. MCLAREN, D. A. & FORSTER, D. F. 1976 The nature of activated sludge flocs. *Water Research* **10**, 25–30.

STERRITT, R. M. & LESTER, J. N. 1980 Interactions of heavy metals with bacteria. *Science of the Total Environment* **14**, 5–17.

STERRITT, R. M. & LESTER, J. N. 1981 The influence of sludge age on heavy metal removal in the activated sludge process. *Water Research* **15**, 59–65.

STERRITT, R. M., BROWN, M. J. & LESTER, J. N. 1981 Metal removal by adsorption and precipitation in the activated sludge process. *Environmental Pollution, Series A* **24**, 313–323.

STOVELAND, S. & LESTER, J. N. 1980 A study of the factors which influence metal removal in the activated sludge process. *Science of the Total Environment* **16**, 37–54.

TEZUKA, Y. 1973 A *Zoogloea* bacterium with gelatinous mucopolysaccharide matrix. *Journal of the Water Pollution Control Federation* **45**, 531–536.

WATER POLLUTION RESEARCH LABORATORY 1971 *Report of the Director*. London: HMSO.

Immobilisation of Microbial Cells and Their Use in Waste Water Treatment

P. S. J. CHEETHAM AND C. BUCKE

Tate & Lyle Group Research & Development, Philip Lyle Memorial Research Laboratory, Whiteknights, Reading, UK

Concern about the hazards of biologically active wastes and other pollutants has been growing throughout the post-war period ever since the publication of "Silent Spring" (Carson 1962). Adherence to and enforcement of regulatory legislation still, however, tends to be variable even in developed countries.

Pollutants include natural compounds, e.g. lignified woods, and artificial ones, e.g. organophosphorus insecticides. Microbial enzymes are a potentially effective means of degrading pollutants. Many have evolved over the course of millions of generations in natural environments, thereby protecting the parent cells from toxic substances, e.g. by utilisation as a carbon or energy source or both. Detoxification may result in either complete mineralisation of the pollutant or partial decomposition associated with the loss of undesirable biological activities. Many enzymes are not completely substrate specific and act on compounds which have been only very recently synthesised by man and released into the biosphere, e.g. surfactants and "builder" compounds (Cain 1981), 3-chlorobenzoate (Dom *et al.* 1974) and *S*-triazines (Cook & Hutter 1981). Such transformations are sometimes co-metabolism reactions which take place only in the presence of a second naturally assimilated carbon source. Thus, despite the variety of potentially toxic substances in the environment, comparatively few are so recalcitrant that they persist for long periods. However, localised large concentrations can have very damaging effects, e.g. oil spoils (Alexander 1973; Faber 1979; Peyton 1984).

Apparently recalcitrant compounds can be made biodegradable by developing microorganisms capable of degrading them, e.g. by chemostat culture, mutation or genetic transfer. For the degradation of wastes the use of enzymes still associated with whole cells, rather than isolated enzymes, appears to be advantageous because the enzymes are at least partially protected from denaturation by being retained inside a substantially intact cell wall and membrane. Also, degradation is often carried out by means of a number of enzymes and co-enzymes working in concert or sequentially. Such systems would be difficult and expensive to reconstruct using enzymes that had been individually extracted, purified and immobilised. Furthermore, the component enzymes are often difficult to extract in a fully active form and tend to decay at different rates; the conditions of use would be sub-optimal for the activity of most of the enzymes, and it would be difficult to reassemble them in their native spatial relationships. Metabolism of xenobiotic substances is often dependent on enzyme induction in microorganisms with little or no previous degradative capability, a process which is only possible when viable cells are used (Leisinger *et al.* 1981).

Additionally, degradation is often only possible by mixed cultures of species, some of which are primary species which can grow in monoculture on the substrate (pollutant) and others which are secondary species which cannot metabolise the primary substrate, but utilise metabolites and lytic products of the primary species. Certainly degradation rates and resistance to toxins and substances are invariably greater with mixed cultures than with pure cultures and isolated enzymes (Bull 1980).

Use of immobilised cells has many advantages over the use of freely suspended cells. Chief among these are the capability of re-using immobilised cells and the ease with which the cells can be separated from the reaction mixture, thus preventing contamination of the product stream. Other advantages include the ability to disperse the cells evenly by immobilisation so as to minimise diffusional restrictions on the rates of reaction and the ease with which immobilised cells can be used to exploit the kinetic features of continuously stirred and packed-bed types of reactor. Minimisation of product inhibition is a feature of the packed-bed reactor, and both types allow fully continuous processes thereby avoiding the changing conditions associated with batch operation and the use of free cells. These advantages also usually result in an increase in the volumetric activity (process intensity) of the system as compared with existing procedures, for instance the activated-sludge method of sewage treatment and anaerobic digestion systems.

In this chapter we have attempted to describe methods of cell immobilisation to survey the use of deliberately immobilised biocatalysts for waste treatment. We have not tried to review the use of naturally immobilised cells for the treatment of sewage etc. This topic has been thoroughly reviewed elsewhere (Cooper & Atkinson 1981; Switzenbaum 1983). Detoxification of pesticides by enzymes and microbial biodegradation in general have been reviewed by Munnecke (1978) and Bull (1980), respectively. Cell and enzyme immobilisation have been reviewed by Mosbach (1977), Cheetham (1980) and Bucke & Wiseman (1981).

Immobilisation of Microbial Cells in Calcium Alginate

If immobilisation is defined as the limitation of the free diffusion of material within a solid, liquid or gaseous environment, it is probably accurate to state that the majority of microorganisms exist in nature in an immobilised state. Many bacteria are either naturally entrapped in extracellular slimes or adsorbed on inert material or growth substrate. Many enzymes occur naturally adsorbed onto soil particles and many algae produce extracellular polysaccharides which allow them to adhere to surfaces and to survive periods of drought. To immobilise microorganisms deliberately does not, therefore, necessarily place them in an environment grossly different from their natural circumstances. However, it may be necessary to eliminate certain enzyme systems to prevent the further metabolism of commercially useful products or the formation of side products. In such cases immobilised cells become convenient sources of single enzymes or sequences of enzymes.

Many methods of immobilising cells have been developed, some mimicking nature by entrapping organisms in natural polymers, some improving on nature by increasing the physical stability of the preparation and others by combatting nature by covalently modifying cells and thus eliminating many of their biochemical activities. Entrapment within calcium alginate is one of the simplest methods of immobilising cells and has found widespread use in laboratory and pilot-scale studies, including the construction of reactors for waste degradation. It is not the perfect immobilisation method but when compared with other methods it appears to be one of the best.

Alginate

Alginic acid is a block co-polymer of D-mannuronic and L-guluronic acids which is synthesised by all algae of the class Phaeophyceae and by various bacteria, principally *Azotobacter* and *Pseudomonas* species. Commercially available alginates are isolated from *Laminaria*, *Macrocystis* and a few other genera of algae. Alginic acid is not a defined substance of constant composition because the proportion of D-mannuronic and L-guluronic acid units varies greatly from species to species and from one part of an organism to another. Furthermore, a typical alginic acid molecule consists of homopolymeric blocks of D-mannuronic acid, homopolymeric blocks of L-guluronic acid and sequences of alternating D-mannuronic acid and L-guluronic acid. There is a wide range of values for the molecular weight of alginate, and as manufacturers "tailor" their products to give desired viscosities, a wide range of properties is available.

For the purpose of entrapping cells and using them in reactors on a commercial scale the most important characteristic is that of gel strength. Thus not all grades of alginic acid are suitable for cell immobilisation. Alginic acid in itself forms weak gels but at undesirably low pH values. The best gelling agents are divalent or trivalent ions, of which calcium is generally the most effective. Calcium cross-links only L-guluronic acid units, which restricts the types of alginic acid suitable for immobilisation because *Macrocystis* alginic acid has a low L-guluronic acid content. In our experience the most useful alginic acid is that isolated from *Laminaria* species; this has a large L-guluronic acid content and forms a strong gel but has a comparatively low solution viscosity which allows comparatively large concentrations to be used.

Cell Immobilisation

Most alginic acids are sold as soluble sodium salts. Dissolution of sodium alginate in water can be achieved only by experience since lumps of solid material become surrounded by a layer of concentrated gel which severely inhibits further dissolution; prolonged stirring overcomes this problem. If large quantities of solution are required, it is wise to add the alginate powder to a stirred solution which is as warm as possible and to stir the preparation overnight to ensure complete dissolution. It is essential to use distilled or de-ionised water. Aqueous solutions of sodium alginate may be sterilised by autoclaving, which reduces the viscosity a little but has no marked effect on the gel strength.

Choice of the concentration of sodium alginate depends on the following factors: viscosity of the alginate solution, gel strength required and the size of the substrate molecules. These factors govern the extent to which diffusion of substrate to the cells limits the rate of subsequent reaction.

A final calcium alginate concentration of 2–5% (w/v) has been found to be suitable for laboratory studies with *Laminaria* alginate (Bucke & Cheetham 1981). Other workers have used concentrations up to 8% (w/v) (Linko *et al.* 1980).

The cells to be immobilised are harvested as a concentrated paste and stirred into the sodium alginate solution which may include osmotica, e.g. sucrose or sorbitol, or substrates (Fig. 1). Problems with premature gelling will be encountered if the cell growth medium contains appreciable levels of calcium. The cell–alginate suspension is then extruded dropwise into a bath containing a solution of a soluble calcium salt, e.g. 0.1- to 1.0-M $CaCl_2$ which may also contain an osmoticum or substrate(s) or both. Note that the outer layer of each bead consists of highly cross-linked alginate which is precipitated instantaneously when the cell suspension first enters the $CaCl_2$ solution, but that the inside of the bead is much more porous, being true calcium

```
Grow cells
   ↓
Harvest cells as a
concentrated paste
   ├──────────────────── Dissolve alginate powder
Prepare cell slurry in
sodium alginate solution
   ↓
Extrude slurry into
CaCl₂ solution
   ↓
Allow time for complete
gelling
   ↓
Recover immobilised cells
by draining off CaCl₂
solution
```

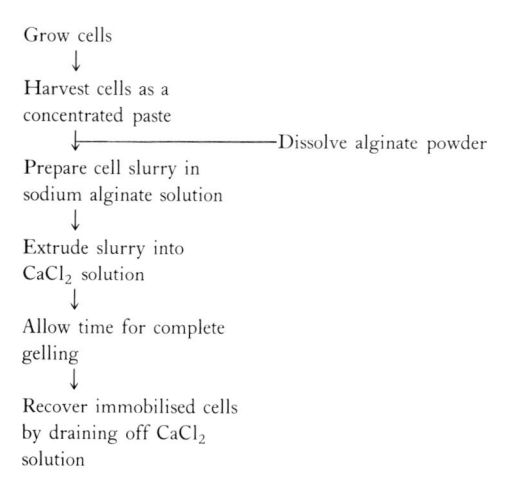

FIG. 1. Flow diagram for the immobilisation of cells in alginate.

alginate gel formed during the slow diffusion of Ca^{2+} into the bead, a process which requires several hours for completion. It is most simple to extrude the suspension dropwise into the $CaCl_2$ solution to produce beads, but spaghetti-like strands may be produced if care is taken to keep the extruded gelled material away from the extruder nozzle. On a small scale a syringe is a convenient extruder, but on a larger scale a pump may be used. The size of the beads or strands is out of the control of the user, being determined principally by the viscosity and surface tension of the alginate–cell suspension. Normally the diameter is about 5 mm, larger than is ideal for the optimal combination of reactor flow conditions and ready diffusion of substrate to, and product from, the immobilised cells (Cheetham *et al.* 1979). It is also possible to form alginates into sheets which may be used in "Swiss roll" configurations (Rosevear 1982).

The relation of bead strength and cell concentration was studied in detail by Cheetham (1979). Gels were shown to be weakened by large concentrations of cells and deformation of the beads, which results in compaction, and eventual blockage of columns was shown to occur continuously through the lifetime of the column. This is a "creep" phenomenon familiar to civil engineers who encounter it in many structural materials (Fig. 2). It is possible to feed growth media to entrapped cells and allow "secondary" growth within the beads. The influence of increased cell concentration produced by secondary growth on gel strength appears not to have been examined but can be expected to cause an appreciable weakening of the gel structure particularly on the outside of the bead which is the predominant site of cell proliferation.

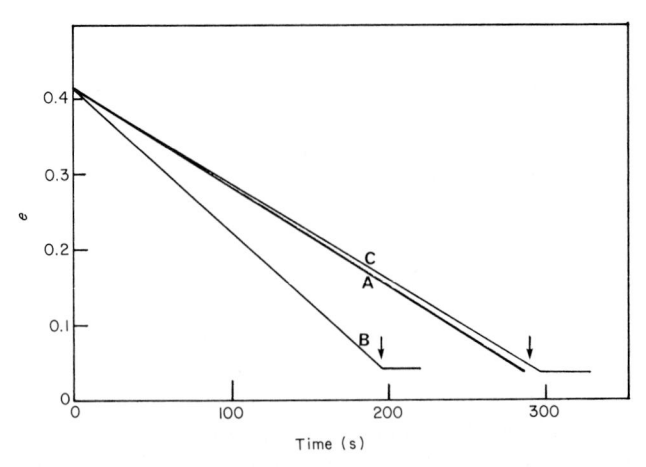

FIG. 2. Measurement of the rate of compression of alginate bead columns with time. (A) Column containing beads incubated in distilled water at a pressure of 13,158 kg/m²; (B), columns of beads incubated in 55% (w/v) sucrose solution at a pressure of 3061 kg/m²; (C), similar to (B) except that the sucrose solution was continually pumped up the column; arrow indicates point at which blockage of the column occurred; e is the void volume of the column and is expresed as a percentage of the packed column volume. (From Cheetham 1979.)

Performance of Calcium Alginate–Cell Complexes

Cells subjected to the alginate immobilisation process have been only briefly exposed to the calcium ions (and a counter anion). Therefore, it is not surprising that the stationary effectiveness factor (n) of alginate-immobilised cells is usually very large:

$$\text{Effectiveness factor } (n) = \frac{\text{Initial activity after immobilisation}}{\text{Initial activity before immobilisation}}$$

Cells retain their viability for longer periods than in liquid media and the activity of immobilised cells may be increased by feeding a complete growth medium to allow cell growth.

When immobilised cells are supplied with a substrate which does not allow growth, no cells are released from calcium alginate beads after the first 24–36 h. When supplied with a growth medium, cells first fill the bead and then spill out into the liquid phase so that the immobilised cell preparation acts as a large inoculum. In such circumstances the alginate beads retain their structure well because there is no destruction of bead integrity as cells emerge. Similarly, calcium alginate beads are not disrupted by CO_2 produced by immobilised yeast cells during ethanolic fermentation (Kierstan & Bucke 1977). The major hazard to the integrity of calcium alginate beads is chelating agents with a greater affinity for calcium ions than L-guluronic acid. Thus alginate is unsuitable for use when phosphate is an important constituent of a growth medium or substrate stream. The effects of small amounts of a calcium chelators may be overcome by including calcium ions in the process stream. The gross effects of calcium chelation have been avoided either by substituting strontium ions for calcium ions as cross-linking agent (Paul & Vignais 1980) or by further cross-linking the alginate chains by using polyethylenimines (Birnbaum *et al.* 1981). The latter method appears adversely to affect the viability and activity of the cells. However, neither of these approaches is ideal if the immobilised cell preparation is intended to produce a material for food use, but they appear to be acceptable for non-food processes.

Although various microorganisms are capable of using alginate as a sole carbon source, since they possess the enzyme alginate lyase, we have encountered no problems of microbial degradation of alginate, even in columns which have been in constant use with unsterilised substrate streams for several months.

Comparison with Alternative Methods of Cell Immobilisation

The earliest attempts at cell immobilisation by entrapment in a gel used polyacrylamide, a method that has proved sufficiently successful to be used in commercial processes producing L-malic acid and L-aspartic acid from fumarate. In both cases, however, polyacrylamide has been replaced by κ-carrageenan in recent years. Also, polyacrylamide has been superseded by calcium alginate for fungal mycelia conducting steroid biotransformations (Larsson *et al.* 1976).

The drawbacks to the use of polyacrylamide are the possible toxicity to microbial cells and certain toxicity to human beings, production of free radicals during the polymerisation reaction and, almost certainly less significantly, heat produced during polymerisation. These factors contribute to a reduction in activity of polyacrylamide-entrapped cells, but this is of little significance where the preparation is to be given a complete growth medium because usually sufficient viable cells remain to allow the bead to be filled with cells by secondary growth.

Probably the primary reason for the preference for natural polysaccharides over polyacrylamide is the unexpected and largely unexplained increase in the longevity of biochemical activities observed on the immobilisation of cells in the neutral polymers. Like alginate, κ-carrageenan is an algal polysaccharide extracted from membranes of the Rhodophyceae. It is a co-polymer of β-D-galactose sulphate and 3,6-anhydro-α-D-galactose which forms gels in the presence of potassium or ammonium salts. Unlike alginate it forms gels only when warm solutions are mixed with the appropriate metal ions and cooled. Nevertheless, the conditions required for cell entrapment are mild, i.e. cell suspension and κ-carrageenan solution are warmed to 37–60°C and mixed, and potassium or ammonium salts are added. Beads of gel are produced easily, and the fact that cells are taken briefly to elevated temperature does not markedly affect their activity and viability. κ-Carrageenan, like alginate, has the property of prolonging the biochemical activities immobilised within it.

κ-Carrageenan, alginate and polyacrylamide have proved to be the most widely used media for cell entrapment. Other media have been used with varying degrees of success. Agar produces a gel that is often too soft for practical use. Many chemically produced polymers have few if any advantages over polyacrylamide; polyurethane foams have been used successfully on occasions. An improvement on conventional polyacrylamide has been the introduction of "pre-polymerised" material which allows cell entrapment in polyacrylamide in milder conditions (Freeman & Aharonowitz 1981).

For some purposes, entrapment in cellulose acetate has proved very successful. A suspension of cells is dispersed into a solution of cellulose acetate which is extruded into a setting bath to form filaments of immobilised cells. Although contact with organic solvents is detrimental to cell activity and viability, under such conditions immobilised cells remain active for long periods of time (Marconi *et al.* 1976).

Entrapment between collagen fibres has proved successful in many cases (Vieth & Venkatasubramanian 1976). This is more than a simple entrapment because, for stability, the complex of cells and collagen must be tanned using a cross-linking agent such as glutaraldehyde. Thus collagen immobilisation falls between entrapment in a polysaccharide or polymer gel and covalent modification. In practice the reactive groups on the collagen so outnumber those on the cells that inactivation of the cells by the cross-linking agent seems not to be a major problem. Entrapment in collagen fibres produces membranes rather than beads which may be of advantage in some cases.

The most gentle means of immobilising cells is by adsorption onto a surface such as stainless-steel knitted wire balls. This process is perhaps better described as semi-immobilisation since cells are readily released by passage of media over the cell–support complex. The use of adsorbed cells is well known in the water industry (Atkinson *et al.* 1980, 1981) and it will not be discussed here.

The immobilisation of cells by covalent binding to each other or to a support material is widely practised but almost invariably when only a single enzyme activity is being used, e.g. glucose isomerase (Poulsen & Zittan 1976).

The Use of Immobilised Cells or Enzymes as Detoxifying Agents

A considerable amount of research has been carried out recently into the enzymic and microbiological degradation of pesticides. For instance, *Pseudomonas putida* has been found to possess dehalogenases which will degrade mono- and dichloroacetates and propionates derived from pesticides and herbicides; especially important is the observation that Dalapon-degrading activity was stable over 20,000 h in a chemostat (Berry *et al.* 1979; Slater *et al.* 1979). An enzyme extract obtained from a mixed microbial population adapted to grow on Parathion was found to hydrolyse a number of organophosphate insecticides. This enzyme could also be successfully immobilised to facilitate its use with an operational half life of 280 days (Munnecke 1977, 1978).

Detection of wastes can often be achieved using immobilised cells or enzymes as quantitative monitors of pollutants, for instance Karube *et al.* (1977) have developed a biochemical oxygen demand (BOD) monitor (reviewed by Guilbault 1982). These analytical devices usually work by measuring the loss of enzyme activity caused by exposure to the pollutant, the degree of inhibition of the enzyme being proportional to the concentration of pollutant present in the sample under test.

Use of α-amylase has been proposed for the treatment of waste waters from the wheat starch industry (Wieg 1984). α-Amylase has also been used with considerable success to clarify colloidal starch–clay suspensions, commonly referred to as "white waters," that are produced in large volumes by paper mills. After enzyme treatment the solids can then be flocculated easily by the addition of alum. It was found to be advantageous to immobilise the enzyme so that it could be re-used, and to avoid the water having to be treated to destroy the enzyme before recycling the clarified water for re-use in paper manufacture (Smiley 1974). Other applications include the removal of carcinogenic aromatic amines, including benzidine, naphthylamines and aminophenyl, from waste waters derived from the coal, resin, plastics and textile industries by using hydrogen peroxide and peroxidase to cross-link the pollutants (Klibanov & Morris 1981).

The brown, partially chlorinated lignin derivatives commonly known as "Kraft lignins" present in pulp mull effluents are normally treated in aerated lagoons or activated sludge systems which, although reducing BOD and COD, fail to reduce colour. Kraft lignins can be decolourised by using either the white-rot fungus *Coriolus versicolor* entrapped in calcium alginate beads and supplemented with various carbon and energy sources (Livernoche *et al.* 1981; Fig. 3) or the lignolytic fungus *Phanerochaete chrysosporium* (Kirk & Yang 1979). In the former study the immobilised cells removed 80% of the colour after incubation for 3 days in the presence of sucrose (Livernoche *et al.* 1983). This biological method is an important alternative to the chemical

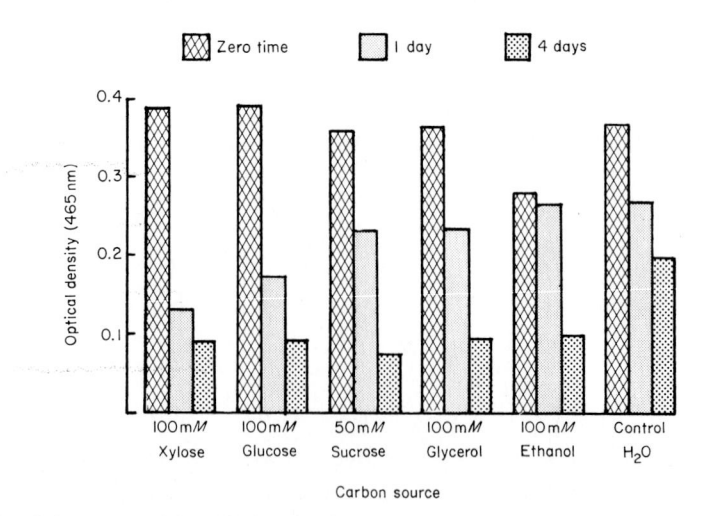

FIG. 3. Colour removal from Kraft mill effluent by immobilised *Coriolus versicolor* using various carbon sources. (From Livernoche *et al.* 1981.)

bleaching of lignin, especially as the chlorinated lignins produced during bleaching have been shown to be carcinogenic (Klibanov & Morris 1981).

Nitrosomonas europaea cells immobilised in alginate have been used to oxidise ammonia in waste waters to NO_2^- and NO_3^- so as to reduce the BOD and so that agal growth is not promoted (van Ginkel *et al.* 1983). The activity of the immobilised cells increased as a result of growth under conditions that would have caused wash-out of the free cells. Further degradation of the resulting soluble nitrates has been achieved by *Pseudomonas denitrificans* immobilised in alginate which have been used to denitrify drinking water (Fig. 4). The cells are provided with an exogenous carbon source and reduce nitrates and nitrites completely to gaseous products. When operated in continuous reactors the activity of the cells decayed with a half-life of 30 days.

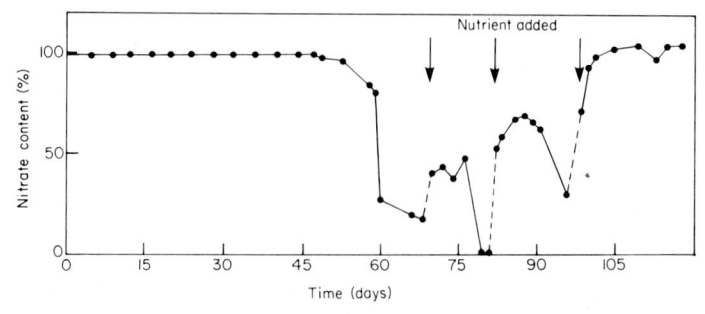

FIG. 4. Denitrification of water using immobilised *Pseudomonas denitrificans*. (From Nilsson & Ohlson 1982.)

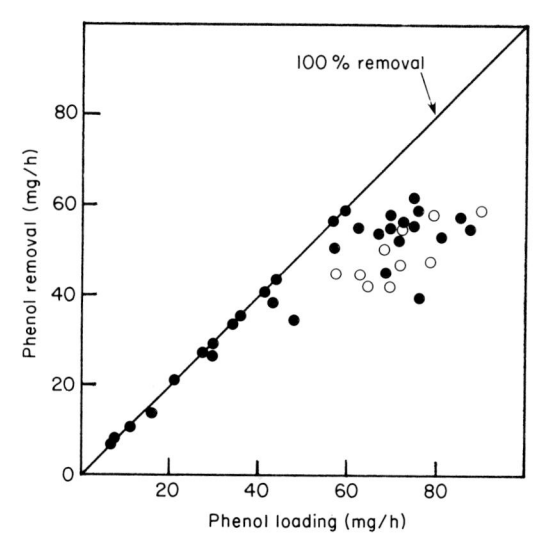

FIG. 5. Effect of phenol concentration on the capacity of immobilised *Aureobasidium pullulans* to remove phenol from waste waters. (●) Influent phenol concentration not greater than 1200 µg/ml; (○) influent phenol concentration of 1700 µg/ml. (From Takahashi *et al.* 1981.)

After addition of nutrients at intervals, fresh cell growth occurred and increases in the rate of reaction were observed (Nilsson *et al.* 1980; Mattiasson *et al.* 1981; Nilsson & Ohlson 1982) (see also Cabane and Vergnault 1978).

Degradation of phenols remaining in the waste waters from hospitals and laboratories and after coal processing to coke and removal of H_2S and NH_3 can be achieved either by using the fungus *Aureobasidium pullulans* adsorbed to fibrous asbestos (Takahashi *et al.* 1981; Fig. 5) or by using cells of *Pseudomonas* sp. either adsorbed to anthracite coal or entrapped on alginate gel (Hackel *et al.* 1975). Treatment of phenol-containing wash waters is advantageous in that it avoids damage to conventional biological treatment systems.

Formation of CH_4 continuously for over 90 days from waste waters using a population of methanogenic cells entrapped in agar, collagen or polyacrylamide membranes has been achieved (Karube *et al.* 1980; see also Kirsop *et al.*, this volume). Karube *et al.* (1980) used a microbial population isolated from a sewage sludge digester to degrade the waste waters from an alcohol fermentation factory to CH_4. Agar gel was found to be the immobilisation support of choice and cell division during operation helped to maintain the activity of the immobilised preparations.

Cyanide present in aqueous wastes can be effectively detoxified using immobilised mycelia of *Stemphylium loti* in which cyanide hydralase has been induced, the cyanide being converted to formamide. Flocculation with polyelectrolytes was the preferred method of immobilisation. Immobilisation

TABLE 1

Activity and stability of Erwinia rhapontici *cells immobilised on various supports**

Immobilisation technique[†]	Activity (g product/g wet cells/h)	Half-life (h)
Calcium alginate	0.325	8500
DEAE cellulose	0.583	400
Polyacrylamide	0.13	570
Glutaraldehyde-aggregated cells	0.153	40
κ-Carrageenan–locust bean gum	0.263	37.5
Bone char	0.01	25
Agar	0.34	27
Xanthan–locust bean gum	0.10	8

* From Bucke & Cheetham (1981).
† For details of the immobilisation methods see Mosbach (1976).

stabilised the enzyme and allowed it to be used continuously. The immobilised cells were, however, inhibited by nickel and so cannot be used for treating wastes from electroplating (Nazaly & Knowles 1981).

Production of glucose from waste cellulose can be regarded both as a means of disposing of a waste material and of producing a valuable product with potential uses as a food and as a substrate for the generation of fuels such as ethanol. Unfortunately, despite the undoubted potential of this project and much research effort, no practicable process using immobilisation techniques has yet emerged. A more feasible process uses immobilised enzymes to hydrolyse the lactose in whey which is a large-volume high-BOD waste derived from the cheese-making industry, the disposal of which gives severe pollution problems. Lastly, several recent studies have shown the potential of microorganisms for the removal of toxic metals, including uranium, from solution. For instance, Norberg & Rydin (1984) have used *Zoogloea ramigera* to accumulate copper continuously.

Discussion

Over the past 6 years the authors have worked on developing cell immobilisation methodologies, especially entrapment in calcium alginate gels, and on a number of immobilised cell reactions on a laboratory scale. On such process, the synthesis of isomaltulose from sucrose using *Erwinia rhapontici*, has been successfully developed to a pilot-plant scale using 20-liter columns and producing several tonnes of pure product (Cheetham *et al.* 1982). Little enzyme activity or cell viability was lost during immobilisation and unexpectedly large yields of product were obtained. The cell-associated enzyme responsible for the reaction was stabilised by using structurally intact, non-growing, but not necessarily viable, cells in preference to isolated enzyme, disrupted cells or growing cell preparations. Also, by using concentrated substrate and achieving high degrees of conversion, an operational half-life of 1 year was obtained (Table 1). These and other studies, such as the use of very large columns of yeast cells to produce ethanol from molasses, e.g. Wada *et al.* (1980), have considerable implications for the large-scale use of immobilised cells for waste treatment. They include choice of the optimal type of alginate, the best strains of cells to use in an immobilised, non-growing state and the conditions of use to maximise activity and stability. Mechanical factors such as the measurement of the rates of cell leakage from their supports, substrate diffusion into the support matrices and pressure drops through packed-bed columns together with the consequent compression of the columns, which occur slowly but continuously throughout the lifetime (several months) of the column, have also been studied.

In conclusion, it seems plausible that immobilised cells will be used increasingly provided that cells possessing the appropriate enzyme activities can be selected and the process can be made sufficiently cheap and convenient to use on a large scale; they are especially suitable for processes where a microbial product as opposed to biomass is required. Where pollutant removal is desired, immobilised cells can be extremely valuable and it is possible that many environmental protection processes using this approach will be revealed in the near future.

References

ALEXANDER, M. 1973 Non-biodegradable and other recalcitrant molecules. *Biotechnology and Bioengineering* **15**, 611–647.

ATKINSON, B., BLACK, G. M. & PINCHES, A. 1980 Process intensification using cell support systems *Process Biochemistry* **15**, 4, 24–32.

ATKINSON, B., BLACK, G. M. & PINCHES, A. 1981 In *Biological Fluidised Bed Treatment of Water and Waste Water* eds. Cooper P. P. & Atkinson, B., p. 75. Chichester: Horwood.

BERRY, E. K. M., ALLISON, N. & SKINNER, A. J. 1979 Degradation of the selective herbicide 2,2-dichloropropionate (Dalapon) by a soil bacterium. *Journal of General Microbiology* **110**, 39–45.

BIMBAUM, S., PENDLETON, R., LARSSON, P. & MOSBACH, K. 1981 Covalent stabilization of alginate gel for the entrapment of living whole cells. *Biotechnology Letters* **3**, 393–400.

BUCKE, C. & CHEETHAM, P. S. J. 1981 Production of isomaltulose. UK Patent Application No. A 2 063 268 A.

BUCKE, C. & WISEMAN, A. W. 1981 Immobilised cells and enzymes. *Chemistry and Industry* **7**, 234–240.

BULL, A. T. 1980 Biodegradation: some attitudes and strategies of micro-organisms and microbiologists. In *Contemporary Microbial Ecology* eds. Ellwood, D. C., Hedger, J. N., Latham, M. J., Lynch, J. M. & Slater, J. M., pp. 107–136. London: Academic Press.

CABANE, B. & VERGNAULT, J. 1978 US Patent Application 4,209,390.
CAIN, R. B. 1981 Microbial degradation of surfactants and "builder" compounds. In *Microbial Degradation of Xenobiotics and Recalcitrant Compounds* eds. Leisinger, T., Cook, A. M., Hutter, R. & Nuesch, J., pp. 325–370. London: Academic Press.
CARSON, R. 1962 *Silent Spring.* Harmondsworth: Pelican Books.
CHEETHAM, P. S. J. 1979 Physical studies on the mechanical stability of columns of calcium alginate gel pellets containing entrapped microbial cells. *Enzyme and Microbial Technology* **1**, 183–188.
CHEETHAM, P. S. J. 1980 Developments in the immobilization of microbial cells and their applications. *Topics in Enzyme and Fermentation Biotechnology* **4**, 189–238.
CHEETHAM, P. S. J., BLUNT, K. W. & BUCKE, C. 1979 Physical studies on cell immobilization using calcium alginate cells. *Biotechnology and Bioengineering* **21**, 2155–2168.
CHEETHAM, P. S. J., ISHERWOOD, J. & IMBER, C. I. 1982 The formation of isomaltulose by immobilized *Erwinia rhapontici. Nature London* **299**, 628–631.
COOK, A. M. & HUTTER, R. 1981 Degradation of *S*-triazines. In *Microbial Degradation of Xenobiotics and Recalcitrant Compounds* eds. Leisinger, T., Cook, A. M., Hutter, R. & Nuesch, J., pp. 287–349. London: Academic Press.
DINELLI, D., MARCONI, W. & MORISI, F. 1976 Fibre-entrapped enzymes. In *Methods in Enzymology* ed. Mosbach, K., Vol. 44, pp. 227–243. London: Academic Press.
DOM, E., HELLWIG, M., REINEKE, W. & KNACKMUSS, H.-J. 1974 Isolation and characterization of a 3-chlorobenzoate degrading pseudomonad. *Archives of Microbiology* **99**, 61–70.
FABER, M. D. 1979 Microbial degradation of recalcitrant compounds and synthetic aromatic polymers, *Enzyme and Microbial Technology* **1**, 226–232.
FREEMAN, A.& AHARONOWITZ, Y. 1981 Immobilization of microbial cells in cross-linked, prepolymerised, linear polyacrylamide gels: antibiotic production by immobilized *Streptomyces clavuligerus* cells. *Biotechnology and Bioengineering* **23**, 2747–2759.
van GINKEL, C. G., TRAMPER, J., LUYBEN, K. C. A. M. & KLAPWIJK, A. 1983 Characterisation of *Nitrosomonas europaea* immobilised in calcium alginate. *Enzyme and Microbial Technology* **5**, 297–303.
GUILBAULT, G. G. 1982 Immobilized enzymes as analytical devices. *Applied Biochemistry and Biotechnology* **7**, 85–98.
HACKEL, U., KLEIN, J., MEGNET, R. & WARNER, F. 1975 Immobilization of cells in polymeric matrices. *European Journal of Applied Microbiology* **1**, 291–294.
KARUBE, J., MITSUDA, S., MATSUNAGA, T. & SUZUKI, S. 1977 A rapid method for the estimation of BOD by using immobilized cells. *Journal of Fermentation Technology* **55**, 243–248.
KARUBE, I., KURIYAMA, S., MATSUNAGA, T. & SUZUKI, S. 1980 Methane production from waste waters by immobilized methanogenic bacteria. *Biotechnology and Bioengineering* **22**, 847–857.
KIRK, T. K. & YANG, M. H. 1979 Partial dilignification of unbleached Kraft pulp with lignolytic fungi. *Biotechnology Letters* **1**, 347–352.
KIERSTAN, M. & BUCKE, C. 1977 The immobilization of cells, subcellular organelles and enzymes in calcium alginate gels. *Biotechnology and Bioengineering* **19**, 387–397.
KLIBANOV, A. M. & MORRIS, E. D. 1981 Removal of carcinogenic aromatic amines from waste waters derived from coal, resin, plastics and textile industries using hydrogen peroxide and peroxidase. *Enzyme and Microbial Technology* **3**, 119–122.
LARSSON, P. O., OHLSON, S. & MOSBACH, K. 1976 New approach to steroid conversion using activated immobilized micro-organisms. *Nature London* **263**, 796–797.
LEISINGER, T., COOK, A. M., HUTTER, R. & NUESCH, J., eds. 1981 *Microbial Degradation of Xenobiotics and Recalcitrant Compounds.* London: Academic Press.

LINKO, Y.-Y., WESKSTROM, C. & LINKO, P. 1980 Sucrose inversion by immobilized *Saccharomyces cerevisiae* yeast cells. In *Food Process Engineering* eds. Linko, P. & Arinkari, J. Vol. 2, pp. 81–91. Applied Science Publishers.

LIVERNOCHE, D., JURASEK, L., DESROCHERS, M. & VELIKY, I. A. 1981 Decolourisation of a Kraft mill effluent with fungal mycelia immobilized in calcium alginate gel. *Biotechnology Letters* **3**, 701–706.

LIVERNOCHE, D., JURASEK, L., DESROCHERS, M., DORICA, J. & VELIKY, I. A. 1983 Removal of colour from kraft mill waste waters with cultures of white-rot fungi and with immobilized mycelia of *Coriolus versicolor*. *Biotechnology and Bioengineering* **25**, 2055–2065.

MATHIASSON, B., RAMSTORP, M., NILSSON, I. & HAHN-HAGERDAL, B. 1981 Comparison of the performance of a hollow-fibre microbe reactor with a column containing alginate entrapped cells. *Biotechnology Letters* **3**, 561–566.

MOSBACH, K., ed. 1976 Immobilized enzymes. *Methods in Enzymology*, Vol. 44. London: Academic Press.

MUNNECKE, D. M. 1977 Properties of an immobilized pesticide-hydrolysing enzyme. *Applied & Environmental Microbiology* **33**, 503–507.

MUNNECKE, D. M. 1978 Detoxification of pesticides using soluble or immobilized enzymes. *Process Biochemistry* **13**, 14–17.

NAZALY, N. & KNOWLES, C. J. 1981 Cyanide degradation by immobilised fungi. *Biotechnology Letters* **3**, 363–368.

NIJPELS, H. H. 1981 Lactases and their applications. In *Enzymes and Food Processing* eds. Birch, G. G., Blakebrough, N. & Parker, K. J., pp. 89–104. London: Applied Science.

NILSSON, I. & OHLSON, S. 1982 Columnar denitrification of water by immobilized *Pseudomonas denitrificans* cells. *European Journal of Applied Microbiology and Biotechnology* **14**, 86–90.

NILSSON, I., OHLSON, S., HAGGSTROM, L., MOLIN, N. & MOSBACH, K. 1980 The use of *Pseudomonas denitrificans* cells to denitrify drinking water. *European Journal of Applied Microbiology and Biotechnology* **10**, 261–274.

NORBERG, A. & RYDIN, S. 1984 Development of a continuous process for metal accumulation by *Zoogloea ramigera*. *Biotechnology and Bioengineering* **26**, 265–268.

PAUL, F. & VIGNAIS, P. M. 1980 Photophosphorylation in bacterial chromatophores entrapped in alginate gel: improvement of the physical and biochemical properties of gel beads with barium as gel-inducing agent. *Enzyme and Microbial Technology* **2**, 281–297.

PEYTON, T. O. 1984 Biological disposal of hazardous waste. *Enzyme and Microbial Technology* **6**, 147–154.

POULSEN, P. & ZITTAN, L. 1976 Continuous production of high-fructose syrup by cross-linked cell homogenates containing glucose isomerase. In *Methods in Enzymology* ed. Mosbach, K., Vol. 44, pp. 809–821. London: Academic Press.

ROSEVEAR, A. 1982 Biologically active composites. UK Patent Application 2 083 827 A.

SLATER, H. J., LOVATT, D., WEIGHTMAN, A. J., SENIOR, E. & BULL, A. T. 1979 The growth of *Pseudomonas putida* on chlorinated aliphatic acids and its dehalogenase activity. *Journal of General Microbiology* **114**, 125–136.

SMILEY, K. L., BOUNDY, J. A., HOFREITER, B. T. & ROGOVIN, S. P. 1974 Use of α-amylase for clarification of colloidal starch–clay suspensions from paper mills. In *Immobilized Enzymes in Food and Microbial Processes* eds. Ohlson, A. C. & Cooney, C. L., pp. 133–147 New York: Plenum Press.

SWITZENBAUM, M. S. 1983 Anaerobic fixed film waste water treatment. *Enzyme and Microbial Technology* **5**, 242–250.

TAKAHASHI, S., ITOH, M. & KANEKO, Y. 1981 Treatment of phenolic wastes by *Aureobasidium pullulans* adhered to fibrous supports. *European Journal of Applied Microbiology and Biotechnology* **13**, 175–178.

VIETH, W. R. & VENKATASUBRAMANIAN, K. 1976 Process engineering of glucose isomerization by collagen-immobilized whole microbial cells. In *Methods in Enzymology* ed. Mosbach, K., Vol. 44, pp. 768–776. London: Academic Press.

WADA, M., KATO, J. & CHIBATA, I. 1980 Continuous production of ethanol using immobilized growing yeast cells. *European Journal of Applied Microbiology* **10**, 275–287.

WEIG, A. J. 1974 Enzymic treatment of waste-waters from the wheat starch industry. *Starch* **36**, 135–140.

Isolation and Growth of Sulphate-Reducing Bacteria

B. N. Herbert and P. D. Gilbert

Shell Research Limited, Sittingbourne Research Centre, Sittingbourne, Kent, UK

The importance of the dissimilatory sulphate-reducing bacteria has been widely recognized for many years. Whilst their role in the sulphur cycle is fundamental in maintaining our environment, the adverse economic consequences of their activities can be devastating in industrial processes. Within the oil-winning industry the generation of H_2S by these bacteria can result in health hazards, corrosion of equipment and impairment of oil-bearing reservoirs. Consequently, steps have to be taken to minimise their growth. Expertise has been largely developed in the United States under the auspices of the American Petroleum Institute (API), which culminated in procedure No. 38 (API 1965). Parallel to this most of the academic investigations into the physiology of these bacteria were carried out in the UK by Postgate and Miller and their co-workers.

As a result of the advent of the UK offshore oil industry there has been a considerably renewed interest in these bacteria. This is exemplified by the use of seawater, which contains large amounts of sulphate (2800 mg/litre), for a number of purposes, including water injection into reservoirs for pressure maintenance and the use of water to ballast oil in storage tanks.

An important development during the last few years has been the work of Pfennig and his co-workers who isolated new genera of sulphate-reducing bacteria. This demonstrated not only the importance of careful medium selection but also that there are more genera of sulphate-reducing bacteria than have been previously recognised. Whether any of these new genera have a significant role to play in industrial and environmental situations needs to be carefully assessed.

Methods for the isolation and growth of sulphate-reducing bacteria were last reviewed by Pankhurst (1971) in a volume of this series. The purpose of this chapter is to update these data with reference to practical details of procedures commonly adopted in the oil industry.

The term "sulphate-reducing bacteria" refers to those bacteria which use sulphate as a terminal electron acceptor. In fact, other electron acceptors, i.e. sulphite, thiosulphate, bisulphite, tetrathionate and dimethyl sulphoxide, can also be used. Perhaps a more correct collective name would be the "inorganic oxidised sulphur-compound-reducing bacteria." Even this is inadequate following the isolation of genera that can use elemental sulphur as a terminal electron acceptor. In addition, some strains can reduce nitrite to ammonia (Steenkamp & Peck 1981). The classification of sulphate-reducing bacteria into two genera by Campbell & Postgate (1965) and Postgate & Campbell (1966) needs re-assessing. Pfennig *et al.* (1981) suggested that all dissimilatory sulphate-reducing bacteria be classified in one physiological–ecological group, as has been done for the phototrophic and methanogenic bacteria.

Those bacteria that have been described as sulphate reducing are listed in Table 1. The sulphur-reducing bacteria are also listed for completeness.

Requirements for Growth

Historical Development of Media

It is of interest to summarise the history of improvements in media composition as this parallels the development of our understanding of the nutritional requirements of sulphate-reducing bacteria. A selected number of key media are listed in the Appendix. From the beginning Beijerinck (1895) identified the requirement for sodium lactate as a carbon source and ferrous salt as an indicator of sulphide production. However, it was not until some 50 years later that Butlin *et al.* (1949) incorporated yeast extract and Grossman & Postgate (1953) recognised the need for redox-poising agents. The importance of medium salinity was pointed out by the American Petroleum Institute (1965) and this has been extended by Herbert (1976) by the inclusion of source water, i.e. water from which the sulphate-reducing bacteria are to be isolated.

Principal Requirements for Growth

Sulphate-reducing bacteria obtain the carbon and energy necessary for cell growth by various routes. Chemo-organotrophic growth may be at the expense of single organic carbon compounds, such as lactate, which provide a common carbon and energy source. Alternatively, the carbon and energy

TABLE 1

List of sulphate-reducing and sulphur-reducing bacteria

Sulphate-reducing bacteria
  Genus *Desulfovibrio*
    D. desulfuricans
    D. vulgaris
    D. salexigens
    D. africanus
    D. gigas
    D. baculatus
    D. thermophilus
    D. sapovorans
  Genus *Desulfotomaculum*
    D. nigrificans
    D. orientis
    D. ruminis
    D. acetoxidans (lactate negative)
  Genus *Desulfomonas*
    D. pigra
  Fatty-acid-utilising sulphate reducers
    Desulfobulbus propionicus
    Vibrioid species (sapovorans group)
    Desulfobacter postgatei
    Desulfococcus multivorans
    Desulfonema limicola
    Desulfonema magnum
    Desulfosarcina variabilis
    Desulfotomaculum acetoxidans

Sulphur-reducing bacteria
  Desulfuromonas acetoxidans
  Desulfuromonas acetexigens
  Campylobacter sp. (saprophytic)

sources may be separate, and organic carbon compounds that are not assimilated for growth, e.g. formate or isobutanol, can serve as electron donors for energy generation whilst other carbon compounds are assimilated for growth. This mode of growth has been termed "mixotrophic" (Postgate 1979). Hydrogen may also serve as an electron donor and in this instance growth should strictly be termed "chemolithotrophic". Recently there have been fresh claims of autotrophic growth (Pfennig *et al.* 1981), although assimilation of CO_2 to provide a proportion of the carbon requirement of the cells has been known for some time (Sorokin 1966a,b; Rittenberg 1969). The capacities for mixotrophic growth and for growth on a common carbon and energy source are not mutually exclusive (Sorokin 1966a,b), but the media required to elicit the separate modes of growth differ in composition.

Historically, the most widely employed carbon source in media for the isolation of sulphate-reducing bacteria has been lactate although pyruvate and malate are common alternatives. The use of lactate is in line with the classical view that the range of combined carbon and energy sources for sulphate-reducing bacteria is narrow. Such sources include short-chain substituted fatty acids, simple alcohols and glycerol which are incompletely oxidised. Acetate and CO_2 are examples of end products. However, a certain proportion of the acetate may be reassimilated (Sorokin 1966b). Recently this picture has changed with the isolation of sulphate-reducing bacteria that grow on a wider range of carbon compounds and those for which lactate will not support growth. Sulphate-reducing bacteria which do not grow on lactate-based media have been isolated using acetate as the sole carbon source. These include sporing (Widdel & Pfennig 1977) and non-sporing types (Widdel & Pfennig 1981a) which oxidise acetate completely to CO_2. Other carbon sources for these strains include ethanol and butyrate. Some of the strains that were isolated on acetate also grew on lactate but growth was slow. The converse does not appear to be true because of nine marine strains isolated on lactate, none grew on acetate; however, enrichments from the same source with acetate produced five acetate-utilising strains which could not use lactate (Laanbroek & Pfennig 1981). The situation for propionate is different as strains isolated from marine and freshwater sediments on lactate could grow on propionate and vice versa. On balance, therefore, it would seem that acetate-utilising sulphate-reducing bacteria are best isolated on media containing acetate as the sole carbon source but the situation for propionate is not so critical. The range of carbon sources for these newly isolated sulphate-reducing bacteria also includes higher carbon number fatty acids such as palmitic. Pfennig et al. (1981) have proposed two groups for sulphate-reducing bacteria on the basis of their oxidative metabolism. The first group encompasses those strains that cannot completely oxidise their growth substrates and form acetate as an end product; members of the second group carry out a complete oxidation to carbon dioxide.

General Conditions for Growth

Sulphate-reducing bacteria are strict anaerobes and as such require strictly anaerobic conditions for the initiation of growth. As growth continues in sulphate-containing media, the production of sulphide ensures the absence of O_2 and the maintenance of a reduced environment. With growth in "sulphate-free" media (sulphate content sufficient only for assimilatory sulphate reduction) this is not the case and a negative redox potential has to be maintained in other ways. The addition of redox-poising agents is one method commonly used for the exclusion of oxygen and the establishment of reducing conditions. Postgate (1979) recommends a negative potential of −100 mV (Eh) for successful growth. These agents include thiol compounds such as cysteine or thioglycollate as well as sodium sulphide. The presence of thioglycollate may be a disadvantage because it could interfere with biocide testing. More recently titanium(III) citrate has been proposed as a combined non-toxic redox-poising agent and redox dye (Zehnder & Wuhrmann 1976). The inclusion of redox dyes such as resazurin or nile blue gives a visible indication of the redox state of the medium. Once a negative redox potential has been obtained, entry of O_2 may be prevented by purging with N_2 or using a N_2 head space (Fig. 1). Alternatively, alkaline pyrogallol plugs to absorb O_2 or entirely filled, sealed vessels can be used. In the case of solid media, redox-poising agents are sometimes added to the medium, and agar plates are incubated in the conventional manner using an anaerobic atmosphere, e.g. N_2 (100%), H_2 (95%) + CO_2 (5%), N_2 (80%) + H_2 (10%) + CO_2 (10%). Gas mixtures containing H_2 have the added advantage of

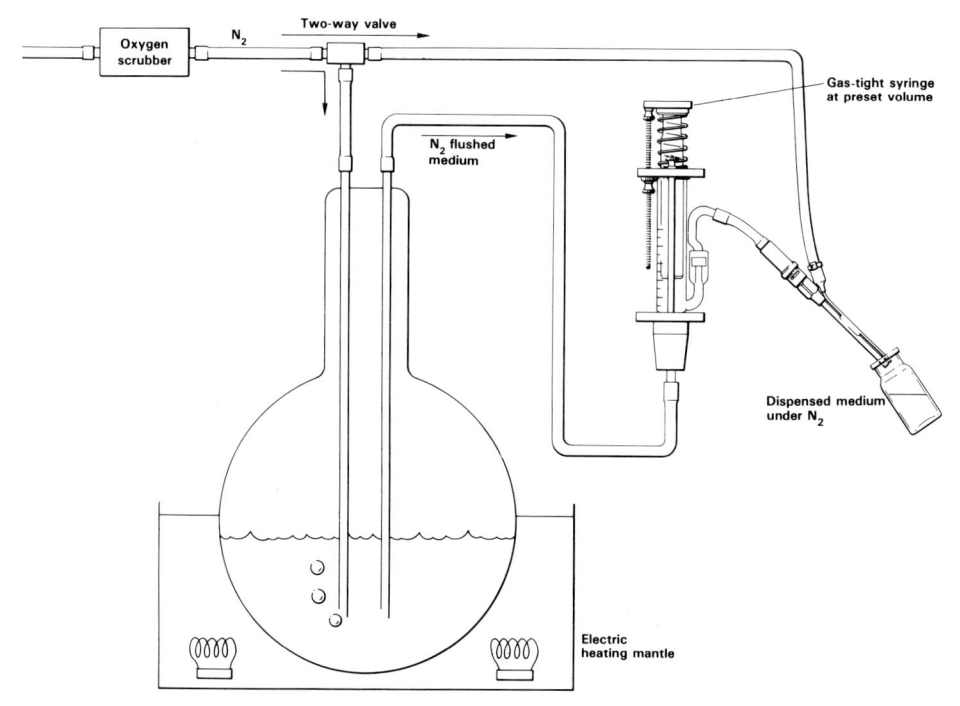

FIG. 1. Assembly for introducing nitrogen cap into media bottles during the dispensing of medium.

reacting with the palladium catalyst supplied with most anaerobic gas jars thereby removing the last traces of O_2. The catalyst is poisoned, however, by the presence of H_2S and therefore it should be replaced or regenerated after exposure.

The pH range most commonly used in media for the growth of sulphate-reducing bacteria is 7.2–7.6. This may exclude the possibility of isolating sulphate-reducing bacteria found in acid environments (Satake 1977). The pH value of the medium can alter during growth; alteration may arise owing to the formation of bicarbonate (Abd-el Malek & Rizk 1963) or due to the evaporation of H_2S from the medium.

The temperature of incubation used depends upon the strain of sulphate-reducing bacterium. Thermophilic *Desulfotomaculum* species are generally incubated at 55°C and mesophilic *Desulfovibrio* species are incubated at 30°C. Temperatures greater than 55°C may be required for sulphate-reducing bacteria isolated from oil-bearing reservoirs.

Sulphate-reducing bacteria that are found in freshwater and marine environments have different salinity requirements. Desulfotomacula are not native to saline environments and do not have a high salinity requirement. Media for the isolation and growth of marine strains are typically supplemented by the addition of 2.5% (w/v) NaCl; however, most marine strains show a wide tolerance to different salinities (Kimata *et al.* 1955; Ockynski & Postgate 1963; Hardy 1981). Where there is an obligate requirement for salinity, it appears to be the sodium ion and not the chloride ion that is required (Trüper *et al.* 1969).

Most of the common media for sulphate-reducing bacteria have a similar inorganic nutrient status providing the elements essential for growth as well as an excess of sulphate for dissimilatory reduction. The sulphate-reducing bacteria have a large requirement for iron and this is reflected in the media.

Very large concentrations of iron (97 mg/litre in Postgate medium B) are included for the diagnostic formation of black ferrous sulphide. The solubility of iron may be increased by adding chelating agents such as citrate. The buffering capacity of most media is low and is provided by KH_2PO_4. When lactate is used as the carbon source the pH value of the medium may decrease on autoclaving owing to the dissociation of lactic acid dimers. This can be lessened by pre-autoclaving the lactate. The final pH should be checked after autoclaving and adjustments made if necessary.

Although sulphate-reducing bacteria can grow in chemically defined media, their growth is stimulated by the addition of yeast extract (Macpherson & Miller 1963) which may have a chelating function thereby increasing the concentration of dissolved iron (Postgate 1951), in addition to providing alternative carbon and nitrogen sources and growth factors. The isolation and growth of some of the more novel types of sulphate-reducing bacteria on non-classical carbon sources may require the addition of a vitamin mixture. In the absence of yeast extract a trace element solution such as that formulated by Macpherson & Miller (1963) may be required.

Isolation Procedures

General Considerations

Media for the isolation of sulphate-reducing bacteria described by Pankhurst (1971) and Postgate (1979) are lactate based and have been used successfully to isolate *Desulfovibrio* and *Desulfotomaculum* from a wide variety of environments. Variations on these media have been used for the isolation of more novel types of sulphate-reducing bacteria.

Members of the sulphur-reducing genus *Desulfuromonas* have been isolated from fresh water and marine sediments although they are more prevalent in the latter. Media for their isolation have been well described by Pfennig & Biebl (1981). These media differed from conventional formulations by the inclusion of elemental sulphur as the terminal electron acceptor although this may be substituted by L-malate, fumarate or organic disulphide compounds such as cysteine or oxidised glutathione. Freshwater strains use only acetate as a carbon and energy source and this is completely oxidised to CO_2. However, marine strains have a wider substrate specificity and can use ethanol, propanol or pyruvate as an alternative to acetate. Acetate was included in the medium at small concentrations (0.05%, w/v) and a vitamin mixture or biotin alone was used as a growth supplement. Sodium sulphide was the redox-poising agent which could not be substituted by thioglycollate. One interesting method of enrichment for *Desulfuromonas* species has been the successful use of co-culture with the marine green sulphur bacterium *Prosthecochloris aestuarii*. The latter provides elemental sulphur as a terminal electron acceptor and also prevents the accumulation of inhibitory concentrations of sulphide by the removal of H_2S produced by *Desulfuromonas*.

Media for the isolation of sulphate-reducing bacteria that grow on a wider range of carbon sources or on acetate alone are presented by Pfennig *et al.* (1981). These media differed from conventional ones in several respects. Smaller concentrations of inorganic nutrients, nitrogen and phosphorus were used. A trace element mixture, CO_2-saturated bicarbonate solutions and a vitamin mixture were included. However, yeast extract was omitted and in some cases growth factors were provided by a mixture of organic acids. Small amounts of carbon sources were added, e.g. acetate 0.20% (w/v), and either sodium sulphide or sodium dithionite was used as the redox-poising agent.

Using such media and an incubation temperature of 36°C, spore-forming sulphate-reducing bacteria which utilise acetate have been successfully isolated from marine sediments (Widdel & Pfennig 1977) and from rumen contents, animal dung and dung-contaminated fresh water (Widdel & Pfennig 1981b). The strains could also utilise butyrate which could be used as the sole carbon source for isolation purposes. The non-sporing counterparts of the acetate utilisers have been isolated from marine sediments at the lower incubation temperature of 28°C (Widdel & Pfennig 1981a). When the growth factor supplement was included in the medium, filamentous, gliding sulphate-reducing bacteria of the type *Desulfonema limicola* were isolated. With benzoate as the sole carbon and energy source, coccoid sulphate-reducing bacteria were isolated from freshwater and marine sediments. Vibrioid sulphate reducers were isolated from the same environments using palmitate as the carbon source; these strains also grew on butyrate. In addition, enrichment with propionate yielded "lemon-shaped" cells similar to *Desulfobulbus propionicus* (Pfennig et al. 1981).

Sulphate-reducing bacteria have been successfully isolated on media formulated for mixotrophic growth. Badziong et al. (1978) used a mineral medium supplemented with acetate and an $H_2 + CO_2$ gas mixture to isolate two *Desulfovibrio* spp. from freshwater sediment and sewage sludge. Both strains used H_2 as an electron donor and assimilated carbon from both acetate and CO_2. Isolations were not successful when acetate was omitted from the medium. Sorokin (1966a) reported the isolation of a *Desulfovibrio* spp. from stratum water of an oil deposit mixed with a minimal medium and incubated under an atmosphere of H_2. Acetate was not added but the organic nutrient status of the stratum water was unclear. Once purified, the strain grew under an atmosphere of H_2, assimilating carbon from CO_2 as well as from acetate that was subsequently included in the medium.

Sulphate-reducing bacteria are reported to include psychrophilic, mesophilic and thermophilic strains and therefore the incubation temperature of isolates appears to be important when enriching for a desired strain. However, Trüper et al. (1969) isolated from saline environments mesophilic strains possessing a wide temperature range for growth, e.g. 13–38°C. Furthermore, the optimum growth temperature was not affected despite isolation either above or below these values, a finding also reported by Hardy (1981) for strains isolated from North Sea waters. Reports of pyschrophilic as opposed to psychrotolerant sulphate-reducing bacteria in the literature are rare, although sulphate-reducing bacteria have been isolated from Antarctic environments (Barghoorn & Nichols 1961). Zobell (1958) reported that most sulphate-reducing bacteria from cold environments grow better at 20–30°C than at lower temperatures. It seems therefore that temperatures used for growing mesophilic bacteria are suitable for isolating sulphate-reducing bacteria from cold environments. Thermophilic sulphate-reducing bacteria have been isolated from high-temperature environments such as oil and sulphur wells (Zobell 1958) and hot springs and geothermal areas (Kaplan 1956). The optimum temperature for isolation remains unclear but that for growth is often quoted as 55°C.

Practical Aspects

When attempting to isolate sulphate-reducing bacteria from an environment, it is important to take all practicable means to ensure that the sample is representative. In many cases, e.g. pipelines, the only available sampling points may be a considerable distance from the location where growth of the bacteria is suspected. In most instances the design of an installation does not permit any direct sampling. Therefore any conclusions based on the numbers of sulphate-reducing bacteria isolated would probably be spurious without other data. The estimation of numbers in oil field operations is usually based on serial dilution in liquid media using the most probable number (MPN) method (see Pankhurst 1971). Unfortunately, in most instances sufficient replicates are not used—owing to both ignorance and limitation in media supplies—and hence numbers quoted in oil field operations are often wildly inaccurate.

There is considerable controversy as to whether samples should be inoculated into the medium of choice immediately after collection or whether a delay is permitted. The API (1965) procedure No. 38 recommends that inoculation takes place within 24 h of sampling. Unpublished studies in which we compared immediate inoculation versus delayed inoculation from water samples obtained from the North Sea were inconclusive. On balance we prefer to inoculate media immediately after sampling. Unfortunately, this can present problems of medium supply. It is often declared that the redox-poising capacity of media is short-lived and that media must be used within hours of preparation. Many oil fields are hours and more usually days from a suitably equipped microbiology laboratory. Media are usually supplied by a number of service companies and may be several months old before being used. Additionally, storage arrangements may not be ideal. Nevertheless, there is no evidence that the capacity of these media to isolate sulphate-reducing bacteria has deteriorated. In our experience the choice, rather than the age, of the medium is of most importance. In all oil field operations, lactate-based media are used as described in the API (1965) procedure No. 38. Other carbon sources are not incorporated into media and despite the findings of Pfennig and his co-workers (Pfennig et al. 1981) there is no reason to suspect that failure to isolate sulphate-reducing bacteria from a troublesome region has been due to incorrect choice of carbon source.

Probably of most importance is the water in which the medium is prepared. The addition of 2.5% (w/v) NaCl to accommodate marine strains may not be sufficient. When attempting to isolate sulphate-reducing bacteria from a "new" environment we initially compare a formulation based on Postgate medium B prepared in 75% source water + 25% distilled water (Herbert 1976) with media that are supplied by service companies. In many cases the former performs more satisfactorily than the commercially available media. Although isolation of sulphate-reducing bacteria may require media prepared in natural waters. subsequent subculture can often be successfully achieved in media prepared using artificial waters or, in the case of marine strains, 2.5% (w/v) NaCl. Ideally, a range of media should be compared, possibly including some with different carbon sources, before a final choice is made for future isolations and monitoring.

The inclusion of ferrous ions in the media, either as a soluble ferrous salt or as a nail, is extremely convenient as an indicator of sulphide production. This permits the non-microbiologist to use these media for monitoring sulphate-reducing bacteria by recording the development of a black precipitate of ferrous sulphide in the medium. However, this technique can give erroneous results. Growth and sulphide generation may occur without a black precipitate, or a black precipitate may develop without sulphide generation. In the latter case the black precipitate has been identified by Iverson (1968) as a phosphide, but may be other, as yet unidentified, products. Another potential disadvantage of the inclusion of iron in the medium occurs when the water being sampled contains sulphide. If this is not previously removed the medium will turn black immediately on inoculation. This problem can be solved by bubbling the H_2S out of the water by the effervescent evolution of CO_2 before inoculation. This is commonly achieved by dissolving an Alka-Seltzer tablet (Miles Laboratories) in the water and waiting until all the CO_2 has been evolved. Hydrogen sulphide can be removed from water to be used for medium preparation by boiling the water for 5 min. It is commonly recommended that the boiled water is passed through a Millipore membrane to remove any precipitate [API (1965) procedure No. 38] but we do not do this. Postgate medium B has a precipitate which in our view favours isolation of sulphate-reducing bacteria—possibly by providing an extensive protected surface area to which the bacteria can attach. The development of bacterial colonies within these deposits, seen as discrete black regions, is often seen before more general blackening of the overlying medium. Some waters that contain dissolved CO_2, e.g. water associated with oil in oil-bearing strata, may present problems in medium preparation. If care is not taken the CO_2 is driven off during autoclaving and the final pH of the medium is too high.

For many of the media to be described most of the ingredients are autoclaved together but thioglycollic acid is sterilised separately by filtration. However, this may not be practicable and in media prepared for oil field operations it is more usual to autoclave all the ingredients together. This may be open to criticism but we have no evidence to suggest that the media so prepared are less effective in isolating sulphate-reducing bacteria.

Perhaps one of the more important practical requirements for the isolation of sulphate-reducing bacteria as part of a monitoring programme in an oil field environment is the speed at which positive results are obtained. In addition to the suitability of the chosen medium, the inoculum size affects the rapidity with which a black precipitate develops. The API (1965) procedure No. 38 recommends that the media should be incubated for up to 28 days. Clearly such long incubation periods would be a hindrance to the monitoring of the success or failure of a control programme. However, the temptation to incubate for shorter periods could result in failure to detect the bacteria if either the numbers are very small or there is the carry-over of an inhibitory concentration of a biocide. The API (1965) procedure No. 38 indicates that the isolation of even one sulphate-reducing bacterium represents a potential problem. Nevertheless, practical considerations dictate shorter incubation periods and 10 days is the maximum period that we recommend. The oil industry would favour even shorter incubations and any improvements in this respect would be welcomed. In some cases ATP photometry has been adopted to give a quick assessment of numbers of viable bacteria. However, this method is not specific for sulphate-reducing bacteria and the results can be affected by interference of sulphide and biocides with the luciferin/luciferase complex.

Other rapid methods that have been suggested are radiorespirometry (Jorgensen 1978; Hardy & Syrett 1983), immunofluorescence (Smith 1982) and electrical impedance (Oremland & Silverman 1979).

Purification of Isolates

Purification of isolates of sulphate-reducing bacteria from a mixed culture is generally done by obtaining discrete colonies on solid media. These are then cultivated in liquid media and the culture is checked for purity by replating onto solid media and by microscopy. Microscopical examination may be inapplicable to strains exhibiting pleomorphism but purity checks on solid media incubated aerobically and anaerobically should reveal any contaminants. Purity checks are especially important in the light of recent reports of close syntrophic relationships between sulphate-reducing bacteria and other species (Boone & Bryant 1980).

Single colonies can be obtained either in the depths of an agar medium or on the surface of an agar plate although the latter is somewhat more erratic and gives poorer recovery. Agar shake and roll tubes cultures are the two main methods of obtaining single colonies within an agar medium. For shake cultures 9 ml agar medium, e.g. Postgate medium E, in test tubes are kept molten at 45°C. A dilution series is prepared in the molten medium using 1 ml culture, mixed quickly, allowed to cool and incubated in air. A tube containing discrete colonies of sulphate-reducing bacteria, recognisable by their black colour, is selected and broken, and single colonies are transferred to liquid medium with a sterile Pasteur pipette or loop. Purity checks are made on cultures in liquid medium. The procedure may also be used for enumeration of sulphate-reducing bacteria. For roll tube cultures the method is similar except that molten medium and culture mixture are distributed around the circumference of a thin-walled, revolving tube where it is allowed to solidify.

Single colonies can be obtained on the surface of a suitable nutrient medium, such as Postgate medium E. In this case 0.1–0.3 ml of a dilution of the culture is spread over the surface of the agar plates which are then incubated in an anaerobic atmosphere. A wick soaked in lead acetate can be included to absorb the H_2S produced; this prevents blackening of the whole agar plate which can hinder identification of black colonies. Single colonies are transferred to liquid media and the culture is checked for purity.

Culture Media

The media referred to below may be sterilised by autoclaving at 121°C for 15 min unless otherwise stated.

Liquid Media

Liquid media will be considered on the basis of the mode of growth that they elicit and their function.

Diagnostic Media

These media contain a large concentration of iron which, in the presence of H_2S produced during growth, forms a precipitate of black ferrous sulphide. This reaction indicates the presence of sulphate-reducing bacteria; samples containing large concentrations of dissolved sulphide will require pre-treatment before inoculation to avoid false positive results.

Postgate medium B (Grossman & Postgate 1953; Postgate 1979):

KH$_2$PO$_4$	0.5 g
NH$_4$Cl	1.0 g
CaSO$_4$	1.0 g
MgSO$_4$·7H$_2$O	2.0 g
FeSO$_4$·7H$_2$O	0.5 g
sodium lactate	3.5 g
yeast extract	1.0 g
ascorbic acid	0.1 g
thioglycollic acid	0.1 g
tap water	1000 ml

The pH is adjusted to 7–7.5 and for marine strains 2.5% (w/v) NaCl may be added or the medium may be prepared with either aged seawater or artificial seawater. It is recommended that the thioglycollic acid is prepared and autoclaved separately.

API medium RP 38 (API 1965):

K$_2$HPO$_4$	0.01 g
MgSO$_4$·7H$_2$O	0.1 g
Fe(SO$_4$)(NH$_4$)$_2$·SO$_4$·6H$_2$O	0.2 g
NaCl	10.0 g
sodium lactate	4.0 g
yeast extract	1.0 g
ascorbic acid	0.1 g
distilled water	1000 ml

The pH is adjusted to 7.3.

Media for Large-Scale Culture

A medium containing sufficient lactate and sulphate for maximum growth and including citrate to prevent precipitation has been described by Postgate (1979).

Postgate medium C:

KH$_2$PO$_4$	0.5 g
NH$_4$Cl	1.0 g
CaCl$_2$·6H$_2$O	0.06 g
MgSO$_4$·7H$_2$O	0.06 g
FeSO$_4$·7H$_2$O	0.004 g
Na$_2$SO$_4$	4.5 g
trisodium citrate (Na$_3$C$_6$H$_5$O$_7$·2H$_2$O)	0.3 g
sodium lactate	6.0 g
yeast extract	1.0 g
distilled water	1000 ml

The pH is adjusted to 7.5. For marine strains 2.5% (w/v) NaCl should be added. The medium contains no redox-poising agent; this may be remedied by adding sterile sodium sulphide to a final concentration of 1 mM (Postgate 1966). Alternatively, a large sulphide-containing inoculum may be used.

Media for Isolation and Cultivation Using Various Carbon Sources

As media are described in detail by Pfennig *et al.* (1981), a brief description only will be given.

Solution 1 (mineral salts base):

KH_2PO_4	0.2 g
NH_4Cl	0.3 g
$CaCl_2 \cdot 2H_2O$	0.15 g
$MgCl_2 \cdot 6H_2O$	0.4 g
Na_2SO_4	3.0 g
NaCl	1.2 g
KCl	0.3 g
distilled water	970 ml

For marine strains use 20 g NaCl and 3 g $MgCl_2 \cdot 6H_2O$.

Solution 2 (trace element stock solution):

$FeCl_2 \cdot 4H_2O$	1.5 g
H_3BO_3	60 mg
$MnCl_2 \cdot 4H_2O$	100 mg
$CoCl_2 \cdot 6H_2O$	120 mg
$ZnCl_2$	70 mg
$NiCl_2 \cdot 6H_2O$	25 mg
$CuCl_2 \cdot 2H_2O$	15 mg
$Na_2MoO_4 \cdot 2H_2O$	25 mg
HCl (25% v/v)	6.5 ml
distilled water	993 ml

Solution 3 (selenite stock solution):

NaOH	0.5 g
Na_2SeO_3	3.0 mg
distilled water	1000 ml

Solution 4 (CO_2-saturated sodium bicarbonate solution):

$NaHCO_3$	8.5 g
distilled water	100 ml

Sterilise by filtration.

Solution 5 (sodium sulphide stock solution):

$Na_2S \cdot 9H_2O$	12 g
distilled water	100 ml

Sterilise by filtration.

Solutions 1–5 are sterilised separately and combined aseptically in the following proportions to form the complete basal medium: solution 1 (970 ml) + solution 2 (1 ml) + solution 3 (1 ml) + solution 4 (30 ml) + solution 5 (3 ml). The pH is adjusted with HCl or Na_2CO_3 to 7.2 for enrichment cultures.

Carbon source (stock solutions):

sodium acetate	($CH_3COONa \cdot 3H_2O$)	20 g
propionic acid	(CH_3CH_2COOH)	7 g
n-butyric acid	($CH_3(CH_2)_2COOH$)	8 g
n-palmitic acid	($CH_3(CH_2)_{14}COOH$)	5 g
benzoic acid	(C_6H_5COOH)	5 g

Solutions are prepared in 100 ml distilled water and the pH adjusted to 9 with NaOH. Add 1 ml of stock solution to 100 ml complete basal medium.

Vitamin stock solution:

biotin	1 mg
p-aminobenzoic acid	5 mg
vitamin B_{12}	5 mg
thiamine	10 mg

The solution is prepared in 100 ml distilled water and sterilised by filtration. Add 1 ml to 1 litre complete basal medium.

Growth factors stock solution:

isobutyric acid	0.5 g
valeric acid	0.5 g
2-methylbutyric acid	0.5 g
3-methylbutyric acid	0.5 g
caproic acid	0.2 g
succinic acid	0.6 g

The solution is prepared in 100 ml distilled water, the pH adjusted to 9.0, and sterilised by filtration. Add 1 ml to 1 litre complete basal medium.

Sodium dithionite stock solution:

O_2-free distilled water	100 ml
$Na_2S_2O_4$	3 g

Prepare freshly, sterilise by filtration, add 1 ml to 1 litre complete basal medium.

The choice of carbon source and inclusion of vitamins, growth factors and sodium dithionite is varied depending on the requirements of the isolates to be studied.

Chemically Defined Media for Nutritional Studies

The chemically defined medium of Macpherson & Miller (1963; cited by Pankhurst 1971) is a medium that supports growth of several strains of *Desulfovibrio* species but does not contain any growth factors, however, addition of amino acid mixtures or yeast extract was reported to stimulate growth.

KH_2PO_4	0.34 g
NH_4Cl	0.53 g
$CaCl_2$	0.06 g
$MgSO_4 \cdot 7H_2O$	0.06 g
$FeSO_4 \cdot 7H_2O$	0.007 g
Na_2SO_4	7.10 g
Na_2S	0.08 g
lactic acid	9.01 g
trace elements B, Co, Cu, Mn, Mo, Zn	0.0005 g of each ion

The medium is prepared without $FeSO_4 \cdot 7H_2O$ and Na_2S; the pH is adjusted to 6.5. Solutions of $FeSO_4 \cdot 7H_2O$ and Na_2S are sterilised by filtration and added aseptically. The final pH of the medium should be adjusted to 7.2–7.4.

"Sulphate-Free" Media

"Sulphate-free" media contain very small concentrations of sulphate, sufficient only for assimilatory sulphate reduction. Growth in sulphate-free medium may proceed by respiratory transport phosphorylation in which an organic electron acceptor replaces sulphate. Alternatively, ATP may be generated by substrate level phosphorylation resulting from phosphoroclastic cleavage of pyruvate. Miller & Wakerley (1966) investigated the metabolism of fumarate in a sulphate-free medium. One of the media used was a modification of that of Macpherson & Miller (*vide supra*) with the replacement of $MgSO_4 \cdot 7H_2O$ with 0.25-mM $MgCl_2 \cdot 6H_2O$ and the omission of the sodium and magnesium sulphates. Growth occurred in media when fumarate (50 mM) or fumarate + lactate (each 50 mM) replaced the lactate in the original medium. In the former fumarate acted simultaneously as electron donor and acceptor, whereas in the latter it acted only as an electron acceptor. The other medium used by Miller & Wakerley (1966) was a modified "medium C" (presumably that of Butlin *et al.* 1949) which contained yeast extract and is similar to Postgate medium C. Malate can replace fumarate in a dismutation medium (Miller *et al.* 1970).

An alternative sulphate-free medium is medium D of Postgate (1979) which contains sodium pyruvate or choline chloride as a carbon source.

Postgate medium D:

KH_2PO_4	0.5 g
NH_4Cl	1.0 g

CaCl$_2$·2H$_2$O	0.1	g
MgCl$_2$·6H$_2$O	1.6	g
FeSO$_4$·7H$_2$O	0.004	g
yeast extract	1.0	g
sodium pyruvate	3.5	g or
choline chloride	1.0	g and
distilled water	1000	ml

The pH is adjusted to 7.5 and the medium sterilised by filtration.

Media for Mixotrophic Growth

The medium of Badziong *et al.* (1978) has been used successfully for the isolation and cultivation of strains growing on H$_2$ with assimilation of acetate and CO$_2$ as carbon sources.

KH$_2$PO$_4$	0.5	g
(NH$_4$)$_2$SO$_4$	5.3	g
CaCl$_2$·2H$_2$O	0.1	g
MgSO$_4$·7H$_2$O	0.2	g
NaCl	1.0	g
Na$_2$CO$_3$ (8%, w/v, in H$_2$O)	50	ml
HCl (25% v/v)	5.5	ml
Na$_2$S$_2$O$_4$ (0.5 M in H$_2$O)	1.0	ml
resazurin (0.2%, w/v, in H$_2$O)	1.0	ml
trace elements mixture	10	ml
sodium acetate	2.0	g
distilled water	1000	ml

The stock solutions of Na$_2$CO$_3$, HCl and Na$_2$S$_2$O$_4$ are prepared separately and sterilised by filtration. They are added to the bulk medium aseptically and the final pH of the medium should be 7.2. The trace elements mixture stock solution has the following composition:

FeCl$_2$·4H$_2$O	0.3	g
H$_3$BO$_3$	0.01	g
MnCl$_2$·4H$_2$O	0.1	g
CoCl$_2$·6H$_2$O	0.17	g
ZnCl$_2$	0.1	g
CuCl$_2$	0.02	g
Na$_2$MoO$_4$·2H$_2$O	0.01	g
nitrilotriacetic acid (pH 6.5 with NaOH)	12.8	g
distilled water	1000	ml

The medium is incubated under an H$_2$ (80%) + CO$_2$ (20%) gas mixture. Alternative electron donors, e.g. sodium formate, may be added to the medium to give a final concentration of 50 mM. For an alternative medium employing a wider range of electron donors Sorokin (1966a) should be consulted.

Media for the cultivation of sulphur-reducing bacteria lie beyond the scope of this chapter; Pfennig & Biebl (1981) should be consulted for further information.

Solid Media

Solid media are generally employed for the purification and enumeration of sulphate-reducing bacteria (Pankhurst 1971). They are often prepared by the addition of agar to an appropriate liquid formulation (see *Liquid media*).

Postgate medium E (Postgate 1979):

KH_2PO_4	0.5 g
NH_4Cl	1.0 g
$CaCl_2 \cdot 6H_2O$	1.0 g
$MgSO_4 \cdot 7H_2O$	2.0 g
$FeSO_4 \cdot 7H_2O$	0.5 g
Na_2SO_4	1.0 g
sodium lactate	3.5 g
yeast extract	1.0 g
ascorbic acid	0.1 g
thioglycollic acid	0.1 g
agar	15.0 g
tap water	1000 ml

The pH is adjusted to 7.6 after boiling. Unfortunately in the formulations of Postgate (1966, 1979), $MgSO_4 \cdot 7H_2O$ appears as $MgCl_2 \cdot 7H_2O$, an error that has been corrected by Pankhurst (1971) who also recommends that the thioglycollic acid should be sterilised by filtration. For marine strains NaCl (2.5% w/v) may be added.

Mara & Williams (1970) evaluated a number of solid media for enumeration using the shake tube method. They recommended a modified iron sulphite medium.

$MgSO_4 \cdot 7H_2O$	2.0 g
$FeSO_4 \cdot 7H_2O$	0.5 g
sodium lactate (70% w/v)	5.0 ml
ascorbic acid	0.75 g
sodium thioglycollate	0.75 g
iron sulphite agar (Oxoid)	23.0 g
distilled water	1000 ml

The pH is adjusted to 7.5 after autoclaving. The ascorbic acid and sodium thioglycollate are prepared as a concentrated ($\times$ 100) stock solution, adjusted to pH 7.5, autoclaved separately, and combined aseptically with the bulk medium after autoclaving.

Commercially Available Media

Water-treatment companies that provide a service to the oil companies provide ready-prepared liquid media based on the API formula. Such a medium is also available in dehydrated form (Difco) and in a prepared form (Easicult S, Orion). It should be clear from the previous discussion that these media have only limited usefulness.

Conclusions

The enumeration and isolation of sulphate-reducing bacteria in the oil industry is used as a means of assessing the extent of associated problems and the effectiveness of control measures. For the most part procedures follow the API (1965) procedure No. 38 or modifications of this. This procedure employs media based on lactate as a carbon source and takes no account of species having different nutritional requirements.

The procedures used in the oil industry need re-assessing in light of our understanding of the role of these bacteria in corrosion mechanisms, their ecology in oil field systems, and the identification of new species that may not be detected by the traditionally used media. There is a growing need for sensitive procedures that can detect sulphate-reducing bacteria activity quickly to allow control measures to be effectively and economically employed.

Appendix

Historical Development of Culture Media

Beijerinck (1895):

potassium phosphate	0.1–0.2 g
$CaSO_4·2H_2O$ or $MgSO_4·7H_2O$	0.5–2.0 g
$FeSO_4(NH_4)_2·SO_4·6H_2O$	trace
Na_2CO_3	1.0 g
sodium or potassium malate or lactate	0.05–0.1 g
asparagine or peptone	0.05–0.1 g
ditch water	1000 ml

van Delden (1903):

K_2HPO_4	0.5 g
$MgSO_4·7H_2O$	1.0 g
$FeSO_4·7H_2O$	trace
sodium lactate	5.0 g
asparagine	1.0 g
tap water	1000 ml

Baars (1930):

K$_2$HPO$_4$	1.0 g
NH$_4$Cl	0.5 g
CaSO$_4$	1.0 g
MgSO$_4$·7H$_2$O	2.0 g
FeSO$_4$(NH$_4$)$_2$·SO$_4$·6H$_2$O	trace
sodium lactate	3.5 g
tap water	1000 ml

Starkey (1938):

K$_2$HPO$_4$	0.5 g
NH$_4$Cl	1.0 g
CaCl$_2$·2H$_2$O	0.1 g
MgSO$_4$·7H$_2$O	2.0 g
FeSO$_4$(NH$_4$)$_2$·SO$_4$·6H$_2$O	trace
Na$_2$SO$_4$	0.5 g
sodium lactate	3.5 g
tap water	1000 ml

Butlin *et al.* (1949):

K$_2$HPO$_4$	0.5 g
NH$_4$Cl	1.0 g
CaCl$_2$·2H$_2$O	0.1 g
MgSO$_4$·7H$_2$O	2.0 g
FeSO$_4$·7H$_2$O	0.002 g
Na$_2$SO$_4$	1.0 g
sodium lactate	3.5 g
yeast extract	1.0 g
distilled water	1000 ml

Grossman & Postgate (1953) (Postgate medium B):

KH$_2$PO$_4$	0.5 g
NH$_4$Cl	1.0 g
CaSO$_4$	1.0 g
MgSO$_4$·7H$_2$O	2.0 g
FeSO$_4$·7H$_2$O	0.5 g
sodium lactate	3.5 g
yeast extract	1.0 g
ascorbic acid	0.1 g
thioglycollic acid	0.1 g
tap water	1000 ml

API Medium RP 38 (API 1965):

K$_2$HPO$_4$	0.01	g
MgSO$_4$·7H$_2$O	0.2	g
FeSO$_4$(NH$_4$)$_2$·SO$_4$·6H$_2$O	0.1	g
NaCl	10.0	g
sodium acetate	4.0	g
yeast extract	1.0	g
ascorbic acid	0.1	g
distilled water	1000	ml

Modified Postgate medium B (Herbert 1976):

KH$_2$PO$_4$	0.5	g
NH$_4$Cl	1.0	g
CaSO$_4$	1.0	g
MgSO$_4$·7H$_2$O	2.0	g
FeSO$_4$·7H$_2$O	0.5	g
sodium lactate	3.5	g
yeast extract	1.0	g
ascorbic acid	0.1	g
thioglycollic acid	0.1	g
distilled water	250	ml
source water	750	ml

References

ABD-EL MALEK, Y. & RIZK, S. G. 1963 Bacterial sulphate reduction and the development of alkalinity. III. Experiments under natural conditions. *Journal of Applied Bacteriology* **26**, 20–26.

AMERICAN PETROLEUM INSTITUTE (API) 1965 Recommended Practice for Biological Analysis of Subsurface Injection Waters, 2nd ed., API RP No. 38. Dallas: Division of Production.

BARGHOORN, E. S. & NICHOLS, R. L. 1961 Sulphate-reducing bacteria and pyritic sediments in Antarctica. *Science* **134**, 190.

BADZIONG, W., THAUER, R. K. & ZEIKUS, J. G. 1978 Isolation and characterisation of *Desulfovibrio* growing on hydrogen plus sulphate as the sole energy source. *Archives of Microbiology* **116**, 41–49.

BAARS, J. K. 1930 Over sulfaat reductive door bacteriën. Dissertation, University of Delft, Holland.

BEIJERINCK, M. W. 1895 Über *Spirillum desulfuricans* als Ursache von Sulfat Reduktion. *Zentralblatt für Bakteriologie, Parasitenkunde, Infektionskrankheiten und Hygiene. Abterlung II.* **1**, 104–114.

BOONE, D. R. & BRYANT, M. P. 1980 Propionate-degrading bacterium, *Syntrobacter wolinii* sp. nov. gen. nov. from methanogenic ecosystems. *Applied and Environmental Microbiology* **40**, 626–632.

BUTLIN, K. R., ADAMS, M. E. & THOMAS, M. 1949 The isolation and cultivation of sulphate-reducing bacteria. *Journal of General Microbiology* **3**, 46–58.

CAMPBELL, L. L. & POSTGATE, J. R. 1965 Classification of the spore-forming sulphate-reducing bacteria. *Bacteriological Reviews* **29**, 359–362.

GROSSMAN, J. P. & POSTGATE, J. R. 1953 Cultivation of sulphate-reducing bacteria. *Nature, London* **171**, 600–602.

HARDY, J. A. 1981 The enumeration, isolation and characterisation of sulphate-reducing bacteria from North Sea waters. *Journal of Applied Bacteriology* **51**, 505–516.

HARDY, J. A. & SYRETT, K. R. 1983 A radiorespirometric method for evaluating inhibitors and sulphate-reducing bacteria. *European Journal of Applied Microbiology and Biotechnology* **17**, 49–52.

HERBERT, B. N. 1976 The effect of hydrostatic pressure on bacteria intended for injection into oil formations. *Journal of Applied Bacteriology* **41**, 12.

IVERSON, W. P. 1968 Corrosion of iron and formation of iron phosphide by *Desulfovibrio desulfuricans*. *Nature, London* **217**, 1265–1267.

JORGENSEN, B. B. 1978 A comparison of methods for the quantification of bacterial sulphate reduction in coastal marine sediments. 1. Measurement with radiotracer techniques. *Geomicrobiology Journal* **1**, 11–27.

KAPLAN, I. R. 1956 Evidence of microbiological activity in some of the geothermal regions of New Zealand. *New Zealand Journal of Science and Technology* **37**, 639–662.

KIMATA, M., KADOTA, H., HATA, Y. & TAJIMA, T. 1955 Studies on the marine sulphate-reducing bacteria. II. Influence of various environmental factors on sulphate-reducing activity of marine sulphate-reducing bacteria. *Bulletin of the Japanese Society of Scientific Fisheries* **21**, 109–112.

LAANBROEK, H. J. & PFENNIG, N. 1981 Oxidation of short-chain fatty acids by sulphate-reducing bacteria in freshwater and marine sediments. *Archives of Microbiology* **128**, 330–335.

MACPHERSON, R. & MILLER, J. D. A. 1963 Nutritional studies on *Desulfovibrio desulfuricans*. *Journals of General Microbiology* **31**, 365–373.

MARA, D. D. & WILLIAMS, D. J. A. 1970 The evaluation of media used to enumerate sulphate-reducing bacteria. *Journal of Applied Bacteriology* **33**, 543–552.

MILLER, J. D. A. & WAKERLEY, D. S. 1966 Growth of sulphate-reducing bacteria by fumarate dismutation. *Journal of General Microbiology* **43**, 101–107.

MILLER, J. D. A., NEUMANN, D. M., ELFORD, L. & WAKERLEY, D. S. 1970 Malate dismutation by *Desulfovibrio*. *Archives für Mikrobiologie* **71**, 214–219.

OCHYNSKI, F. W. & POSTGATE, J. R. 1963 Some biochemical differences between fresh water and salt water strains of sulphate-reducing bacteria. In *Symposium on Marine Microbiology* ed. Oppenheimer, C. H., pp. 426–441. Illinois: Thomas.

OREMLAND, R. S. & SILVERMAN, M. P. 1979 Microbial sulfate reduction measured by an automated electrical impedance technique. *Geomicrobiology Journal* **1**, 355–372.

PANKHURST, E. S. 1971 The isolation and enumeration of sulphate-reducing bacteria. *Society for Applied Bacteriology Technical Series 5. Isolation of Anaerobes* eds. Shapton, D. A. & Gould, G. W., pp. 223–240. London: Academic Press.

PFENNIG, N. & BIEBL, H. 1981 The dissimilatory sulphur-reducing bacteria. *The Prokaryotes. Handbook on Habitats, Isolation and Identification of Bacteria* eds. Starr, M. P., Stolp, H., Trüper, H. G., Belows, A. & Schlegel, H. G., Vol. 1, pp. 941–947. Berlin: Springer-Verlag.

PFENNIG, N., WIDDEL, F. & TRÜPER, H. G. 1981 The dissimilatory sulphate-reducing bacteria. *The Prokaryotes. Handbook on Habitats, Isolation and Identification of Bacteria* eds. Starr, M. P., Stolp, H., Trüper, H. G., Belows, A. & Schlegel, H. G., Vol. 1, pp. 927–940. Berlin: Springer-Verlag.

POSTGATE, J. R. 1951 On the nutrition of *Desulfovibrio desulfuricans*. *Journal of General Microbiology* **5**, 714–724.

POSTGATE, J. R. 1966 Media for sulphur bacteria. *Laboratory Practice* **15**, 1239–1244.

POSTGATE, J. R. 1969 Media for sulphur bacteria. Some amendments. *Laboratory Practice* **18**, 286.

POSTGATE, J. R. 1984 *The Sulphate-Reducing Bacteria*. 2nd Edition. Cambridge: University Press.

POSTGATE, J. R.& CAMPBELL, L. L. 1966 Classification of *Desulfovibrio* species, the non-sporulating sulphate-reducing bacteria. *Bacteriological Reviews* **30**, 732–738.

RITTENBERG, S. C. 1969 The roles of exogenous organic matter in the physiology of chemolithotrophic bacteria. *Advances in Microbial Physiology* **3**, 159–196.

SATAKE, K. 1977 Microbial sulphate reduction in a volcanic acid lake having pH 1.8 to 2.0. *Japanese Journal of Limnology* **38**, 33–35.

SMITH, A. D. 1982 Immunofluorescence of sulphate-reducing bacteria. *Archives of Microbiology* **133**, 118–121.

SOROKIN, Y. I. 1966a Sources of energy and carbon for biosynthesis in sulphate-reducing bacteria. *Mikrobiologiya* **35**, 761–766.

SOROKIN, Y. I. 1966b Investigation of the structural metabolism of sulphate-reducing bacteria with ^{14}C. *Mikrobiologiya* **35**, 967–977.

STARKEY, R. L. 1938 A study of spore formation and other morphological characteristics of *Vibrio desulfuricans*. *Archives of Microbiology* **9**, 268–304.

STEENKAMP, D. J. & PECK, H. D., JR. 1981 Proton translocation associated with nitrite respiration in *Desulfovibrio desulfuricans*. *The Journal of Biological Chemistry* **256**, 5450–5458.

TRÜPER, H. G., KELLEHER, J. J. & JANNASCH, H. W. 1969 Isolation and characterisation of sulphate-reducing bacteria from various marine environments. *Archives für Mikrobiologie* **65**, 208–217.

VAN DELDEN, A. 1903 Beitrag zur Kenntnis der Sulfatreduktion durch Bakterien. *Zentralblatt für Bakteriologie, Parasitenkunde, Infektionskrankheiten und Hygiene*. Abteilung II. **11**, 113–119.

WIDDEL, F. & PFENNIG, N. 1977 A new anaerobic, sporing, acetate-oxidising, sulphate-reducing bacterium, *Desulfotomaculum* (amend.) *acetoxidans*. *Archives of Microbiology* **112**, 119–122.

WIDDEL, F. & PFENNIG, N. 1981a Studies on dissimilatory sulphate-reducing bacteria that decompose fatty acids. I. Isolation of new sulphate-reducing bacteria enriched with acetate from saline environments. Description of *Desulfobacter postgatei* gen. nov. sp. nov. *Archives of Microbiology* **129**, 395–400.

WIDDEL, F. & PFENNIG, N. 1981b Sporulation and further nutritional characteristics of *Desulfotomaculum acetoxidans*. *Archives of Microbiology* **129**, 401–402.

ZEHNDER, A. B. J. & WUHRMANN, K. 1976 Titanium (III) Citrate as a non-toxic oxidation–reduction buffering system for the culture of obligate anaerobes. *Science* **194**, 1165–1166.

ZOBELL, C. E. 1958 Ecology of sulphate-reducing bacteria. *Producers Monthly* **22**, 12–29.

Estimation and Control of Microbial Activity in Landfill

J. M. GRAINGER, K. L. JONES AND P. M. HOTTEN

Department of Microbiology, The University, Reading, Berkshire, UK

J. F. REES*

Environmental Safety Group, Harwell Laboratory, Oxfordshire, UK

In the United Kingdom some 20 million tonnes of municipal refuse is generated per annum of which approximately 85% is disposed of to landfill. This means of disposal can present problems such as pollution of ground water by leachate containing carboxylic acids and inorganic ions (Zanoni 1973; Rees & Viney 1982) and risk of explosion or fire from landfill gas (Bromley & Parker 1979; Pacey 1980). However, successful control of microbial activity in landfill could lead to progress in containing the problems, using landfill sites for biogas production (Marchant 1981), and improving predictions of the period of activity of completed landfills with respect to use in reclamation projects.

An improved understanding of the microbiological processes involved is a necessary preliminary to successful control of hazards and maximisation of benefits because landfill microbiology is still in its infancy. Until the late 1970s, reports on microorganisms in landfill were limited to enumeration of faecal indicator bacteria; such studies were not relevant even to environmental issues. Then Filip & Kuster (1979) described studies on microbial activity in landfill in Germany and Jones & Grainger (1980) reported work in the UK.

Microbial activity and interaction in other anaerobic environments such as sediments (Sorensen *et al.* 1981; Jones *et al.* 1982), the rumen (Hungate 1966) and anaerobic digestion of sewage sludge and agricultural waste (Hobson *et al.* 1974) are well documented. Hydrolysis and fermentation of polymers such as starch, protein, cellulose and hemicellulose result in the pro-

---

*Present address: BioTechnica Ltd., 5 Chiltern Close, Cardiff CF4 5DL, UK.

duction of volatile fatty acids (Hobson *et al.* 1974; Bryant 1977; Miller *et al.* 1979; Wolfe & Higgins 1979). Further microbial transformations occur via reduction of carboxylic acids to acetate, H_2 and CO_2 by obligate proton reducers, fixation of CO_2 and H_2 and methanogenesis (Wolfe & Higgins 1979). The processes in landfill are assumed to be similar (see Senior & Balba, this volume, Fig. 5). However, for much of its depth landfill, unlike the other environments mentioned, is an example of solid substrate fermentation with poor mixing characteristics.

Landfills are physically, chemically and biologically heterogeneous environments. For this reason and the large areas involved, it is not easy to imagine a more difficult habitat to study microbiologically. Given these problems and a lack of established procedures for landfill microbiology, an attempt was made to assess aspects of overall microbial activity which might be of use in monitoring the effects of different landfill engineering practices. The chosen approach was to study "extracellular" (*sensu lato*) enzyme activity with respect to biologically important refuse polymers, i.e. cellulose, starch, protein and lipid. Procedures were tested in laboratory and field experiments by monitoring the effect of moisture content, a parameter that is relevant to landfill engineering practice. In this chapter the emphasis is on the methods used; fuller accounts and discussion of the results have been published elsewhere (Jones & Grainger 1983; Jones *et al.* 1983; Rees & Grainger 1982).

Enzyme Activities

Extraction of Enzymes

For proteases, amylases and lipases 10-g samples of refuse were extracted by vigorous shaking for 2 min with 30 ml distilled water containing 0.2% (v/v) Triton X-100 and 0.75-μM $MgSO_4$. The resulting slurry was compressed, and the supernatant fluid was decanted and then, after centrifugation at 4000 g in a bench top centrifuge, used for enzyme activity measurements. For cellulases extraction was by shaking 50 g refuse with 200 ml distilled water containing 0.2% (v/v) Triton X-100 for 5 min. The slurry was centrifuged at 2500 g for 10 min.

Assay Procedures

Heat-denatured (100°C for 1 min) controls were incubated with all enzyme assays, and the assay incubation temperature and the pH value used for each enzyme were chosen after preliminary experiments with extracts of pulverised refuse (see p. 263). As the enzyme assays were not done with sterilised extracts, some of the recorded enzyme activities may have been affected by the presence of microorganisms. The reason for using non-sterile extracts was that membrane filtration caused reduction of activity. Similar reductions in activity that were seen on filtration of commercial enzyme preparations added to autoclaved refuse extracts suggested that the observed effects were not solely due to the removal of microbial cells. Sterilisation by methods other than filtration was not attempted.

Protease, Amylase and Lipase Assays

Protease activity was determined by adding 1 ml enzyme extract to 10 mg dyed, powdered cow hide (Azocoll, Calbiochem) in 1 ml 0.2-M Tris buffer, pH 9.0, followed by incubation at 37°C until dye release was detectable. The reaction was terminated by addition of 2.5 ml 5% (v/v) trichloroacetic acid (TCA) and absorbance was measured at 520 nm. Standard curves were prepared by addition of protease from *Bacillus subtilis* (Sigma) to 0–10 mg Azocoll under the conditions described above and incubated until the substrate protein was degraded. The weight of protein degraded for a given absorbance was then determined from a standard curve of weight of Azocoll degraded against absorbance at 520 nm. Activity was expressed as mg Azocoll degraded per hour per gram dry-weight refuse.

Amylase was assayed by addition of 1 ml enzyme extract to 1 ml 1% (w/v) amylopectin-free starch (Sigma) in 0.2-M phosphate buffer, pH 7.0. Incubation was at 37°C for 20 min. The reducing sugars produced were reacted with 0.5 ml freshly prepared solution of 1.7% (w/v) dinitrosalicylic acid, 2.62% (w/v) KOH and 1.55% (w/v) NaOH for 10 min at 100°C. Absorbance was measured at 500 nm after addition of 4.0 ml water. The weight of starch degraded for a given absorbance was determined from standard curves prepared by adding amyloglucosidase (Sigma) to 0–10 mg starch under optimum conditions (pH 4.5, 37°C, 24 h) and incubating until all the substrate was degraded. Activity was expressed as milligrams of starch degraded per hour per gram dry-weight refuse.

Lipase assays were carried out using a serum lipase determination kit (Sigma), which involved the titrimetric measurement of fatty acids released from triglycerides.

Cellulase Assays Using Cellulose-Azure

Cellulase activity was determined by adding 1 ml enzyme extract to 10 mg cellulose-Azure (Sigma) in 0.2-M phosphate buffer, pH 7.0, or acetate buffer, pH 5.0. Incubation was continued at 37°C until dye release was detectable, i.e. from 2 h for strong activity to 24 h for weak activity. The reaction was terminated by addition of 2.5 ml 5% (v/v) TCA and the colour produced was measured at 595 nm. Standard curves of absorbance against weight of cellulose degraded were prepared by addition of cellulases from *Aspergillus niger* (Sigma) to different weights of cellulose-Azure under the conditions described above, followed by incubation until a strong blue colour was detected. The residual weight of cellulose was determined by filtration using desiccated, pre-weighed Oxoid membrane filters (APD 0.2 μm), dried to constant weight at 80°C after filtration. Activity was expressed as milligrams of cellulose degraded per 24 hours per gram dry-weight refuse.

The dye release method was chosen in preference to other techniques that involve disappearance of substrate, reduction in viscosity of carboxymethylcellulose or sugar release. The advantages lie in the length of time involved in the overall procedure and in avoiding the necessity of either already knowing the approximate levels of cellulase activity or using a range of incubation times. The method has been found to be satisfactory when compared with sugar release determinations and has been used as the basis for developing a mathematical model of cellulose degradation using fungal cellulase and cultures of *Clostridium thermocellum* (Hotten et al. 1983).

Cellulase Assays Using Dyed Cellophane Strips

Measurement of *in situ* cellulase activity in an environment is not possible when the method depends on recovery of dye that has been released from dyed substrate. However, Moore et al. (1979) developed a method whereby dyed substrate (Cellophane) was examined after degradation and the extent of cellulolysis estimated from the fraction of the weight of substrate remaining undegraded, calculated from the amount of dye retained in the substrate.

In the original method dye that had become loosely bound by degradation was first removed by autoclaving in water, and then the residual dye remaining after degradation, representing undegraded substrate, was extracted by boiling in alkali at 100°C (see Appendix). Cellulolytic activity was estimated by comparing the amount of residual dye remaining in the degraded strip with that in a control, undegraded strip and expressing the results in terms of weight loss.

Although the method is suitable for circumstances where there is strong cellulolytic activity, e.g. with fungi, the method was not found to be appropriate for laboratory experiments with degrading landfill under either aerobic or anaerobic conditions where cellulolytic activity was weak, i.e. $\leq$ 6–7% weight loss of substrate. In a preliminary report, Hotten et al. (1981) indicated that two factors together account for the problem. (1) Not all of the residual dye from undegraded substrate is removed in alkali at 100°C. (2) Dye that has become "loosened" by weak cellulolysis remains too firmly associated with the substrate to be removed by autoclaving in water, the stage which precedes the extraction of residual dye, but it is released together with dye from undegraded substrate by the subsequent alkali treatment. Consequently, an overestimate of the amount of substrate remaining undegraded is obtained. The contribution of (2) becomes less important if the limitations of (1) are reduced by increasing the proportion of residual dye that can be extracted from undegraded substrate. An improvement in this respect may have contributed to the successful application of the method to lake sediments (Jones & Simon 1981) where residual dye was extracted by autoclaving at 121°C in alkali instead of heating at 100°C (J. G. Jones, personal communication); this possibility was explored by Moore *et al.* (1979) but not used by them routinely because of occasional instances of discolouration of the extract.

Some Factors Affecting Enzyme Activity Estimations

Effect of Assay pH Value

Refuse samples of different pH values were obtained from laboratory experiments (p. 267) at different stages of fermentation. The samples used for investigating protease activity were of pH 6.0, 6.5 and 7.0 and those used for examining amylase activity were of pH 4.8, 6.0 and 6.9. Assays of enzyme in refuse extracts were done as previously described at pH 5.0, 6.0, 7.0, 8.0 and 9.0 in 0.1-M Tris or acetate buffer. An intended similar experiment with cellulases was limited by rapid loss of activity during degradation in the laboratory experiments (p. 267); it was decided therefore to assay for cellulolytic activity at pH 5.0 and 7.0. Results for proteases (Fig. 1) indicate that of the pH values tested, greatest protease activity occurred at pH 9.0 irrespective of the pH value of the refuse from which the extract was prepared. In contrast, amylases (Fig. 2) showed optimal activity when assayed at a pH value similar to that of the respective refuse sample.

Although the activity of extracted proteases was estimated at pH 9.0, refuse pH values of <7 are recorded at some stages during fermentation and therefore *in situ* activity may be less than that determined under assay conditions. Similarly, as the optimal assay pH value for amylase varied with refuse pH value, data on activity obtained at pH 7.0, the chosen value for assay, may differ from activity in landfill sites.

Effect of Assay Temperature

Refuse proteases and amylases from laboratory experiments were assayed at 20, 25, 30, 37, 40 and 45°C. Since temperature profiles showed optimal activity of protease and amylase at an assay temperature of 37°C (Fig. 3), this temperature was chosen for routine use. A similar effect was shown for

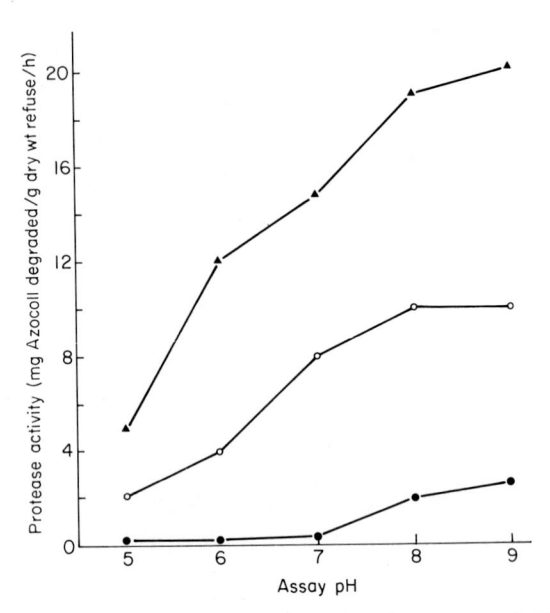

FIG. 1. Effect of assay pH value on protease activity in refuse extracts of pH 6.0 (●), 6.5 (○) and 7.0 (▲).

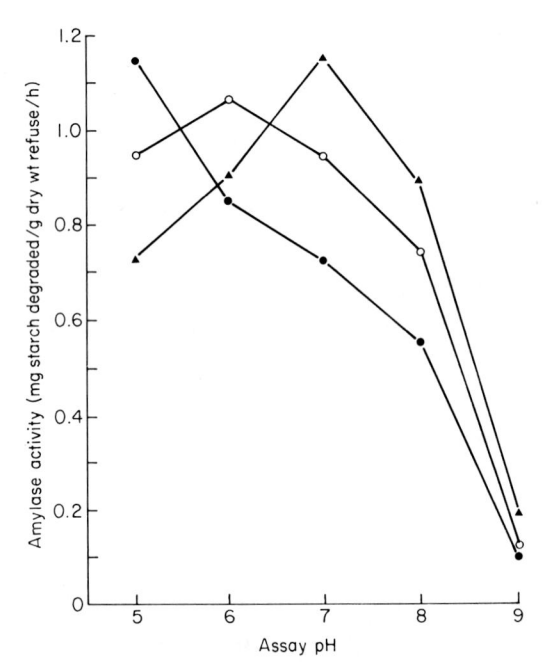

FIG. 2. Effect of assay pH value on amylase activity estimation in refuse extracts of pH 4.8 (●), 6.0 (○) and 6.9 (▲).

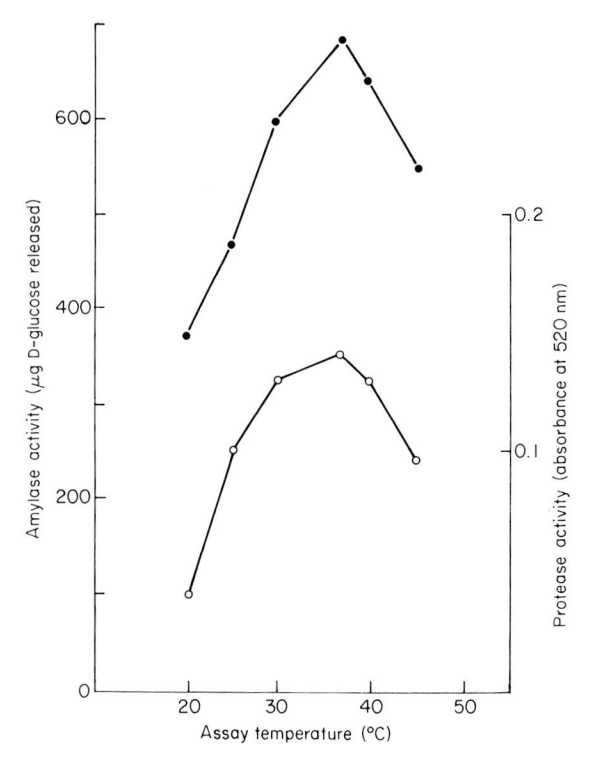

FIG. 3. Effect of assay incubation temperature on protease (○) and amylase (●) activity estimations in refuse extracts.

cellulases. However, it should be noted that laboratory refuse degradation experiments were done at room temperature but the temperature of the landfill site chosen for field experiments ranged from 28°C in the surface layers to 45–48°C at a depth of 10 m.

Recovery of Cellulase Activity

In an experiment to investigate the recovery of cellulases from refuse, a range of concentrations of *Trichoderma reesei* cellulase was added to saturated non-sterile and sterile (autoclaved) refuse. After 24 h at room temperature only the extracts from the sterilised refuse showed satisfactory recovery of activity (Fig. 4). Therefore, although added enzymes may be more sensitive to inhibitory effects than those produced *in situ* according to observations with some soil enzymes by Pettit *et al.* (1976, 1977), there are problems with enzyme inactivation, perhaps by proteases, and extraction that should not be overlooked.

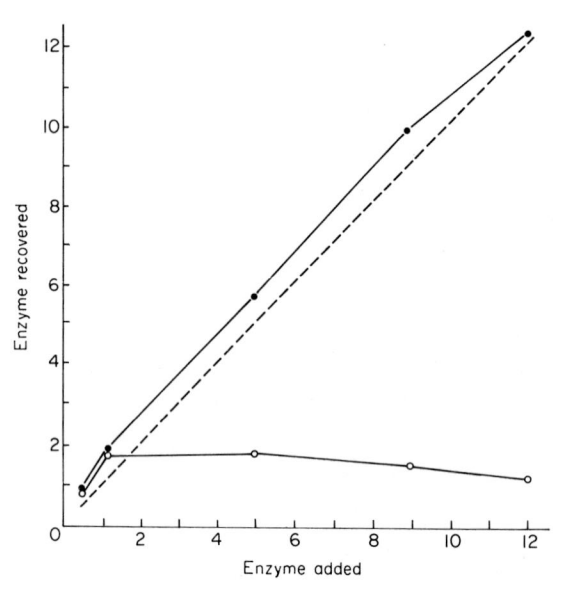

FIG. 4. Recovery of *Trichoderma reesei* cellulase from refuse after addition to non-sterilised (○) and autoclaved (●) samples. Cellulase activity (milligrams cellulose degraded per gram dry weight g refuse per day) results given recorded at assey pH 5.0; broken line represents theoretical recovery (after Jones & Grainger 1983).

Chemical Analysis

Refuse Composition

Refuse samples were dried for 4 days at 105°C and hammer milled before analysis. Cellulose, lignin, starch and hemicellulose were determined gravimetrically (Goering & Van Soest 1970). Protein was estimated by the Kjeldahl–Nesslerization method (Herbert *et al.* 1971).

Gas and Leachate Production and Composition

Gas production in field experiments was measured by incubation of refuse samples in 1.5-litre Kilner jars and displacement of acidified 12% (w/v) NaCl solution to reduce solubility. Composition was determined by gas chromatography using a 2.5 m × 1.5 mm stainless steel column packed with 60–80 mesh molecular sieve 5A (Chromatography Services, Lower Bebington, Wirral) at 100°C for H_2, N_2, O_2 and CH_4; for CO_2, a similar column packed with 60–80 mesh silica gel was used at 100°C. Helium (40 ml/min) was used as a carrier gas and detection was by hot wire. Samples of leachate for analysis were obtained by centrifugation of core samples at 4000 g for 45 min. Carboxylic acid concentrations were determined using 2.1 m × 4 mm glass columns containing 5% free fatty acid phase (FFAP) on 80–100 mesh chromosorb (treated with dimethyl chorosilane) at 135°C for 6 min programmed at 5°C/min to final temperature of 200°C. Helium (50 ml/min) was used as a carrier gas and detection was by flame ionisation.

Laboratory Experiments

Pulverised refuse was chosen for study owing to the lack of homogeneity of unprocessed refuse. Samples were obtained from High Heavens Landfill Site, High Wycombe, Buckinghamshire. Freshly pulverised material [moisture content 65% (w/v), referred to as "dry", i.e. unsaturated refuse] was compressed to a density of 0.7 t/m^3 either in carboys (5 liters) or plastic confectionery jars (3.5 liters) containing a 1- to 2-in. layer of gravel to aid drainage. Excess water was added to some samples (i.e. saturated refuse) to investigate the effects of using saturated conditions. The lids of the containers were fitted with bungs and outlets for monitoring the volumes of gases produced by venting through acidified 12% (w/v) NaCl solution to reduce solubility. The containers were incubated at room temperature (22–25°C). Samples were taken at 5- to 10-day intervals and the containers resealed as quickly as possible. Liquid samples for leachate quality analysis were obtained through a tap at the bottom of the container.

Changes in protease activity over a period of 6 months showed that saturation with water induced a threefold increase in activity to approximately 70 mg Azocoll degraded per hour per gram dry-weight refuse after 15 days, which then rapidly declined to less than the initial value after 40 days. In dry refuse as received from the pulveriser (65% moisture), activity was unchanged for 50 days and then declined. Amylase activity was similarly affected by refuse moisture contact. In saturated refuse, activity increased for the first 50 days to a level of 55 mg starch degraded per hour per gram dry-weight refuse, some 1000 times greater than the initial value. In contrast, activity in dry refuse was detected only in microgram amounts of starch degraded per hour per gram dry-weight refuse. Cellulase activity decreased rapidly within 10 days in saturated and dry samples. However, this effect was less marked with refuse degrading in a shallow layer in a tray (Fig. 5). Lipase activity was not detected in refuse extracts.

Enzyme activity determinations were used in small-scale (boiling tube) experiments to quantify further the effect of refuse moisture content on microbial activity; values in the range 3–8 g H$_2$O per gram dry weight were used. Protease and amylase activities were determined after 20 and 40 days, respectively, following incubation of 10-g samples of pulverised refuse in

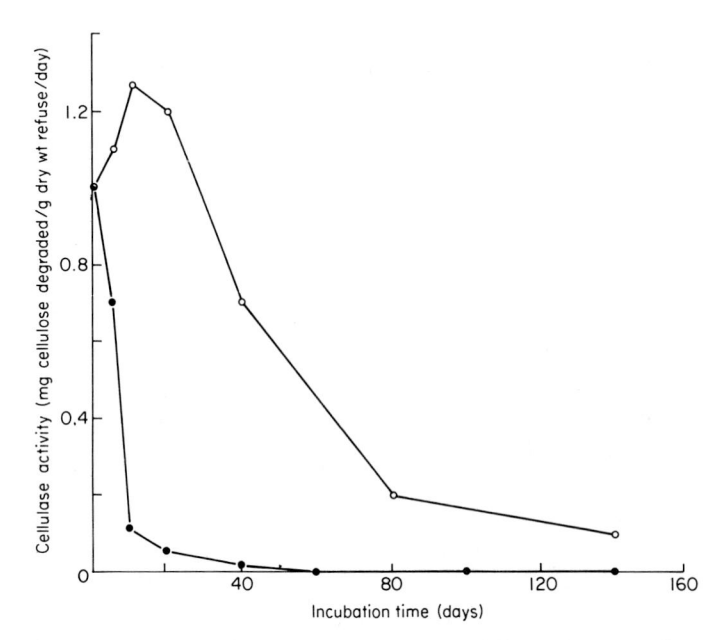

FIG. 5. Cellulase activity in refuse degrading under aerobic (○) and anaerobic (●) conditions. Enzyme activity results given recorded at assay pH 5.0.

test tubes sealed under N_2 with Suba Seals (Gallenkamp). Greatest activity was observed with moisture content of 6–8 g H_2O per gram dry-weight refuse. [See Jones and Grainger (1983) for further details.]

Field Experiments

Following the exploratory work in the laboratory, the approach was evaluated in field experiments. The site chosen (Aveley Landfill Site, Essex) had been partially characterised (Rees 1980a,b) and was appropriate for the study because a gradually rising water table presented the opportunity to examine the effect of different moisture regimes on microbial activity.

Sampling was done with a percussion drilling rig fitted with a 25-cm-diameter drill. Samples were taken at 1-m intervals, stored in air-tight containers (5-litre capacity) and processed in the laboratory as soon as possible after collection. The pH and temperature profiles of bore holes were recorded.

Enzyme activity profiles are summarised in Fig. 6. Measurements of amylase and protease activity showed a stimulatory effect of increased moisture, i.e. position of the water table, thereby confirming observations in

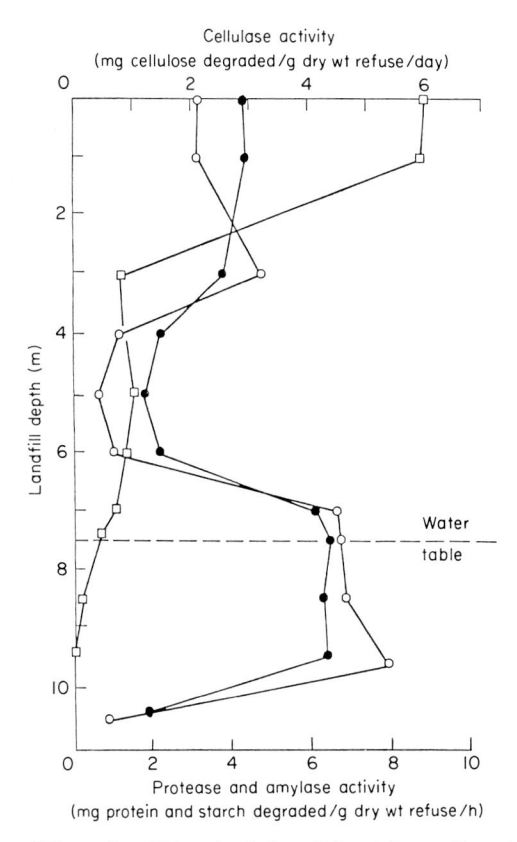

FIG. 6. Protease (○), amylase (●) and cellulase (□) activity profiles with depth in a landfill. (From Jones et al. 1983.)

the laboratory experiments. Starch and protein content of core samples was less below the water table than above it. Cellulase activity was detected only in assays done at pH 5.0, suggesting that either the enzymes were of fungal origin or bacterial cellulases had not been extracted successfully, or both. It was also noted that maximum cellulase activity in the landfill was detected in surface samples where O_2 permeation would allow fungal growth. No cellulase activity was detected below the water table which could be explained in part in terms of destruction of cellulases by proteolytic enzymes, the activity of which was stimulated within the water table. However, there had been cellulolytic activity in the depths of the landfill because cellulose levels and the cellulose:lignin ratio decreased with depth, the latter from 4.5 in the top 5 m to 0.5 below that depth. Lignin degradation is not expected to occur anaerobically and therefore would remain unchanged. Records of the age of

270   J. M. GRAINGER ET AL.

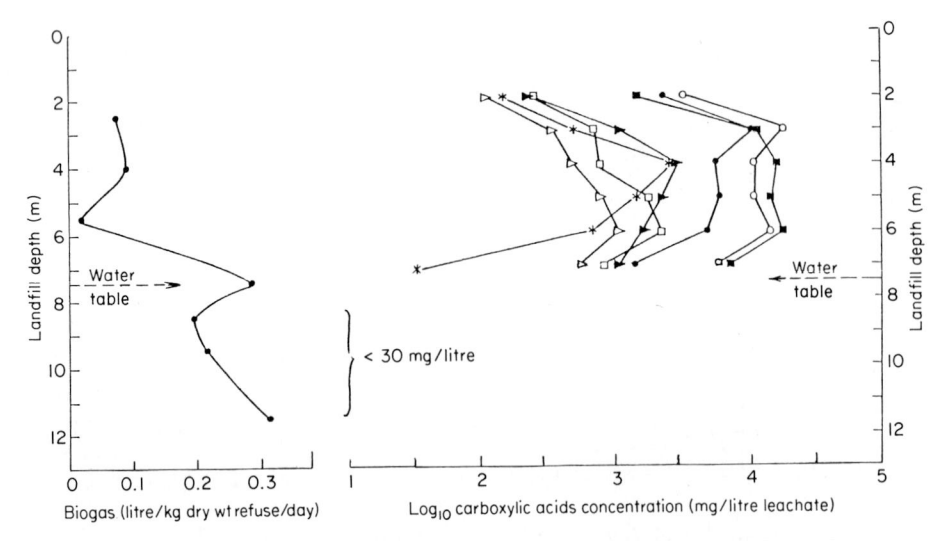

FIG. 7. Biogas and carboxylic acids production profiles with depth in a landfill (after Jones et al. 1983). (○), Acetate; (●), propionate; (□), isobutyrate; (■), butyrate; (▷), isovalerate; (▶), valerate; (*), caproate.

refuse at the various depths would have helped to assess the influence of time on the enzyme activities detected.

Results of some analyses of gas and leachate are summarized in Fig. 7. Decrease in carboxylic acid content of leachate from a mean value of ca. 4000 mg/litre at 3–6 m below the surface to <30 mg/litre below the water table coincided with increase in rates of production of biogas (60% CH_4, 40% CO_2).

Conclusions

"Extracellular" enzyme activity measurements, as used by soil microbiologists (e.g. Burns 1978), appear to be of value in studying the biodegradation of polymers in municipal refuse when used in conjunction with chemical and physical analyses. The effects of changing environmental factors on microbial activity can be examined and some interactions between different physiological groups of organisms can be studied. That the trends observed here differ from those recorded elsewhere by most probable number colony counts of equivalent physiological groups is not surprising as the two approaches are based on different principles (Jones & Grainger 1983; Jones et al. 1983).

The trends observed in laboratory experiments on the effect of moisture content on enzyme activity were also shown in field experiments, where the position of the water table controlled moisture levels. In the field experiments the correlations observed between polymer degradation, leachate quality and biogas generation show the value of using microbiological methods to study the initial stages in the degradation process in conjunction with chemical analyses of the products of the subsequent stages.

Several points of technique need further development in the enzyme activity approach. For example, much preliminary work is needed to ascertain the most appropriate conditions for the assay, and there may be difficulties with enzyme inactivation and extraction. Although the choice of conditions for enzyme assays may place constraints on absolute measurements of rates of refuse decomposition, the approach is valuable for assessing trends in changes in microbial activity and the effects on them of environmental factors and various landfill engineering practices. Therefore, there is promise that the variety of approaches being made to the study of landfill microbiology (e.g. Senior and Balba, this volume) and common factors with work on other environments (e.g. Theodorou *et al.*, Lynch *et al.*, Hobson *et al.*, Kirsop *et al.*, this volume) will enable microbiology to make an increasingly important contribution to efforts to contain the environmental problems and maximise the potential benefits associated with landfill disposal of municipal refuse.

Appendix

Cellulase Assay by Dye Release from Cellophane Strips (after Moore *et al.* 1979)

Cellophane (grade 325P non-moisture-proof, British Cellophane Ltd) is cut into 2 × 5-cm strips and boiled in batches of 100 in two changes of 500 ml of glass distilled water (GDW) to remove plasticisers. A further 500 ml GDW is added and heated to 80°C while stirring. This temperature is maintained for the following additions to be made: 1.5 g Remazol Brilliant Blue-R (Sigma), followed by 30% (w/v) aqueous Na_2SO_4 added as five 20-ml aliquots at 2-min intervals, and then 2.5 g $Na_3PO_4 \cdot 12H_2O$ solution in 15 ml GDW. The temperature is maintained at 80°C for a further 20 min. The dyed strips are then rinsed with hot water until the washings are colourless and autoclaved at 121°C for 15 min in two changes of 1 litre GDW to remove any remaining unbound dye. The strips are air-dried and stored until required.

After cellulolytic action, loosely attached dye is removed from the Cellophane strips by repeated autoclaving at 121°C for 15 min individually in 20 ml GDW until the washings are colourless. Retained dye is extracted by heating each strip to 100°C in 20 ml 0.35% (w/v) KOH for 15 min. The extracts are then made up to 50 ml with GDW and the absorbance at 595 nm measured as soon as possible after cooling to avoid fading. All test and control strips of Cellophane for an experiment should be taken from the same dyeing batch because of variation in intensity of dyeing.

Acknowledgements

Funding for KLJ and JFR was provided through the Harwell Laboratory in association with the Department of the Environment-sponsored Landfill Research Programme; PMH was supported by a Science and Engineering Research Council post-graduate studentship.

References

BROMLEY, J. & PARKER, A. 1979 Methane from landfill sites. *International Environment and Safety* **8**, 9–11.
BRYANT, M. P. 1977 Microbiology of the rumen. In *Dukes' Physiology of Domesticated Animals* ed. Swenson, M. J., pp. 287–304. New York: Cornell University Press.
BURNS, R. G., ed. 1978 *Soil Enzymes*. London: Academic Press.
FILIP, Z. & KUSTER, E. 1979 Microbial activity and the turnover of organic matter in municipal refuse disposed of to landfill. *European Journal of Applied Microbiology and Biotechnology* **7**, 371–379.
GOERING, H. & VAN SOEST, P. J. 1970 *Forage fibre analysis*. Agricultural Handbook No. 379. Washington: U.S. Department of Agriculture.
HERBERT, D., PHIPPS, P. H. & STRANGE, R. E. 1971 Chemical analysis of microbial cells. In *Methods in Microbiology* eds. Norris, J. R. & Ribbons, D. W., Vol. 5B, pp. 209–344. London: Academic Press.
HOBSON, P. N., BOUSFIELD, S. & SUMMERS, R. 1974 Anaerobic digestion of organic matter. *Critical Reviews in Environmental Control* **4**, 131–191.
HOTTEN, P. M., JONES, K. L. & GRAINGER, J. M. 1981 The use of a dye release method for estimating activities of cellulase and numbers of anaerobic cellulolytic bacteria in domestic refuse. *Second European Congress of Biotechnology, Abstracts of Communications*, p. 260. London: Society of Chemical Industry.
HOTTEN, P. M., JONES, K. L. & GRAINGER, J. M. 1983 The application of a mathematical model to an appraisal of the cellulose Azure method for determining cellulase activity. *European Journal of Applied Microbiology and Biotechnology* **18**, 346–349.
HUNGATE, R. E. 1966 *The Rumen and its Microbes*. New York: Academic Press.
JONES, J. G. & SIMON, B. M. 1981 Differences in microbial decomposition processes in profundal and littoral lake sediments with particular reference to the nitrogen cycle. *Journal of General Microbiology* **123**, 297–312.
JONES, J. G., SIMON, B. M. & GARDENER, S. 1982 Factors affecting methanogenesis and associated anaerobic processes in the sediments of a stratified eutrophic lake. *Journal of General Microbiology* **128**, 1–12.
JONES, K. L. & GRAINGER, J. M. 1980 Estimation of cellulolytic, proteolytic and amylolytic activity in domestic refuse. *Journal of Applied Bacteriology* **50**, viii.

JONES, K. L. & GRAINGER, J. M. 1983 The application of enzyme activity measurement to a study of factors affecting protein, starch and cellulose fermentation in domestic refuse. *European Journal of Applied Microbiology and Biotechnology* **18**, 181–185.

JONES, K. L., REES, J. F. & GRAINGER, J. M. 1983 Methane generation and microbial activity in a domestic refuse landfill site. *European Journal of Applied Microbiology and Biotechnology* **18**, 242–245.

MARCHANT, A. J. 1981 Practical aspects of landfill management of landfill gas—a local authority view. In *Landfill Gas Symposium Proceedings* Paper 7, Harwell: Harwell Laboratory.

MILLER, D., BROWN, C. M., PEARSON, T. H. & STANLEY, S. O. 1979 Some biologically important low molecular weight organic acids in the sediments of Loch Eil. *Marine Biology (Berlin)* **50**, 375–383.

MOORE, R. L., BASSET, B. B. & SWIFT, M. J. 1979 Developments in the Remazol Brilliant Blue dye-assay for studying the ecology of cellulose decomposition. *Soil Biology and Biochemistry* **11**, 311–312.

PACEY, J. G. 1980 *Design and Planning for End Use of Sanitary Landfill*. San Jose, California: EMCON Associates.

PETTIT, N. M., SMITH, A. R. J., FREEMAN, R. B. & BURNS, R. G. 1976 Soil urease: activity, stability and kinetic properties. *Soil Biology and Biochemistry* **8**, 479–484.

PETTIT, N. M., GREGORY, L. J., FREEMAN, R. B. & BURNS, R. G. 1977 Differential stabilities of soil enzymes: assay and properties of phosphatase and arylsulphatase. *Biochimica Biophysica Acta* **485**, 357–366.

REES, J. F. 1980a The fate of organic compounds in the landfill disposal of organic matter. *Journal of Chemical Technology and Biotechnology* **30**, 161–175.

REES, J. F. 1980b Optimization of methane production and refuse decomposition in landfills by temperature control. *Journal of Chemical Technology and Biotechnology* **30**, 458–465.

REES, J. F. & GRAINGER, J. M. 1982 Rubbish dump or fermenter? Prospects for the control of refuse fermentation to methane in landfills. *Process Biochemistry* **17**, 41–44.

REES, J. F.& VINEY, I. 1982 *Leachate Quality and Gas Production from a Domestic Refuse Landfill. The Implications of Water Saturated Refuse at Aveley Landfill* (AERE-R 10328). London: HMSO.

SORENSEN, J., CHRISTENSEN, D. & JORGENSEN, B. B. 1981 Volatile fatty acids and hydrogen as sediments for sulphate-reducing bacteria in anaerobic marine sediment. *Applied and Environmental Microbiology* **42**, 5–11.

WOLFE, R. S. & HIGGINS, I. J. 1979 Microbial biochemistry of methane—a study in contrasts. *International Review of Biochemistry* **21**, 267–353.

ZANONI, A. E. 1973 Potential for ground water pollution from the land disposal of solid wastes. *Critical Reviews in Environmental Control* **3**, 225–260.

The Use of Single-Stage and Multi-Stage Fermenters to Study the Metabolism of Xenobiotic and Naturally Occurring Molecules by Interacting Microbial Associations

E. SENIOR

Department of Bioscience and Biotechnology, Applied Microbiology Division, University of Strathclyde, Glasgow UK

M. T. M. BALBA

Biomass International, Milnthorpe, Cumbria UK

World Health Organisation estimates have suggested that before 1980 some 4 million chemicals had been either isolated from natural products or had been synthesised. Of these, some 60,000 are thought to be in use daily and 200 new chemicals are marketed each year. Most of these compounds ultimately appear in the natural enviornment where their degradation and subsequent recycling are natural phenomena mediated through microbial activities. Thus their ultimate fates are of great interest to microbiologists and biochemists and have also attracted considerable public concern with respect to both persistence and toxicity.

Many of these chemicals are aromatic and although studies on biodegradation and persistence under aerobic conditions are well documented, anoxic ecosystems have largely been neglected. This is somewhat surprising since the latter constitute significant components of many environments and are characteristic of aquatic sediments, landfills and the rumen.

From the plethora of studies made on oxygen-linked reactions which lead to aromatic ring hydroxylations as a preliminary to ring fissions, it is apparent that two mechanisms are operative. In one, a single atom of oxygen becomes incorporated into the aromatic ring as a hydroxyl; in the other, both atoms of the oxygen molecule are incorporated and result in the formation of a cis dihydrodiol. In both cases subsequent events lead to the opening of the ring by a cleavage dioxygenase, either of the ortho (intradiol) or meta (extradiol) type. The resulting aliphatic acids then become components of microbial intermediary metabolism. Molecular oxygen is obligatory for these transformations to occur since it becomes incorporated into the catabolic products and also serves as the terminal electron acceptor (Evans *et al*. 1976; Dagley 1978).

In the primitive biosphere which lacked O_2, fermentative biochemical reactions must have afforded the energy for prokaryotic growth and survival. Aromatic compounds of abiogenic or biogenic origin were likely to have been degraded to aliphatic molecules and anaerobic bacteria must have evolved mechanisms to accomplish this. The essential biochemical transformations may be summarised as follows:

(1) preliminary metabolic changes such as decarboxylation (Balba *et al*. 1979) or $-NH_2$ group replacement of $-OH$ (Balba *et al*. 1981);
(2) reduction of the aromatic ring to cyclohexane derivative;
(3) cleavage of cyclohexane ring (where the attached functional group is a carboxyl the mechanism is a CoA-mediated β-oxidation; when the group is a ketone a hydrolase must be operative);
(4) metabolism of the aliphatic acids produced by ring fission, eventually to acetate;
(5) production of CH_4 from acetate and CO_2.

Pertinent Research Areas

Lignin Degradation

Cellulose is the most abundant biopolymer on earth, closely followed by lignin which is composed of aromatic alcohols, e.g. coniferyl, *p*-coumaryl and sinapyl alcohol, linked by a variety of C—C and C-ether bonds. Although cellulose is readily utilised by certain species of anaerobes, particularly those present in the rumen, evidence indicates that lignin is recalcitrant in such conditions, with the exception of limited demethylation of methoxy substituents (Meyer & Bondi 1952). This recalcitrance has long been recognised for its adverse effect on plant utilisation by farm animals. In the natural environment lignin is slowly degraded under aerobic conditions by the white rot fungi and, possibly, some bacterial species although the polymer does not serve as a carbon source.

In order that the utilisation of this natural product may be commercialised, physical and chemical pretreatments are often employed to produce labile substrates for subsequent microbial catabolism. However, hydrolysates so produced are variable in terms of their chemical composition, and thus it is often better to simplify the research problem by selecting specific aromatic moieties for detailed study, an approach which can also be applied to complex industrial effluents. Suitable molecules for study and their functional groups include anthranilate (o-NH$_2$, —COOH); p-amino benzoate (p-NH$_2$, —COOH); cinnamate (—CH = CHCOOH); cinnamyl alcohol (—CH = CHCH$_2$OH); benzyl alcohol (—CH$_2$OH); p-cresol (p-CH$_3$, —OH).

Methanogenic Fermentation

Many simple aromatic compounds such as benzyl alcohol, benzoate, cinnamyl alcohol, cinnamate, ferulate and vanillate are completely degraded to CO$_2$ and CH$_4$ under anoxic conditions by naturally occurring, interacting microbial associations. This ultimate production of only two gases is a unique characteristic of the metabolism and is mediated by an oxidation–reduction sequence in the presence of water:

$$C_nH_aO_b + [n - (a/4 - b/2)] H_2O = (n/2 - a/8 + b/4)CO_2 + (n/2 + a/8 - b/4)CH_4$$

However, when the catabolic potentials of methanogens are examined it can be seen that they are unable to utilise aromatic substrates in monoculture (Barker 1956; Wolfe 1972, 1979).

Many studies have been made on the methanogenic fermentation of aromatic substrates since this phenomenon was first reported by Tarvin & Buswell (1934). Notable amongst these were the studies of Ferry & Wolfe (1976), Healy & Young (1979) and Balba & Evans (1977, 1979, 1980a,b). In all of the reports the responsible organisms have constituted very stable interacting microbial associations whose integrities were retained through repeated subcultures, a commendation for all work in this area to be made with microbial associations.

Microbiological Processes in Landfill

Although new methods of municipal waste utilisation are being explored, landfill is still a common disposal method and is likely to remain so in many countries for the foreseeable future for several reasons: It is relatively cheap; the technology and environmental control measures are reasonably well understood; re-establishment of contours and restoration of low-lying land are permitted; leachates may be collected and treated by anaerobic digestion under controlled conditions to produce high-value chemicals; any CH$_4$ generated may provide a potentially valuable source of energy.

Before 1980 most studies on landfill were made on the engineering aspects with very few on either microbiology or biochemistry. It is difficult to imagine a more heterogeneous environment than landfill and this is probably reflected in the extremely sparse literature (see also Grainger *et al.*, this volume).

Growth Media and Analytical Methods

Growth Media

Basal Mineral Salts Medium

The basal salts medium contained (g/litre glass-distilled water) K_2HPO_4, 1.5; $NaH_2PO_4 \cdot H_2O$, 0.85; NH_4Cl, 0.90; $Na_2SO_4 \cdot 10H_2O$, 0.32; $MgCl_2 \cdot 6H_2O$, 0.20; $NaHCO_3$, 0.5; trace elements, 9 ml (see below); vitamins, 5 ml (see below); Na_2CO_3, 2.5 ml (8%, w/v, solution); $FeCl_3$, 0.03 ml (10%, w/v, solution). The medium, supplemented with the appropriate carbon and energy source, was filter sterilised by passage through a Sartorius membrane filter (0.2 μm SM 11307), under an atmosphere of O_2-free N_2 (OFN). The final pH value of media was 7.0.

Trace Elements

The trace elements mixture contained (g/litre) EDTA, 1.5; $MgSO_4$, 3.0; NaCl, 1.0; $MnSO_4$, 0.5; $FeSO_4$, 0.1; $CaCl_2$, 0.1; $CoCl_2$, 0.1; $CuSO_4$, 0.01; $AlK(SO_4)_2$, 0.01; H_3BO_3, 0.1; $NaMoO_4$, 0.01; $NiCl_2 \cdot 6H_2O$, 0.006.

Vitamins

The following vitamins were added (mg/litre): biotin, 2.0; folic acid, 2.0; pyridoxin HCl, 10; riboflavin, 5.0; thiamin, 5.0; nicotinic acid, 5.0; pantothenic acid, 5.0; cobalamin, 0.1; *p*-amino benzoate, 5.0; thioctic acid, 5.0.

Carbon and Energy Sources

Simple aromatic compounds (2 m*M*) were used as carbon and energy sources, e.g. anthranilic acid, benzyl alcohol, cinnamyl alcohol and *p*-cresol. They were selected as being representative of compounds which contain basic nuclei of common xenobiotic and naturally occurring molecules. In addition, more complex molecules such as lignin alkali hydolysate and solid domestic refuse were used. Hexoic acid (10 m*M*) was selected as a labile compound for comparative studies.

Other Additions

The basal mineral salts medium was amended with a reducing agent (sodium diothionite, 30 mg/litre) when necessary. When the inoculum was taken from saltmarsh sediment, NaCl (25 g/litre) was also included in the medium.

Sources of Inocula

Inocula used in this study were sampled from two sources, i.e. landfill from the Wilderness site of Glasgow City Corporation Waste Disposal Department, and saltmarsh sediment from Colne Point Saltmarsh, Essex.

Analytical Methods

Methane

The CH_4 concentration of the culture gas phase was determined by use of either a Pye Unicam 104 gas chromatograph or a Perkin Elmer Sigma 115 gas chromatograph.

Each gas chromatograph was equipped with a flame ionisation detector. A glass column (length 2 m, internal diameter 4 mm) was filled with 15% SP 1220 + 1% H_3PO_4 on acid-washed chromosorb (mesh size 100–120). The flow rate of the carrier gas (N_2) was set at 40 ml/min and the temperatures of the column and detector were 70°C and 150°C, respectively. The volume of gas sample injected was 50 μl and pure CH_4 (British Oxygen Company) was used as a standard for analysis.

Fatty Acids

All sample preparation method manipulations were made in ice baths.

Volatile fatty acids. For volatile fatty acids analysis, to 1 ml of culture supernatant was added 0.1 ml 6-M HCl followed by 1 ml ether. After shaking for 3 min the sealed vial was allowed to stand overnight in the cold room (+4°C). Samples of the ether layer (2 μl) were then injected into the GLC under the same conditions as those described for CH_4 analysis with the exceptions that the detector and column temperatures were 200 and 150°C, respectively. Standards solutions (10 mM) of acetate, propionate, butyrate, valerate and isovalerate were prepared by the same method.

Long-chain fatty acids. Samples for long-chain fatty acid analysis were also prepared by the same method. The ether extract was then separated and methylated with diazomethane before GLC analysis.

Carbon Dioxide

Chemostat gas-phase CO_2 was trapped in standardised barium hydroxide (0.1 N) and the concentration determined by back titration against 0.1-N HCl with phenolphthalein as indicator.

Sulphate

The standard $BaCl_2$ turbidimetric method for sulphate estimation was used. To a 100-ml culture sample in a 250-ml conical flask were added 5 ml of conditioning reagent (concentrated HCl, 30 ml; distilled water, 300 ml; 95% ethyl or isopropyl alcohol, 100 ml; NaCl, 75 g; glycerol, 50 ml) with mixing. One teaspoonful of $BaCl_2$ crystals (20 to 30 mesh) was added and the resulting solution stirred for exactly 1 min at which point the OD_{420} was read. The reading was then repeated at 30-s intervals for 4 min and the sulphate determined from a similarly prepared calibration curve (0–40 mg/litre).

Sulphide
Soluble sulphide. Sulphide was first liberated as H_2S from the samples, maintained under anoxic conditions by acidification with concentrated HCl and then trapped in zinc acetate (1%, w/v) as zinc sulphide. The standard thiosulphate-iodine back titration method was used to determine the sulphide concentration.

Hydrogen sulphide. Chemostat gas samples were bubbled through zinc acetate traps before assay for H_2S as described above.

Benzyl Alcohol, Cinnamyl Alcohol and Anthranilic Acid
These three carbon and energy sources were determined spectrophotometrically by measuring the absorbances at 258, 250 and 240 nm, respectively.

Identification of Metabolic Intermediates
Thin-layer chromatography (TLC) was used to identify intermediates of aromatic monomer catabolism. Glass plates (20 × 20 cm) were coated with silica gel (40%, w/w) (Kieselgel 60 PF_{254}, Merck, West Germany) to a thickness of 0.25 mm and left to dry at room temperature. The plates were then activated at 100°C for 30 min before use.

Both acidic and neutral fractions of the chemostat culture supernatants were extracted for identification. Extraction of the neutral fractions was made by the addition of 5 ml of diethyl ether to 10 ml of culture supernatant. Each extraction was repeated twice, the resulting solutions batched and the ether layer transferred to a bottle containing anhydrous sodium sulphate to dry the preparation. The acidic fractions were also extracted with diethyl ether after first reducing the pH to 3 by the addition of HCl (50%, v/v). The resulting ether layer was dried as before. Finally the acidic fractions were concentrated by bubbling OFN through, and the resulting residue was used for TLC analysis.

Concentrated extract (10 μl) was spotted on the TLC plates together with the selected standards such as benzoic acid, phenol, aniline, *p*-hydroxybenzoate, cyclohexane carboxylic acid, heptanoic acid, pimelic acid, adipic acid, valeric acid, cyclohexanol, cyclohexanone and 2-hydroxycyclohexanone. The chromatography solvent used contained benzene, dioxan and acetic acid at a volume ratio of 90 : 25 : 10.

Acidic fractions were visualised by spraying the plates, after drying, with a solution of bromocresol green (0.2%, w/v) in ethanol, whereas other organic compounds were visualised by exposing the plates to I_2 vapour in a closed chromatography tank for a few minutes. Finally, unsaturated compounds were detected by their capacity to absorb ultraviolet light. All visualised spots were identified by comparison of R_f values with known standards.

Enrichment for and Isolation of Interacting Microbial Associations and Component Monocultures

Although it has been postulated that stable interacting microbial associations are common in natural ecosystems and monocultures are largely laboratory artefacts (Harrison & Wren 1976), comparatively few associations have been defined, possibly because of the use of inappropriate enrichment and isolation techniques. A number of discussions and literature reviews have been made on enrichment culture techniques; see Hutner (1962) for nutritional aspects, Stanier (1953) for methods, Veldkamp (1970) for a history and review, Pochon & Tardieux (1967) for methods for soil bacteria and Parkes (1982) for laboratory systems analysis. Close examination of the literature shows that the majority of enrichment methods have been developed for aerobic species, whereas anaerobes have largely been excluded. Two notable exceptions have been the Winogradski soil column, which mimics the habitat formed by anoxic muds and the covering natural waters, and the closed culture with gas collection system developed by Balba (1978), which was first used to enrich for benzoate-metabolising species under anoxic conditions.

In the work to be described, the method of Balba (1978) was used together with a simple closed-culture technique; also the use of various forms of open-culture fermenter will be described.

Closed-Culture Fermenters

The simple closed-culture system was a conical flask (250 ml) containing 150 ml appropriate enrichment medium, inoculated with either refuse or salt-marsh sediment, closed with Subaseals (Gallenkamp), overgassed with OFN and incubated unshaken at 25°C. Gas head space samples were aseptically removed at regular intervals, replaced by equal volumes of OFN and analysed by GLC for CH_4 until the associations were actively evolving this gas, at which point the cultures were used as chemostat inocula.

The closed batch enrichment apparatus of Balba (1978) consisted of a Pyrex flask (1 liter) fitted with two side arms with the central neck connected

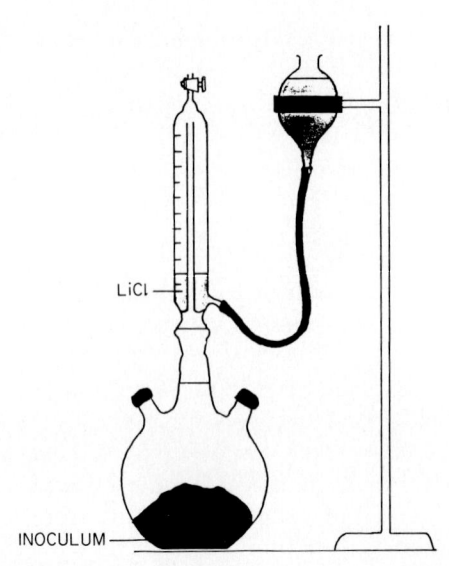

FIG. 1. Closed enrichment apparatus with gas collection system (Balba 1978).

to a graduated gas-collecting chamber (Fig. 1). Fermentation gases were collected by displacement of a solution of either LiCl (saturated solution) or NaCl (20%, w/v) in citric acid (0.5%, w/v). Flasks containing 850 ml required enrichment medium were inoculated with either refuse or saltmarsh sediment (10%, w/v) and OFN flushed over for 20 min with the tap at the upper end of the gas chamber open to establish anaerobiosis. One side arm was closed with a rubber Subaseal through which samples were aseptically taken for analysis by means of plastic syringes fitted with long needles. The vessel was maintained at 25°C and gas samples were taken and analysed as described above.

The system has the advantage that the gas : liquid ratio is maintained at a large value and thus any possible gas overpressure effects are minimised. Also, by application of an influent feed, an OFN gas line and an effluent siphon tube, the system can readily be converted to a chemostat (Fig. 2).

Isolation of the component monocultures of the interacting microbial associations was achieved by repeated subculture in basal mineral salts medium supplemented with a substrate such as benzoate, hexoate, vanillate, cinnamate, coumarate, *m*-anisate, *p*-anisate, phenol and catechol. All manipulations were made under an atmosphere of OFN to maintain anaerobiosis. Where low redox potentials were required, sodium diothionite reducing agent was also added.

Some of the more interesting isolates were the bacteria obtained from enrichments containing benzoate, hexoate, vanillate and cinnamate which

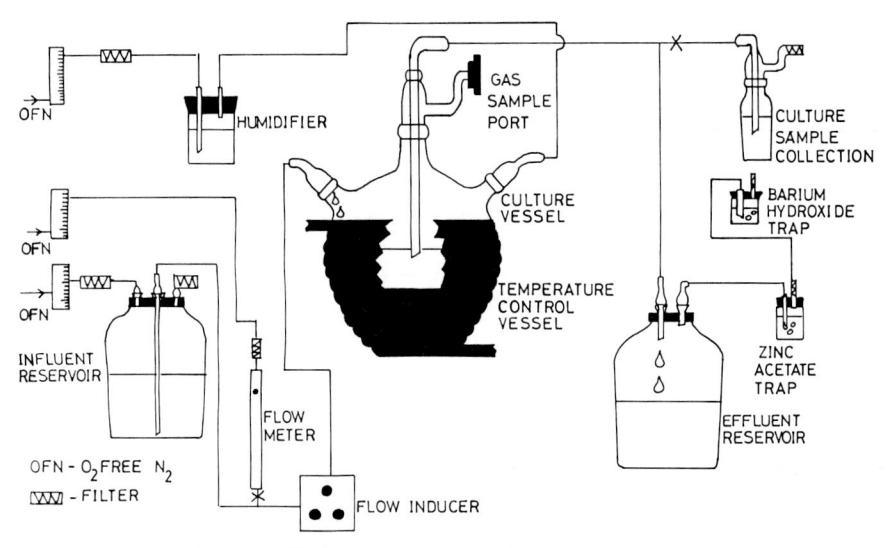

FIG. 2. All-glass continuous anoxic culture apparatus.

were able to reduce sulphate. These features suggest that "new" species had been isolated since sulphate-reducing bacteria are known to be specific for simple substrates such as lactate, acetate and hydrogen. Thus it could be that when appropriate enrichment and isolation methods are applied to other anoxic ecosystems, many other "new" species will be obtained.

In addition, under aerobic conditions, isolations were made of monocultures of bacterial species capable of degrading a wide variety of aromatic monomers such as *m*-anisate, *p*-anisate, phenol and catechol.

Open-Culture Fermenters

Single-Stage Anoxic Systems

Single-stage fermenters have often been used in the past to study the detailed microbial physiology and biochemistry of the aerobic catabolism of various products found in the environment by monocultures, and less frequently, by interacting microbial associations. Only rarely have associations isolated from anoxic ecosystems been studied.

All-Glass Chemostat

The single-stage, all-glass chemostat shown in Fig. 2 is similar to the one described by Veldkamp & Kuenen (1973). A slow stream of OFN (60 ml/h)

TABLE 1

Key intermediary metabolites detected in the methanogenic fermentation of anthranilic acid, benzyl alcohol and p-*cresol*

	Intermediary metabolites identified by TLC	
Substrate	Neutral fraction	Acidic fraction
Anthranilic acid	Aniline, phenol	Phenol
Benzyl alcohol		Benzoic acid
p-Cresol		p-Hydroxybenzoate
		Phenol

was led over the unstirred culture to maintain an overpressure of 0.1 atm. The temperature was maintained at 25°C by means of a MGW Lauda thermocirculator (Lauda, West Germany). Mineral salts medium, supplemented with the appropriate carbon and energy source, was pumped into the chemostat by means of a Gilson Minipuls 2 flow inducer (Anachem, Luton). Samples for gas and liquid analyses were taken as already described.

The single-stage chemostat proved to be an extremely useful tool to investigate the biochemical transformations which occurred, e.g. during the methanogenic fermentations of aromatic substrates, such as anthranilic acid, benzyl alcohol and *p*-cresol, when the key intermediates were aniline, phenol, benzoate and *p*-hydroxybenzoate (Table 1). Other common intermediary metabolites detected and their rates of production are shown in Table 2, and their substrate utilisation and gas production rates are shown in Table 3.

Such data are essential prerequisites for the elucidation of catabolic pathways of aromatic molecules (Fig. 3). The pathway for anthranilic acid is rather different from the one reported previously in the absence of sulphate-reducing bacteria (Balba & Evans 1980b). The sequence for the anaerobic reactions proposed to account for the intermediates detected postulates the

TABLE 2

Steady-state concentrations of common intermediary metabolites detected in the methanogenic fermentation of anthranilic acid, benzyl alcohol and p-*cresol*

	Steady-state concentrations (mM) of intermediary metabolites				
Substrate	Acetic acid	Propionic acid	Butyric acid	Isovaleric acid	Valeric acid
Anthranilic acid	0.550	0.057	0.009	0.022	NT*
Benzyl alcohol	0.510	0.049	0.009	0	NT
p-Cresol	0.370	0.054	0.010	NT	NT

* NT, not tested.

TABLE 3

Substrate utilisation and gas production rates in the methanogenic fermentation of anthranilic acid, benzyl alcohol and p-cresol

Substrate	Substrate utilisation (mmole/day)	Gas production (mmole/day)		
		H_2S	CH_4	CO_2
Anthranilic acid	1.350	0.191	1.001	5.947
Benzyl alcohol	1.134	0.190	0.981	5.120
p-Cresol	0.936	0.186	1.011	4.075

use of a molecule of H_2O for hydroxylation together with the usual addition and/or abstraction of hydrogen. For example, once the decarboxylation of anthranilic acid occurs, then the amino group could possibly be replaced by a hydroxyl group by the addition of H_2O across an imino double bond followed by the elimination of NH_3. The conversion of p-cresol to phenol could be initiated by a methyl hydroxylase similar to that demonstrated in *Pseudomonas putida* (Hopper 1976, 1978).

Such studies are essential for assessing the roles of microorganisms in environmental pollution. Many researchers have shown the efficient detoxification of noxious chemicals by microbial intervention but few have demonstrated, as is the case here, how a toxic chemical (aniline) may be gener-

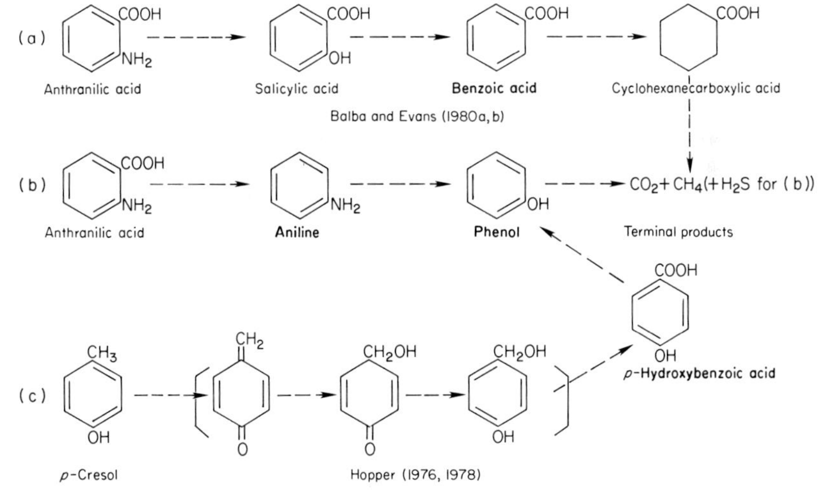

FIG. 3. Proposed pathways for the anaerobic catabolism of anthranilic acid (*a*) in the presence and (*b*) in the absence of sulphate, and (*c*) of p-cresol in the presence of sulphate. Bold type indicates detected intermediates.

ated from an innocuous plant component (anthranilic acid) under anoxic conditions. This finding could have a wide implication in the area of waste disposal biotechnology where it may be postulated that anoxic landfill site conditions create the right niche for this transformation to occur.

Refuse Column

A second type of single-stage cultivation system used was the refuse column (Fig. 4) which was maintained at room temperature. Tap water was perfused through a column (internal diameter 6 cm) packed with fresh refuse to a working volume of 905 ml, at a constant dilution rate (0.01/h) by means of a Gilson Minipuls 2 flow inducer. Anaerobiosis was maintained by an overpressure of OFN and was facilitated by a zinc acetate (1% w/v) and barium hydroxide (18 g/litre) pressure heads which were used to trap evolved H_2S and CO_2, respectively. Samples for culture analysis were withdrawn into sterile syringes via the four sampling ports and samples for gas analysis were taken from the exhaust gas stream.

This type of cultivation system proved to be extremely useful for enrichment studies, for the determination of microbial succession and for detailed examination of leachate generative pathways with respect to important interacting parameters such as different landfill types, materials, tipping practices and climatic conditions. The study of leachate may be regarded as a research priority area as a result of increasing national and international legislation.

An extension of the leachate production study was to link the refuse

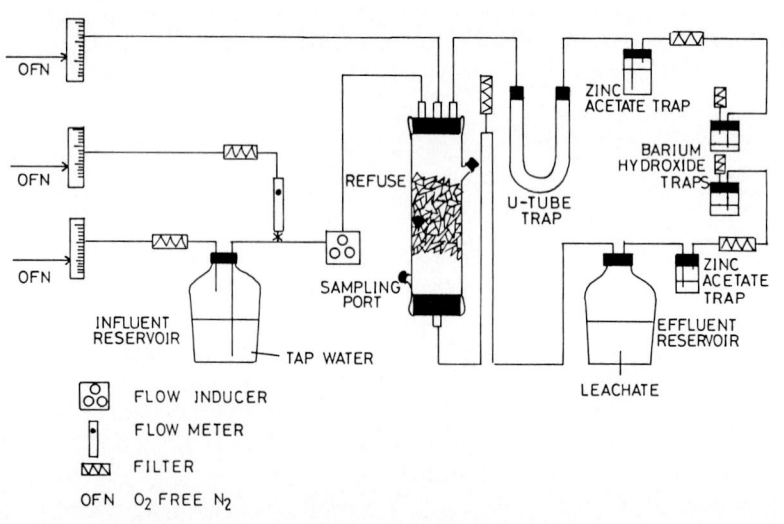

FIG. 4. Refuse column system used in enrichment and leachate studies.

column in series with an anoxic chemostat to examine the anaerobic digestion of leachate with respect to high-value product formation. Research into the generation of specific chemicals from both industrial and agricultural effluents has now reached an advanced state and a number of processes have been commercialised. The feasibility of producing similar products from landfill leachate was also established in the laboratory and is now proceeding to the development stage.

Two-Stage System

Although the single-stage anoxic fermenter is extremely useful for the study of biochemical transformations under anaerobic conditions, it is limited as a laboratory cultivation system since it cannot be used to model the interface between aerobic and anoxic niches. As a consequence of this a two-stage cultivation system was developed, which consisted of an aerobic chemostat linked to an anoxic vessel, to examine the catabolism of neutralised sawdust lignin alkali hydrolysate. In preliminary studies analysis of the chemostat liquid and gas effluent stream showed that the culture, which had previously been enriched and isolated from landfill, rapidly cleaved the rings of the aromatic monomers of lignin in the aerobic stage with the subsequent production of a mixture of carbon dioxide and methane in the anoxic vessel.

This simple two-stage culture system proved to be ideal for feasibility studies of high-value chemical and chemical feedstock production from lignin hydrolysate.

Multi-Stage System

From the foregoing discussions of single and two-stage fermenters it can be seen that much valuable information can be gained of both microbial interactions and biochemical transformations operative in the catabolism of aromatic molecules. However, when very heterogeneous ecosystems are examined, more sophisticated models have to be employed.

Arguably the most complex of all ecosystems is landfill which contains dust and cinders, vegetables and putrescibles, paper and board, metals, textiles and man-made fibres, glass, plastics and industrial wastes (Cooper 1980). Although only few studies have been made on refuse degradation, it is possible to develop a hypothetical pathway (Fig. 5).

Initially, aerobic decomposition predominates. However, this phase is usually quite short due to the limited oxygen availability and the extremely high biochemical oxygen demand (BOD) of the refuse. During this stage heat is generated by microbial metabolism and the landfill temperature is raised above ambient. As O_2 becomes depleted decomposition is continued

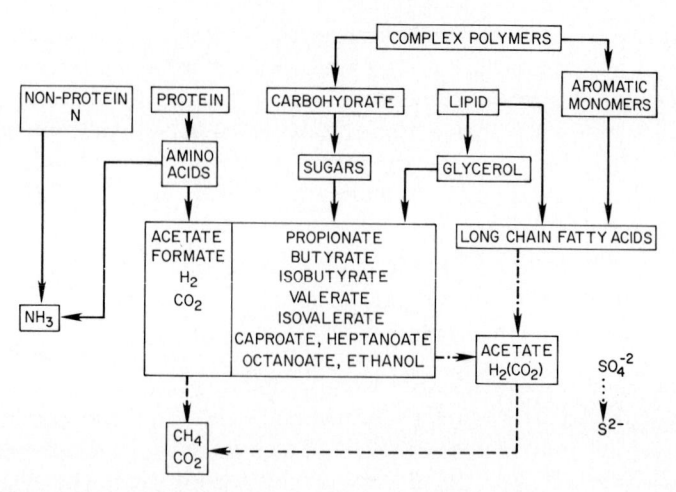

FIG. 5. Hypothetical scheme of reactions occurring in landfill. Hydrolysis and fermentation (→), acetogenesis (–·→), methanogenesis (– –→), sulphate reduction (· · ·→). Adapted from Laanbroek (1978) and Rees (1980).

by facultative anaerobic bacteria followed by strict anaerobes such as sulphate reducers, and this results in the production of large concentrations of volatile fatty acids, particularly acetic acid, and CO_2. The net effect is a reduction in pH value to 4–5 with a concomitant redox potential of 0 mV and below. The terminal stage is the production of CH_4 by obligately anaerobic methanogens and this is characterised by landfill temperatures between 30 and 40°C.

In order to examine the sequential microbial interactions which result in the production of CH_4, a multi-stage model system was developed consisting of aerobic and anaerobic components (Fig. 6).

For the aerobic component the culture vessel consisted of a glass tower fermenter, fitted with side ports, of working volume 13 litres. The base of the tower was filled with stones as a support for the domestic refuse. Mineral salts medium, supplemented with hexoic acid, was pumped up through the tower by means of a Watson Marlow Ltd. (Falmouth) flow inducer (MHRE/12) to maintain a constant dilution rate of 0.01/h.

In the anoxic component a portion of the tower effluent was introduced at a constant dilution rate (0.03/h) into the first vessel of a five-stage, anoxic, unstirred cultivation system by means of a Watson Marlow flow inducer (MHRE/2). Excess effluent from the tower was collected in a waste reservoir by use of a weir overflow system. The culture vessels (Fig. 7) were each of 0.7-litre capacity and were maintained at a constant temperature of 25°C by water jackets in conjunction with a Churchill thermocirculator (05-CTCV)

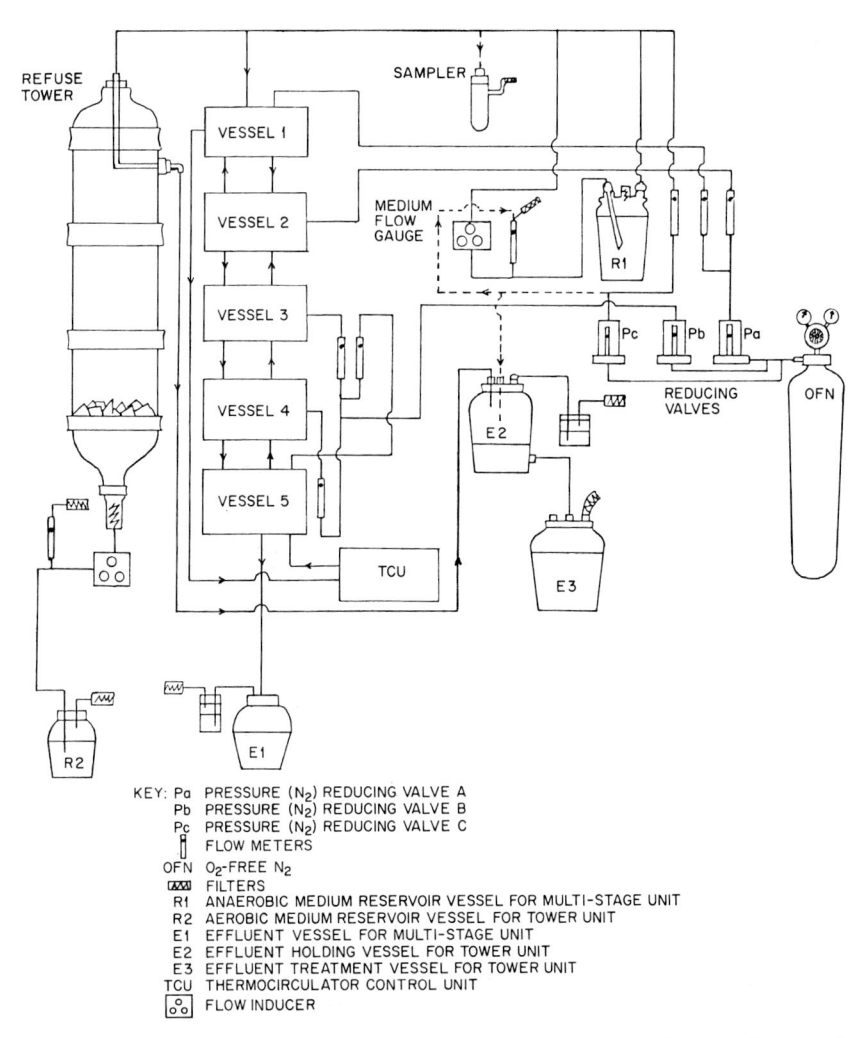

FIG. 6. Landfill multi-stage model consisting of aerobic (refuse tower) and anoxic (five-vessel) components.

(Churchill Instrument Co., Uxbridge). Each vessel was fitted with an angled effluent overflow tube, the height of which was calculated to give the required working volume. The pots were arranged on the gradostat principle so that the entire liquid effluent of the first vessel became the influent of the second and so on. Culture liquid from each pot was introduced via a centrally located medium input tube which allowed influent to be emitted into the vessels at their base only. The design ensured that the bulk flow charac-

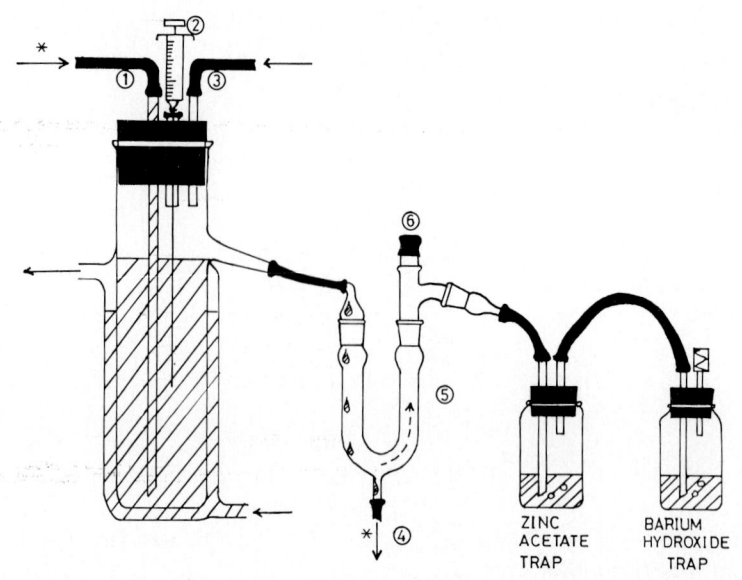

FIG. 7. A vessel of the five-vessel anoxic component of the landfill multistage model. 1, Culture inlet; 2, syringe for culture sampling; 3, OFN inlet; 4, culture for next vessel; 5, effluent OFN and culture gases for sampling; 6, gas sample port.

teristics of the system approximated to a continuous but segmented plug flow. Anaerobiosis was maintained by an overpressure of 0.1 atm OFN in conjunction with pressure heads of zinc acetate (1%, w/v) and barium hydroxide (18 g/litre) which were also used for trapping H_2S and CO_2, respectively. Oxygen-impermeable butyl rubber tubing [Esco (Rubber Ltd.), Teddington] was used throughout.

To facilitate the characterisation of individual fermentation balances within the component stages, independent throughputs and collections of gases were developed. Degasification of the effluent from each stage was effected by the U-tube arrangement so that no gas entered the succeeding vessel. Samples for gas analysis were aseptically withdrawn into gas syringes through the rubber diaphragms and similarly for culture analysis the samples were withdrawn into sterile syringes from the U-tubes. Effluent from the final vessel was collected in a 20-litre reservoir.

When attempting to examine the sequential interactions which result in the production of CH_4 in landfill, the researcher is faced with a somewhat daunting task. The approach employed in this study was first to consider the terminal processes of sulphate reduction and methanogenesis and then to work successively on each preceding step towards the aerobic phases.

The interactions between these two groups of organisms are particularly interesting and are also common to other environments such as anoxic sediments, in which it has been shown that spatial and/or temporal separation in terms of maximum activities may often occur (Barnes & Goldberg 1976; Reeburgh & Heggie 1977; Martins & Berner 1977; Jorgensen 1980; Reeburgh 1980; Senior *et al.* 1980). In these situations methanogenesis was apparently inhibited in the presence of active sulphate reduction. Thus, competition for common precursors, particularly electron donors, would appear to offer a possible explanation. However, it has since been suggested that anaerobic re-oxidation by sulphate-reducing bacteria may also be a contributory factor (Mountfort 1980; Devol & Ahmed 1981; Iversen & Blackburn 1981). In fact, it has been estimated that in anoxic ecosystems 30–75% of the downward sulphate flux may be consumed by CH_4 oxidation (Reeburgh 1980). Oxidation of CH_4 by methanogens could also be included as a minor factor (Zehnder & Brock 1979).

By use of the multi-stage anoxic fermenter described above, it was possible to spatially separate sulphate-reducing bacteria from methanogens, with a combination of carbon source limitation, electron acceptor limitation and specific growth rate (flow rate and culture volume), so that the exact interfaces may be examined in detail.

Conclusion

A range of fermenters from single to multi-stage has been developed to examine problems related to environmental biotechnology. The overall impression gained is that the cultivation system should be developed for the specific problem and that the researcher must be sufficiently versatile to meet this interesting challenge.

Acknowledgements

Grateful thanks are expressed to Professor W. C. Evans and Professor J. G. Kuenen for stimulating discussions and to Elly M. Bonnet, Karawan A. K. Al-Sarraj, Mastura Abdullah and Ann Toman for technical assistance, and to Glasgow City Corporation Waste Disposal Department and Essex Naturalists Trust for sampling permission.

References

BALBA, M. T. M. 1978 Part 1: The methanogenic fermentation of aromatic substrates. Part 2: The origin of hexahydrohippurate in the urine of herbivores. Ph.D. Thesis, University of Wales.

BALBA, M. T. M. & EVANS, W. C. 1977 The methanogenic fermentation of aromatic substrates. *Biochemical Society Transations* **5**, 302–303.

BALBA, M. T. M. & EVANS, W. C. 1979 The methanogenic fermentation of ω-phenyl alkane carboxylic acids. *Biochemical Society Transactions* **7**, 403–404.

BALBA, M. T. M. & EVANS, W. C. 1980a The methanogenic biodegradation of catechol by a microbial consortium: evidence for the production of phenol through *cis*-benzenediol. *Biochemical Society Transactions* **8**, 452–453.

BALBA, M. T. M. & EVANS, W. C. 1980b The methanogenic fermentation of the naturally-occurring aromatic amino acids by a microbial consortium. *Biochemical Society Transactions* **8**, 625–627.

BALBA, M. T. M., CLARKE, N. A. & EVANS, W. C. 1979 The methanogenic fermentation of plant phenolics. *Biochemical Society Transactions* **7**, 1115–1116.

BALBA, M. T. M., SENIOR, E. & NEDWELL, D. B. 1981 Anaerobic metabolism of aromatic compounds by microbial associations isolated from saltmarsh sediment. *Biochemical Society Transactions* **9**, 230–231.

BARKER, H. A. 1956 *Bacterial Fermentations.* New York: John Wiley & Sons.

BARNES, R. D. & GOLDBERG, E. G. 1976 Methane production and consumption in anoxic marine sediments. *Geology* **4**, 297–300.

COOPER, G. H. 1979 Disposal of town's waste. *Symposium on Solid Waste Disposal.* Glasgow: Institute of Public Health Engineers.

DAGLEY, S. 1978 Microbial catabolism, the carbon cycle and environmental pollution. *Naturwissenschaften* **65**, 85–95.

DEVOL, A. H. & AHMED, S. I. 1981 Are high rates of sulphate reduction associated with anaerobic oxidation of methane? *Nature, London* **291**, 407–408.

EVANS, W. C., DAGLEY, S., TRUDGILL, P. W., CHAPMAN, P. J., WILLIAMS, P. A., WORSEY, M. J., ORNSTON, L. N. & PARK, D. 1976 *Biochemical Society Transactions* **4**, 452–473.

FERRY, J. C. & WOLFE, P. S. 1976 Anaerobic degradation of benzoate to methane by a microbial consortium. *Archives of Microbiology* **107**, 33–40.

HARRISON, D. E. F. & WREN, S. J. 1976 Mixed microbial cultures as a basis for future fermentation processes. *Process Biochemistry* **11**, 30–32.

HEALY, J. B., JR. & YOUNG, L. W. 1979 Anaerobic biodegradation of eleven aromatic compounds to methane. *Applied and Environmental Microbiology* **38**, 84–89.

HOPPER, D. J. 1976 The hydroxylation of p-hydroxybenzaldehyde in *Pseudomonas putida*. *Biochemical and Biophysical Research Communications* **69**, 462–468.

HOPPER, D. J. 1978 Incorporation of (^{18}O) water in the formation of p-hydroxybenzyl alcohol by the p-cresol methylhydroxylase from *Pseudomonas putida*. *Biochemical Journal* **175** 345–357.

HUTNER, S. H. 1962 In *This is Life* eds. Johnston, W. H. & Steene, W. C., pp. 109–137. New York: Holt.

IVERSEN, N. & BLACKBURN, T.H. 1981 Seasonal rates of methane oxidation in anoxic marine sediments. *Applied and Environmental Microbiology* **41**, 1295–1300.

JORGENSEN, B. B. 1980 Mineralization and bacterial cycling of carbon, nitrogen and sulphur in marine sediments. In *Contemporary Microbial Ecology* eds. Ellwood, D. C., Hedger, J. N., Latham, M. J., Lynch, J. M. & Slater, J. H., London: Academic Press.

LAANBROEK, H. J. 1978 Ecology and physiology of L-aspartate- and L-glutamate-fermenting bacteria. Ph.D. Thesis, University of Groningen.

MARTINS, C. S. & BERNER, R. A. 1977 Interstitial water chemistry of anoxic Long Island Sound sediments. I. Dissolved Gases. *Limnology and Oceanography* **27**, 10–25.

MEYER, H. & BONDI, A. 1952 Lignin in young plants. *Biochemical Journal* **52**, 95–99.

MOUNTFORT, D. O., ASHER, R.A., MAYS, E. R. & TIEDJE, J. M. 1980 Carbon and electron flow in mud and sandflat intertidal sediments in Delaware inlet, Nelson, New Zealand. *Applied and Environmental Microbiology* **39**, 686–694.

PARKES, R. J. 1982 Methods for enriching, isolating and analysing microbial communities in laboratory systems. In *Microbial Interactions and Communities* eds. Bull, A. T. & Slater, J. H., New York & London: Academic Press.

POCHON, J. & TARDIEUX, P. 1967 In *The Ecology of Soil Bacteria* eds. Gray, T. R. G. & Parkinson, D., pp. 123–138. Liverpool: University Press.

REEBURGH, W. S. 1980 Anaerobic methane oxidation-rate depth distributions in Skan Bay sediments. *Earth and Planetary Science Letters* **47**, 345–352.

REEBURGH, W. S. & HEGGIE, D. T. 1977 Microbial methane consumption reactions and their effect on methane distributions in freshwater and marine environments. *Limnology and Oceanography* **22**, 1–9.

REES, J. F. 1980 Fate of carbon compounds in the landfill disposal of organic matter. *Journal of Chemical Technology and Biotechnology* **3**, 161–175.

SENIOR, E., BANAT, I. M., LINDSTROM, E. B. & NEDWELL, D. B. 1980 Interactions between organotrophic, methanogenic and sulphate-reducing bacteria in anaerobic saltmarsh sediments. I. Field Studies. *Society for General Microbiology Quarterly* **7**, 79.

STANIER, R. Y. 1953 Adaptation, evolutionary and physiological: or Darwinism among microorganisms. In *Adaptation in Micro-organisms* eds. Gale, E. F. & Davies, R., pp. 1–20, Society for General Microbiology Symposium No. 3. Cambridge: University Press.

TARVIN, D. & BUSWELL, M. 1934 The methane fermentation of organic acids and carbohydrates. *Journal of the American Chemical Society* **56**, 1751–1755.

VELDKAMP, H. 1970 Enrichment cultures of prokaryotic organisms. In *Methods in Microbiology* eds. Norris, J. R. & Ribbons, D. W. Vol. 3A, pp. 305–361. New York & London: Academic Press.

VELDKAMP, H. & KUENEN, J. G. 1973 The chemostat as a model system for ecological studies. In *Modern Methods in the Study of Microbial Ecology* ed. Rosswall, T., pp. 347–355. Stockholm: Swedish Natural Science Research Council.

WOLFE, R. S. 1972 Microbial formation of methane. In *Advances in Microbial Physiology* eds. Rose, A. H. & Wilkinson, J. F., Vol. 7, pp. 107–146. New York & London: Academic Press.

WOLFE, R. S. 1979 Microbial biochemistry of methane—a study in contrasts. Part 1: Methanogenesis. *Microbial Biochemistry, International Review of Biochemistry* **21**, 268–300.

ZEHNDER, A. J. B. & BROCK, T. D. 1979 Methane formation and methane oxidation by methanogenic bacteria. *Journal of Bacteriology* **137**, 420–432.

The Investigation and Analysis of Heterogeneous Environments Using the Gradostat

J. W. T. WIMPENNY AND R. W. LOVITT

Department of Microbiology, University College, Cardiff, UK

Faced with the undoubted complexity of microbial ecosystems some microbiologists study them *in situ;* others, usually practicing microbial physiologists, use the chemostat to investigate properties of one or a few species in detail in the laboratory. However, the chemostat is by definition homogenous, a feature not shared by the great majority of microbial ecosystems where heterogeneity and physico-chemical gradients are of paramount importance. Some examples of heterogeneous ecosystems have been reviewed by Wimpenny (1981, 1982) and by Wimpenny *et al.* (1983). They occur in deep, stably stratified oceans and lakes, aquatic sediments, soil profiles and soil crumbs. They appear as microbial films on animal surfaces as dental plaque, on the surfaces of roots and leaves of plants, and on many inanimate surfaces where they may be involved in blocking water pipes, fouling ship hulls and other marine installations and more usefully as the catalytic elements in effluent treatment trickling filters. The bacterial colony is a morphological entity whose structure is dominated by diffusion gradients.

Given the complexity of these ecosystems it seems necessary to simplify the experiments designed to give a better understanding of some of the properties of heterogeneous communities. However, simplification per se is not necessarily commendable since if a system is too simple only relatively trivial questions can be asked of it. Although simplification encompasses the art of model building, we believe that homogeneous laboratory models are too simple. As useful as the chemostat is, it can reveal little of importance concerning the vectorial reactions characteristic of heterogeneous ecosystems.

Until recently there have been very few laboratory models which incorporate heterogeneity. Among the oldest are percolating columns which have been reviewed by Bazin *et al.* (1976). Caldwell & Hirsch (1973) described a two-dimensional gradient plate for the cultivation of bacteria on steady-state gradients in agar plates. Wimpenny (1981) produced a simplified stopped-time-dependent gradient plate which is also capable of developing two-dimensional diffusion gradients. Wimpenny & Whittaker (1979), Wimpenny (1981) and Wimpenny *et al.* (1981) have described a gel-stabilised gradient system capable of generating opposing gradients of O_2 and substrate. Many model film fermenters have been used, predominantly by chemical engineers, to predict the behaviour of trickling filters used in effluent treatment (Atkinson & Fowler, 1974; Characklis, 1980). A number of model systems have been described to replicate dental plaque systems *in vitro*, e.g. Russel & Coulter (1975), and Coombe *et al.* (1982) have described a simple, constant-depth film fermenter designed particularly for the examination of the dental plaque ecosystem.

The most important advantage of homogeneous liquid culture devices such as the chemostat is that they are open systems capable of reaching steady states, a condition that allows unequivocal responses to environmental manipulations. Temporal but not spatial heterogeneity can be investigated in the single-stage chemostat. A good example is the response of photosynthetic bacteria to cycles of light intensity (van Gemerden 1974) or nutrient concentration (Beeftink & van Gemerden 1979). Multi-stage continuous culture systems incorporate spatial heterogeneity but flow is in one direction only and hence reciprocal interactions between neighbouring vessels are impossible. Margalef (1967) suggested the need for bidirectionally linked multi-stage systems and one was developed later for the investigation of estuarine ecosystems by Cooper & Copeland (1973).

Lovitt and Wimpenny have developed a bidirectional compound chemostat, the "gradostat" (Lovitt & Wimpenny 1979, 1981a; Wimpenny 1981, 1982), the construction and some properties of which are described in this chapter.

Description and Operation of the Gradostat

The gradostat (Figs. 1 and 2) consists of a series of usually five fermentation vessels linked together so that culture is transferred between neighbouring vessels in two directions at the same time. Solutes are fed into the system from reservoirs located at each end of the array, and there are receivers for waste culture from each of the end vessels. In the absence of microbial cells, the different solutes present in the two reservoirs will be distributed as opposing gradients across the whole array. Under steady-state conditions the solute gradients will be linear from each reservoir to each receiver provided

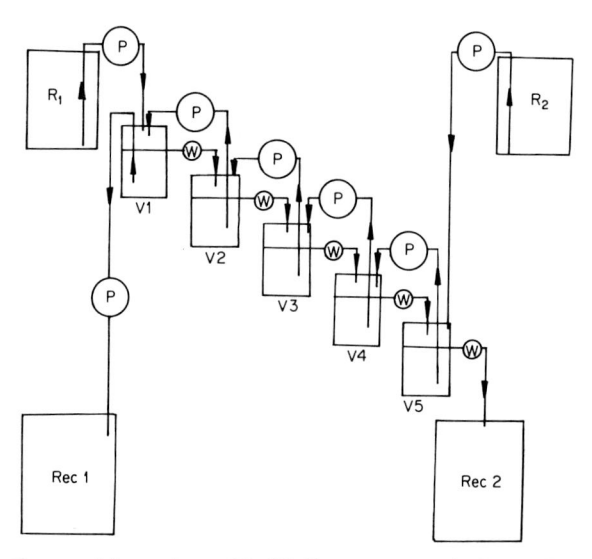

FIG. 1. A flow diagram of the gradostat. V1–V5, Fermenter vessels; R_1 and R_2, reservoirs; Rec 1 and Rec 2, receivers; P, pumped lines for medium; W, overflow weirs for culture.

that the volumes of each vessel and the flow rates in each direction are the same. The presence of cells acting as "sinks" for substrates or as sources of products alters the distribution pattern.

Construction of the Gradostat

The Vessels

Each vessel is basically a single fermentation system. We have used 1-litre QVF (Quickfit & Quartz) all-glass fermenters for four of five vessels and employed an LH Engineering (Slough Bucks UK) series 500 bench top chemostat as the fifth vessel to obtain high O_2 transfer rates where aerobic conditions were needed at one end of the array. Each glass vessel is equipped with a magnetic stirring bar and with a glass weir to allow liquid to flow from vessel to vessel in one direction only. All fermenters have sampling ports and temperature sensors, and are equipped with medium input and output lines which are fitted either with black butyl rubber tubing (Esco Rubber Co., London) or with Tygon tubing (Baird and Tatlock, London). Stainless steel tubing connectors were purpose made. Temperature and agitation rate are controlled in each vessel by purpose-built fermenter bases of standard design.

The Complete System

Each vessel is arranged on a Dexion stand which provides a series of platforms, each higher than the next. Medium flows by gravity from the

FIG. 2. The gradostat.

highest to the lowest vessel over the weirs, whose outlets enter the next lower vessel beneath its liquid surface to ensure that gas is not transferred up the series from the lowest vessel, which is normally aerated. Culture, fresh medium and waste culture are transferred by a multi-channel tubing pump (Technicon Autoanalyser, Technicon Instruments, New York, USA) using matched Technicon pumping tubes. Medium is held in sterilised 20-litre glass aspirator bottles (reservoirs). Waste culture is collected at both ends of the array through outlets connected to unsterile glass bottle receivers containing disinfectant. The gradostat is sterilised empty by autoclaving at 121°C for 20 min.

Operating the Gradostat

After autoclaving the sterile system is assembled on the stand and the pumping lines are inserted into the tubing pump. The system is filled with either inoculated culture or sterile medium by syphoning from the highest vessel and allowing it to flow through the system over the weirs. Where sterile medium is used the system is inoculated by injecting 5 ml inoculum into each vessel. Temperature control, stirring and agitation are started and the system is allowed to develop. Samples are removed using a sterile disposable syringe. The first sample which contains material from the unstirred dead region in the sampling line is discarded.

Transfer of Materials in the Gradostat

Equations Defining the Transfer of Material

To determine the concentration x_n of a solute in the nth vessel the following differential equation applies.

$$V\frac{dx}{dt}n = ux_{n-1} + vx_{n+1} - (u + v)x_n \quad (n = 1, 2 \ldots N) \tag{1}$$

where u and v are media flow rates through the vessels and V is the volume of each vessel.

If $x_0 = a$ and $x_{n+1} = b$, the concentrations of different solute in each reservoir, and the initial conditions are such that $x_n = 0$ at time $t = 0$ ($n = 1, 2 \ldots N$). Then solving the differential equation gives

$$x_n = \left(\frac{u}{v}\right)^{n/2} \frac{1}{N+1} \sum_{k=1}^{N} \frac{C_k}{\alpha_k} (e^{\alpha_k t} - 1) \sin\left(\frac{nk\pi}{N+1}\right) \tag{2}$$

where:

$$\alpha_k = 2\lambda\left(\cos\frac{k\pi}{N+1}\right) + \mu$$

$$C_k = 2\lambda\left[a + b\left(\frac{v}{u}\right)^{\frac{N+1}{2}} (-1)^{k+1}\right] \sin\left(\frac{k\pi}{N+1}\right)$$

$$\lambda = -\frac{\sqrt{u+v}}{V} \quad \text{and} \quad \mu = \frac{-u+v}{V}$$

Equation (2) defines the concentrations of a solute in a particular vessel at any time after pumping commences. However, as t tends to infinity the system tends to a steady state given by

$$x_n = E(u/v)^n + F \qquad (3)$$

where $E=(a-b)/\{1-(u/v)^{N+1}\}$ and $F=\{b-a(u/v)^{N+1}\}$. Where the flow rates u and v are equal, this simplifies to

$$X_n = A + (n+1)B \qquad (4)$$

where $A=\{(N+2)a-b\}/(N+1)$ and $B=(b-a)/(N+1)$.

Computer Predictions on the Effects of Flow Rates on Solute Concentrations

A series of computer simulations was run to determine the effects on solute concentration when the proportions of flow in each direction were varied from 1 (equal flow rates in each direction) to 10. The results (Fig. 3) show that under steady-state conditions, only where this ratio is unity are solutes distributed linearly from sources at each end of the array to receivers at the opposite end. The steady-state solute concentrations at other values show an interesting range. Although unequal flow has not been tested experimentally, it is clear that useful experiments, perhaps in challenging an ecosystem with a xenobiotic substance, could result from the use of a gradostat in this way. Alterations in proportional flow require slight changes in the construction of the gradostat. Thus the weir system would need to be replaced with a

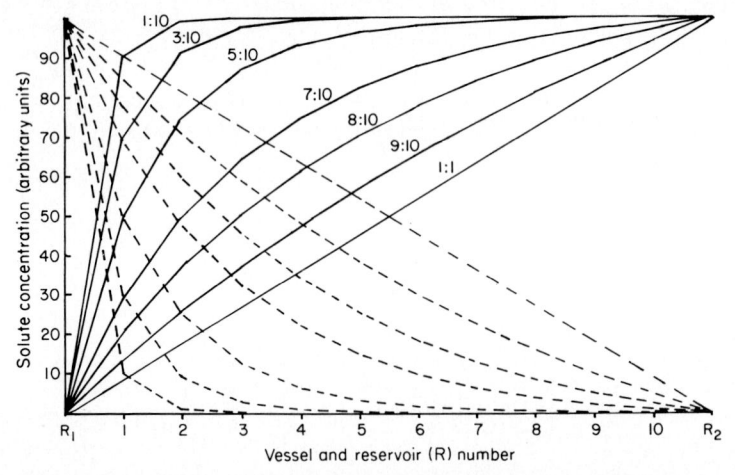

FIG. 3. Computer simulations of the distribution of two solutes in the gradostat at steady state at different flow rates. The flow ratios from the reservoirs range from 1:10 to 1:1.

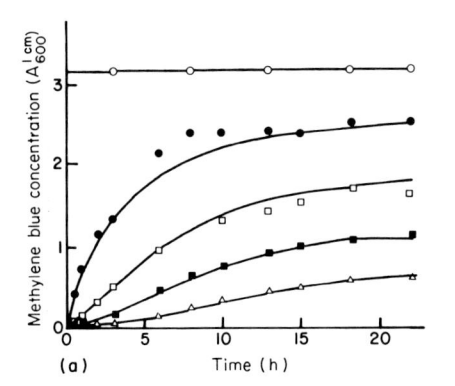

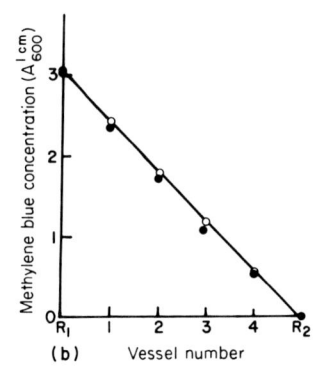

FIG. 4. Distribution of a dye (methylene blue) in each vessel of a four-vessel gradostat and in the reservoir (R). (a) As a function of time. Simulated values (continuous lines) and observed results (symbols). Vessels: 1 (●); 2 (□); 3 (■); 4 (△); R (○). (b) Steady-state conditions: theoretical (○) and observed (●) values.

series of pumped lines. One possible consequence is that slight variations in pumping rates due to variations in tube geometry may lead to changes in vessel volume if care is not taken.

An Experimental Verification of the Distribution Kinetics

A dye marker was used to test the distribution equations. Each vessel and one of the reservoirs contained distilled water. The second reservoir contained methylene blue, the concentration of which was determined in each vessel as a function of time after pumping started. Pumping rates and the volumes in each vessel were maintained as constantly as possible and carefully measured; the slight variations noted almost certainly explain the very slight differences between the theoretical and observed results (Fig. 4). It was concluded from these experiments that the system operates as predicted from theoretical considerations in both steady-state and non-steady-state cases.

Residence Times in the Gradostat

The residence time of any particle in a simple single-stage chemostat is an inverse function of the dilution rate of the vessel. Computer simulations of single- and multi-stage chemostats and of the gradostat (Fig. 5) show considerable differences between each system. In each case the dilution rate for the system as a whole is the same. Each vessel in each system contains the same initial concentration of particles. Individual vessels in a multi-stage array wash out sequentially as particle-free medium is pumped into the

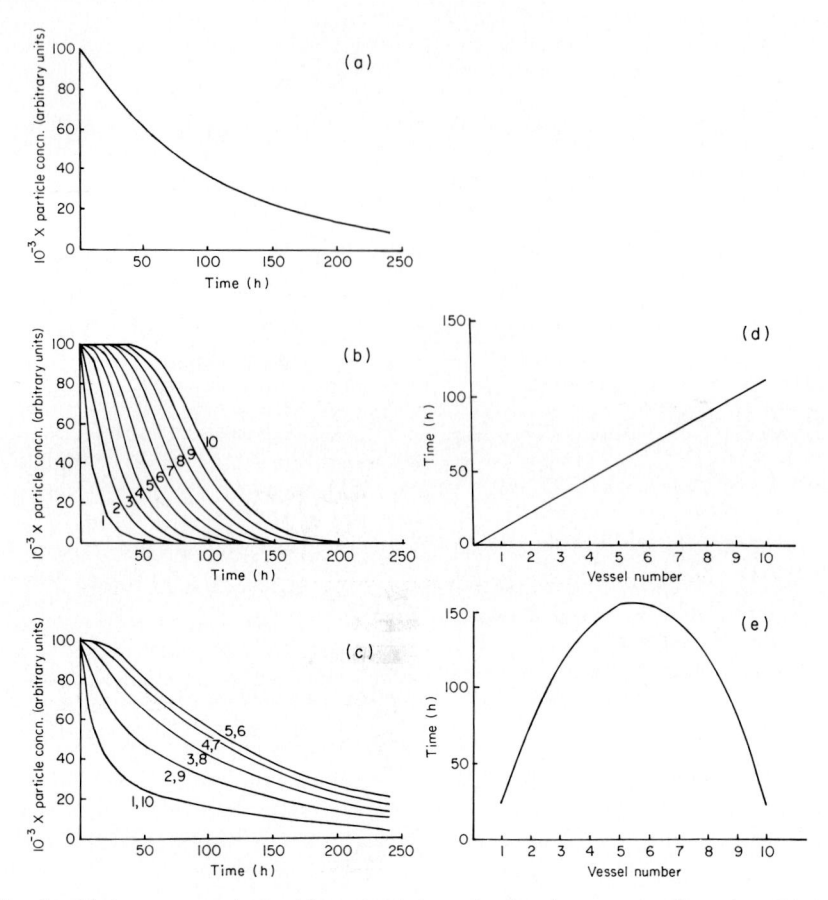

FIG. 5. Wash out curves obtained from simulations of various fermenter conformations: (a) the chemostat; (b) a 10-element multi-stage chemostat; (c) a 10-element gradostat. Each system was operated at an overall dilution rate of 0.01/h and initially contained uniformly distributed particles. Time required for cell concentrations to fall to 40% of their initial value in: (d) a multistage chemostat; (e) a gradostat.

system. The residence time for the system as a whole is approximately the same as for a single vessel having the same total volume as predicted by Herbert (1960). The situation is quite different in the gradostat, where the residence time of any one vessel depends on its position in the array. Residence times of vessels near the ends of the gradostat are shorter than residence times of the more central vessels. Again the average residence time for the gradostat as a whole is approximately the same as that for a single vessel of equivalent volume. This pattern of behaviour has important consequences for biological experiments where organisms may be growing near one end of the system (e.g., see p. 305).

Growth of Organisms in the Gradostat

Growth in Opposing Gradients of Two Essential Nutrients

Lovitt & Wimpenny (1981a) grew *Paracoccus denitrificans* in opposing gradients of a reducing agent plus carbon source (viz. succinate) from one end of the array versus an oxidant (viz. nitrate) from the other. Preliminary experiments had shown that provided that the concentrations of the two nutrients were balanced adequately, the organism grew in the centre vessel in a five-vessel array; cells were distributed away from this position approximately linearly towards each outlet whilst growth in the centre vessel (3) provided a sink for substrates from each reservoir. However, the results suggested that the concentrations of substrates were not symmetrical across the array because growth also occurred in vessel 2 though not in vessel 4.

Computer simulations which were used to predict growth patterns in the array related well to experimentally observed results despite the following limitations to the numerical model: The computer simulation is based on an absolute requirement for each solute, which is only used in the presence of the other to allow growth by simple Monod growth kinetics; the model neither incorporates any deviation from these kinetics, e.g. endogenous respiration or maintenance energy, nor allows for the accumulation of one solute in the absence of the other for the synthesis of endogenous storage materials. The model does not incorporate lag periods or the death of organisms in the absence of substrates.

Factors Determining Distribution of Growth in the Gradostat

Since the majority of microbial habitats are spatially organised and cells are growing often at very small concentrations of solute molecules, it is of considerable importance to determine which kinetic parameters are responsible for the precise spatial location of particular groups of organisms. Computer simulations have been used to examine some of these parameters for a 10-vessel array.

Importance of Growth Yield

The basic strategy for these simulations is that the organism is growing in opposing gradients of two essential nutrients. Different growth yield coefficients are included and all other parameters are constant. When the growth yield coefficients for the two nutrients are equal, growth is best in the centre of the array, i.e. in vessels 5 and 6 (Fig. 6). Alterations in yield coefficient change of position of maximal growth; the greater the cell yield on a given substrate, the further from the source of that substrate can growth take place.

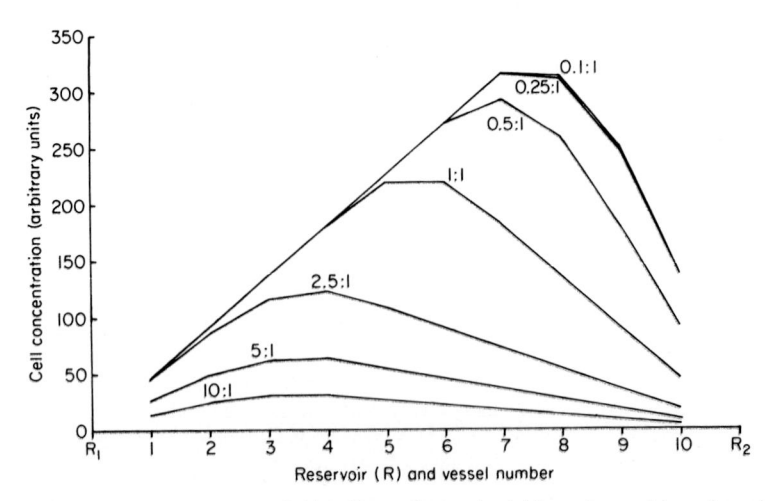

FIG. 6. A computer simulation of the effects of growth yield on the position of maximum growth in the gradostat. Ratios of the amount of each substrate needed to synthesise one unit of cells range from 10:1 to 0.1:1. A value of 1 represents a yield value of 0.02 mmole substrate/mg dry wt cells.

Effects of Substrate Affinity Coefficient

Similar simulations were made where the affinity of the organism for each substrate was varied over a wide range. The affinity coefficient (K_s) is the substrate concentration at which growth rates are half their maximum value (Monod 1949, 1950) and can be very small (of the order of micromolar). Unexpectedly, only a minor effect on the position of growth was observed for wide differences in K_s, i.e. 0.0002 − 2.0 mmole/litre for one substrate and a constant value of 0.02 mmole/litre for the other. This was probably because the cells were substrate limited in all but the centre vessels so that substrate availability rather than substrate affinity was the limiting factor. This is another important observation since opposing solute gradients at limiting concentrations are probably very common in nature. We must stress that these comments apply to a single species growing in opposing solute gradients; however, affinity coefficients are of great importance where two organisms are competing for the same nutrient supplies.

Effect of Growth Rate

Growth rate may be varied in the gradostat, as in a chemostat, by altering the dilution rate. Surprisingly, as the dilution rate is increased, growth in the central region of the gradostat becomes unrestricted by substrate concentration because, as the simulations show, the concentrations of both substrates increase markedly near the centre of the system (Fig. 7). As dilution rate is

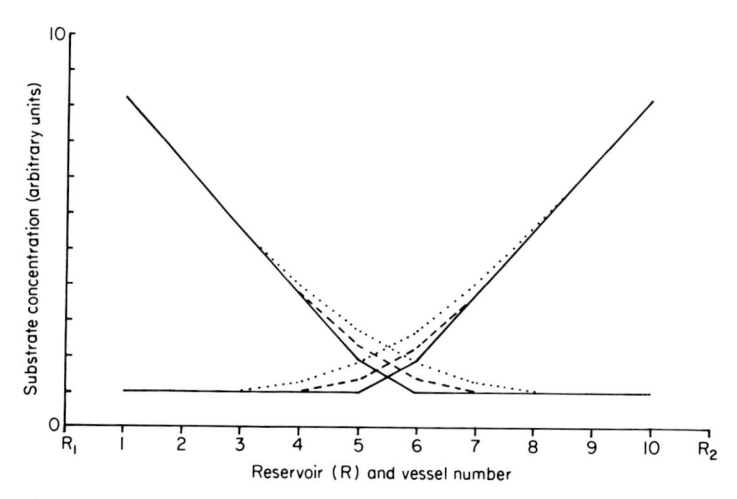

FIG. 7. A computer simulation of the effects of varying dilution (growth) rates in the gradostat on the steady-state concentrations of two essential substrates. Dilution rate (ml/hr): (—) 200; (---), 800; (· · ·), 1000.

increased further this region of unrestricted growth spreads towards the ends of the system. Wash out occurs only when the dilution rate *expressed on a per vessel basis* becomes much higher than the maximum specific growth rate of the organism. The effect described is related to the spread in residence times in the gradostat discussed in page 302. The much shorter residence times at the edges of the system clearly have consequences for persistence in the gradostat. Thus cells which, through competition, grow best at the ends of a system are much more likely to be washed out than those which occupy a more central vessel. Competition near the ends of the array will take place on the basis of growth rate rather than substrate affinity since the more slowly growing cells will be washed out of the gradostat. It has been suggested by Kuenen (1982) that gradients of essential nutrients are not possible in the gradostat since, if the nutrient is limiting, it cannot spread beyond the region of growth. Clearly this is true for any spatially organised natural ecosystem since nutrients diffusing vectorially from a source are unlikely to penetrate beyond a sink where demand exceeds supply. However, the simulations described suggest that gradients are possible over the whole gradostat when dilution rates are adequately high and growth is not nutrient limited.

Some Laboratory Experiments Using the Gradostat

Adaptation

Lovitt & Wimpenny (1981b) grew *Escherichia coli* in the gradostat in opposing gradients of O_2 and nitrate + glucose. The basal medium contained casamino acids and a mineral salts mixture. Although cells, together with culture medium, are transferred at equal rates across the five-vessel gradostat array, many adaptive changes were seen in cells removed from each vessel. Only cells from the most anaerobic vessel possessed the enzyme hydrogenase, which is a good marker for strict anaerobiosis. Cells from the aerobic end of the array possessed nitrate reductase, an enzyme that is normally repressed and inhibited by O_2 (Wimpenny 1969), suggesting that conditions there were oxygen limiting. Energy charge measurements indicated that cells near the centre of the array were either starving or inhibited by the large levels of nitrite which had accumulated. Incorporation of a "transfer" zone in which cells either cannot grow or are subject to toxic agents is another interesting possibility for the gradostat which distinguishes it from the single-stage chemostat. It is not possible to model starvation under steady-state conditions in the chemostat since growth must occur for the system to operate. Transfer zones can be envisaged in natural habitats where groups of organisms cannot grow because they lack one essential nutrient but through which other nutrients diffuse to regions where growth of other organisms is possible.

Growth and Viability

In a number of experiments which involved competitive interactions and enrichments from natural ecosystems (R. W. Lovitt & J. W. T. Wimpenny, unpublished data), it was noted that the cell population decreased exponentially from the vessels in which growth occured to other regions in the gradostat. As the distribution predicted from the equations on page 299 is a linear decrease in cell (or solute) concentration, the experimental results indicate that antagonism and growth inhibition are possible in the gradostat.

Competition

Competitive exclusion has been readily demonstrated in the gradostat. A steady-state culture of *Escherichia coli* in a casamino acids–yeast extract–salts medium in counter gradients of O_2 and glucose was displaced in the aerobic vessels by subsequent inoculation with a culture of *Pseudomonas aeruginosa* and then in the anaerobic vessels by the addition of *Clostridium acetobutylicum* to the system.

Enrichment Culture Technique

The gradostat is well suited to the enrichment culture technique because it provides a range of habitats each having a distinct physico-chemical identity. One such enrichment culture has been established by us to find a consortium of organisms capable of oxidising and reducing inorganic sulphur com-

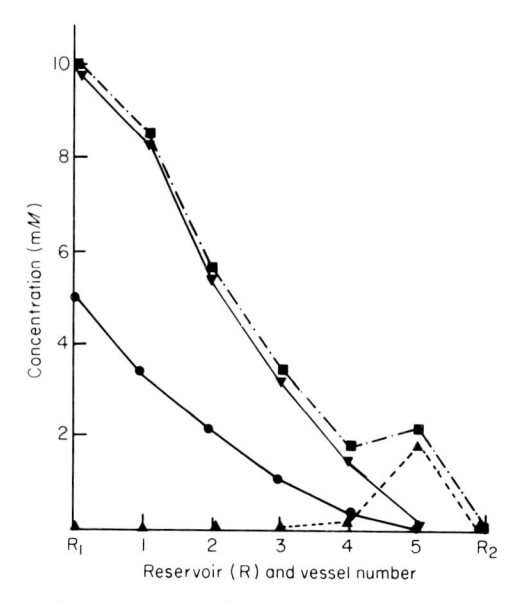

FIG. 8. Enrichment culture technique in the gradostat using lactate and sulphate + nitrate. R_1 contained 10-mM sodium sulphate and 5-mM sodium nitrate; R_2 contained 27-mM sodium lactate. Total sulphur (■); sulphide (▲); sulphate (▼); nitrate (●).

pounds. The basal medium was a salts solution to which lactate was added to the reservoir at one end and sulphate + nitrate to the reservoir at the other. The inoculum was an enrichment culture of water from the base of an oil storage tank. After incubation obligately anaerobic sulphate-reducing bacteria were isolated from vessels near to the lactate (reducing) reservoir and sulphide-oxidising nitrate-reducing bacteria were isolated from those near to the sulphate + nitrate (oxidising) reservoir. The distribution of sulphur compounds across the array gave evidence that a sulphur cycle was operating (Fig. 8).

Transient States

The lactate and sulphate + nitrate system gave an interesting example of the effects of the transient state between two steady states when concentrations of lactate were increased and those of sulphate + nitrate were reduced. The results (Fig. 9) show that concentrations of lactate in vessel 5 increase and of sulphate in vessel 1 decrease as might be expected. There is a large transient increase in sulphide concentration in vessel 4 whilst a smaller decrease in sulphide is recorded in vessel 5 before the new steady state is reached.

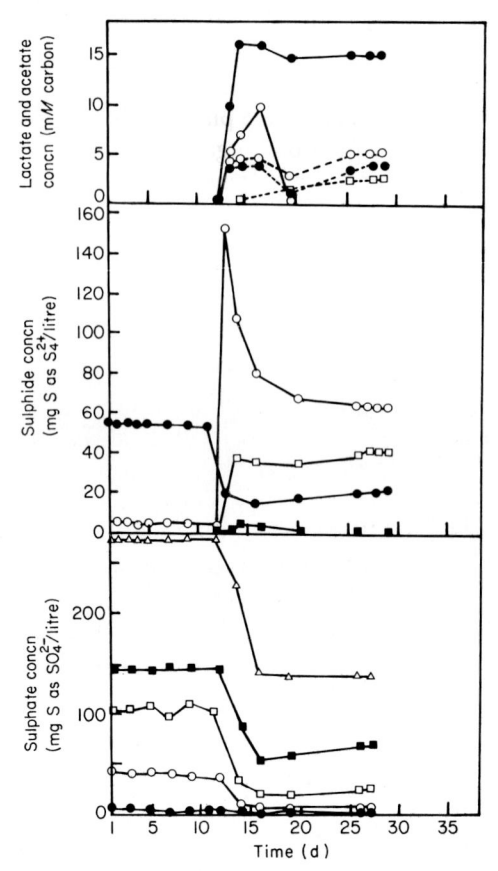

FIG. 9. Transient changes between two steady states in a lactate and sulphate + nitrate enrichment culture in the gradostat. Initially, the limiting concentrations of substrates were: lactate, 27 mM; sulphate, 10 mM; and nitrate, 5 mM. On day 12, they were changed to: lactate, 33 mM; sulphate, 5 mM; nitrate, 1 mM. Vessels 1 (△); 2 (■); 3 (□); 4 (○); 5 (●). Top: (—), lactate; (---), acetate.

Discussion

Possible Applications for the Gradostat

It seems likely that the gradostat will have a number of applications in pure and applied microbiological research. The gradostat can generate gradients of any desired shape, suggesting that it has valuable properties in genetic experiments designed to investigate the selection of mutant strains resistant to particular agents. For instance, if the ratio of flow rates in each direction is greater than unity, large concentrations of the agent will be present only in the first vessel providing a strong selective challenge, whilst the remaining vessels act as "sanctuaries" for sensitive cells. It should be possible to devise experiments using the gradostat where plasmids specifying particular characteristics can be transferred from one organism to another.

For work on, e.g. the sulphur cycle where obligate anaerobes interact with obligate aerobes via inorganic sulphur compounds, and the nitrogen cycle where nitrification and nitrate reduction require the presence and absence of oxygen, respectively, the gradostat provides conditions for the enrichment culture technique where microbial consortia can interact, but constituent species may require mutually exclusive habitats. Such experimental systems could provide a useful model for investigating the effects of xenobiotics on microbial ecosystems, a subject that is of great importance environmentally and one which is in need of more reliable experimental results (see also Senior and Balba, this volume).

The value of the gradostat in investigating the physiological responses of a single species to changes in the environment could be applied to the survival of organisms in habitats where essential nutrients are either absent or present only transiently and in small concentrations. The incorporation of transfer zones in the gradostat is a particularly valuable facility in this context and in learning more about survival in transfer zones in nature through which nutrients pass to other areas where growth may be possible.

In the competition experiment described on page 306 a "generalist" organism, i.e. *E. coli*, occupied a wide range of habitats whilst being the only organism in the system but was eliminated by the addition of "specialists," i.e. an obligate aerobe and an obligate anaerobe. This observation adds weight to a growing view, e.g. expressed by Kuenen & Harder (1982), that specialists will outgrow generalists in their own specific habitat whilst the generalist can dominate where the nature of the habitat permits a choice of metabolic routes. This and other forms of competition, e.g. amensalism (growth inhibition by the production of inhibitory substances) and even predator–prey relationships, can be investigated using the gradostat.

Other Possible Gradostat Conformations

The principle of bidirectional exchange can equally apply to a system where nutrient is entering from one end only. The bacterial colony growing aerobically on the surface of an agar plate and a narrow stratum in a marine sediment containing sulphur-oxidising bacteria being fed with nutrients from below and with oxygen from above can be represented by double-ended gradostat experiments. Bacterial film, however, forms in most cases on the surface of impermeable objects (e.g. ship hulls, rocks, steel structures in the sea, filter beds, human teeth) where solute exchange is through one surface only. Elimination of the reservoir and receiver from one end of the gradostat establishes a laboratory analogue of this type of system (Fig. 10a).

It is theoretically possible to construct a two-dimensional gradostat by connecting stirred vessels in two directions at the same time. The simplest

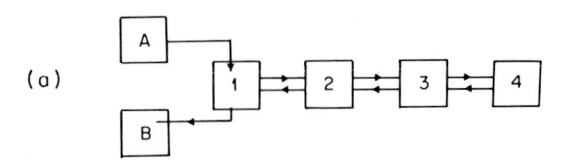

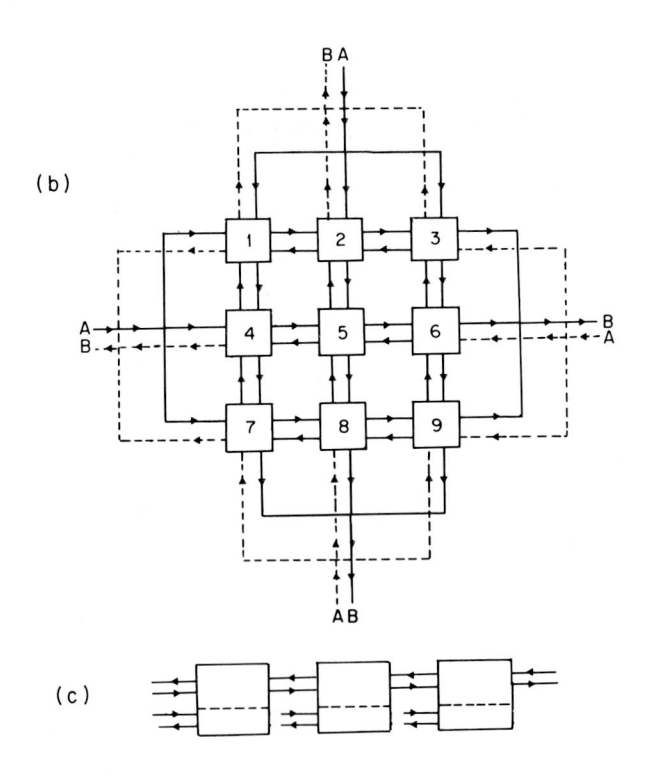

FIG. 10. Different gradostat conformations. (a) Single-ended version with input and output from same end of the array. (b) Two dimensional gradostat consisting of a 3 × 3 array of 9 vessels. (c) Heterogeneous form where cells and solutes are separated by a semi-permeable membrane; the cell chambers can be open or closed.

two-dimensional array consists of four vessels; the system would presumably get unmanagable if arrays larger than say 3 × 3 were needed (Fig. 10b). It ought to be possible to construct a scaled-down version of the gradostat using interconnecting square or circular chambers connected by diffusion couplers as originally used by Cooper & Copeland (1973). It is clear that adding another spatial dimension increases the degree of freedom of an experiment though it also increases its complexity.

It is also theoretically possible to separate cells from substrates using filter membranes (Fig. 10c). Here it is imagined that solutes would be transferred as usual from vessel to vessel by pumping, whilst the cells would remain associated with one vessel. This association either could be permanent in a closed system or the cell chamber itself could be open to solutions passing at right angles to the main gradostat flow. The latter version is intrinsically more attractive an idea since a true steady state is possible. R. A. Herbert (personal communication) has built a model system embracing some of these concepts and has investigated the growth of an association of three bacterial species in a three-vessel system. In the end vessel a clostridium fermented substrate to organic acids and alcohols which diffused into the centre vessel, allowing a sulphate reducer to reduce sulphate to sulphide which diffused into an illuminated chamber where it was re-oxidised to sulphate by an anaerobic photoorganotroph.

Such systems clearly have a major part to play in the analysis of microbial ecosystems that are important in environmental biotechnology.

Acknowledgements

We are grateful to Esso Research Limited, the Natural Environment Research Council and the Science and Engineering Research Council for grants supporting much of this work. It is a pleasure to thank our colleagues for practical help and for many valuable discussions: in particular Mr. S. Jaffe who wrote the gradostat simulation programmes and Mr. D. J. Harries for deriving the transfer equations.

References

ATKINSON, B. & FOWLER, H. W. 1974 The significance of microbial film in fermenters. *Advances in Biochemical Engineering* 3, 224–277.

BAZIN, M. J., SAUNDERS, P. T. & PROSSER, J. I. 1976 Models of microbial interaction in the soil. *CRC Critical Reviews in Microbiology* 4, 463–499.

BEEFTINK, H. H. & VAN GEMERDEN, H. 1979 Actual and potential rates of substrate oxidation and product formation in continuous cultures of *Chromatium vinosum*. *Archives of Microbiology* 121, 161–167.

CALDWELL, D. E. & HIRSCH, P. 1973 Growth of microorganisms in two-dimensional steady-state diffusion gradients. *Canadian Journal of Microbiology* **19**, 53–58.

CHARACKLIS, W. G. 1980 Fouling biofilm development: a process analysis. *Biotechnology and Bioengineering* **23**, 1923–1960.

COOMBE, R. A., TATEVOSSIAN, A. & WIMPENNY, J. W. T. 1982 Bacterial thin films as *in vitro* models for dental plaque. In *Surface and Colloid Phenomena in the Oral Cavity: Methodological Aspects* eds. Frank, R. M. & Leach, S. A., pp. 239–249. London: IRL Press Ltd.

COOPER, D. C. & COPELAND, B. J. 1973 Responses of continuous-series estuarine micro-ecosystems to point-source input variations. *Ecological Monographs* **43**, 213–236.

HERBERT, D. 1960 A theoretical analysis of continuous culture systems. *Society of Chemical Industry Monograph*, No. 12, pp. 21–53.

KUENEN, J. G. 1982 Discussion to paper by J. W. T. Wimpenny. *Philosophical Transactions of the Royal Society of London, Series B* **297**, 514–515.

KUENEN, J. G. & HARDER, W. 1982 Microbial competition in continuous culture. In *Experimental Microbial Ecology* eds. Burns, R. G. & Slater, J. H. pp. 342–367. Oxford: Blackwell.

LOVITT, R. W. & WIMPENNY, J. W. T. 1979 The gradostat: a tool for investigating microbial growth and interactions in solute gradients. *Society for General Microbiology Quarterly* **6**, 8.

LOVITT, R. W. & WIMPENNY, J. W. T. 1981a The gradostat: a bidirectional compound chemostat and its applications in microbiological research. *Journal of General Microbiology* **127**, 261–268.

LOVITT, R. W. & WIMPENNY, J. W. T. 1981b Physiological behaviour of *Escherichia coli* grown in opposing gradients of oxidant and reductant in the gradostat. *Journal of General Microbiology* **127**, 269–276.

MARGALEF, R. 1967 Laboratory analogues of estuarine plankton systems. In *Estuaries: Ecology and Populations* ed. Lauff, G. M., pp. 515–524. Baltimore: Hornshafer.

MONOD, J. 1949 The growth of bacterial cultures. *Annual Review of Microbiology* **1**, 371–394.

MONOD, J. 1950 La technique de culture continue; theorie et applications. *Annales de l'Institut Pasteur, Paris* **79**, 390–410.

RUSSELL, C. & COULTER, W. A. 1975 Continuous monitoring of pH and Eh in bacterial plaque grown on a tooth in an artificial mouth. *Applied Microbiology* **29**, 141–144.

VAN GEMERDEN, H. 1974 Coexistence of organisms competing for the same substrate: an example among the purple sulfur bacteria. *Microbial Ecology* **1**, 104–119.

WIMPENNY, J. W. T. 1969 Oxygen and carbon dioxide as regulators of microbial growth and metabolism. In *Microbial Growth* eds. Meadow, P. & Pirt, S. J., Society for General Microbiology Symposium No. 19, pp. 161–197. Cambridge: University Press.

WIMPENNY, J. W. T. 1981 Spatial order in microbial ecosystems. *Biological Reviews* **56**, 295–342.

WIMPENNY, J. W. T. 1982 Responses of microbes to physical and chemical gradients. *Philosophical Transactions of the Royal Society of London, Series B* **297**, 497–515.

WIMPENNY, J. W. T. & WHITTAKER, S. 1979 Microbial growth in gel stabilised nutrient gradients. *Society for General Microbiology Quarterly* **6**, 80.

WIMPENNY, J. W. T., COOMBS, J.P., LOVITT, R. W. & WHITTAKER, S. G. 1981 A gel-stabilised model ecosystem for investigating microbial growth in spatially ordered solute gradients. *Journal of General Microbiology* **127**, 277–287.

WIMPENNY, J. W. T., LOVITT, R. W., & COOMBS, J. P. 1983 Laboratory model systems for the investigation of spatially and temporally organised microbial ecosystems. In *Microbes in their Natural Environments* eds. Slater, J. H., Whittenburg, R. & Wimpenny, J. W. T., Society for General Microbiology Symposium No. 34, pp. 67–117. Cambridge: University Press.

Computer Control of a Photobioreactor to Maintain Constant Biomass during Diurnal Variation in Light Intensity

M. R. WALACH, H. H. M. BALYUZI,* M. J. BAZIN, Y.-K. LEE AND S. J. PIRT

Department of Microbiology and Department of Physics, Queen Elizabeth College, University of London, London, UK*

This chapter is an example of how biotechnology might be used to overcome some of the limitations of conventional agriculture. Computerised algal bioreactors allow greater control over photosynthesis. They remove the requirement for arable land, dependence on rainfall, large nutrient loss, CO_2 limitation, intermitent crop growth and low photosynthetic efficiency. Pirt (1981) and Pirt *et al.* (1983) described the operation of such a novel algal culture system, the function of which is to increase the rate of solar energy capture over that obtained by conventional agricultural techniques. The reactor consists of 52 m of glass tubing, through which the algae are recirculated, exposed either to the sun or, under the experimental conditions to be described, to tungsten lamps with spectral properties similar to those of natural light. A minimal salts medium is supplied to, and excess culture is removed from, the system by a single pump. Carbon dioxide at a concentration of 5% (v/v) in air is the sole carbon source and surplus gas, including O_2 produced by the photosynthetic activity of the algae, is removed by a degassing device to which the loop of tubing is connected and into which monitoring probes are inserted. The alga used is a *Chlorella*-like coccoid chlorophyte which is grown with three heterotrophic bacterial species. This consortium appears to be ecologically stable and resistant to contamination. The alga grows optimally at 37°C at pH 6.5.

In order to optimise the amount of light trapped by the algae, it is necessary to provide conditions which allow the culture to grow as rapidly as possible while the biomass concentration is large. These conditions should obtain even though environmental parameters may vary, in particular when the light intensity is fluctuating in a diurnal pattern. In order to operate the culture for extended periods of time, it is best to have some automated method of providing these conditions. Therefore, the algal reactor was interfaced with a microcomputer and a feedback control system was designed to regulate the biomass density and the rate of algal growth. This system will be described in this chapter.

The specific goal in controlling the algal culture using a computer was to maximise the rate of biomass formation under conditions of fluctuating light intensity. This was achieved by balancing algal production against the rate of harvest from the bioreactor.

The mass balance for the reactor can be written as

$$dx/dt = \mu(I_0)x - (F/V)x \quad (1)$$

where x is the algal biomass density, F the culture harvest rate, V the volume of the system, and $\mu(I_0)$ the specific growth rate of the algal population, a function of the incident light intensity, I_0. When $F = 0$, the algal population increases in density until an essential nutrient is exhausted or toxic products inhibit growth. By adjusting the rate of harvest so that the dilution rate D (equivalent to F/V) equals the specific growth rate of the culture, no net change in population density occurs, i.e.

$$\mu(I_0) = F/V \quad \text{and} \quad dx/dt = 0 \quad (2)$$

Under such conditions the steady-state biomass density, x', is independent of F and $\mu(I_0)$ and depends only on how long the culture was allowed to grow before being harvested. Therefore x' can be set by the operator of the system and the biomass density maintained at a specified value as long as production and harvest are balanced.

In order to use this strategy, a Research Machines Ltd (RML, P.O. Box 75, Oxford) 380Z microcomputer was attached, via an appropriate interface, to sensors and control devices in the reactor. Details of the electronics and programmes associated with computer control are given by Walach *et al.* (1983) and a diagrammatic representation of the system is given in Fig. 1.

Computation of Specific Growth Rates

In order to determine the specific growth rate (μ), which is the rate of growth per unit of biomass, it is necessary to be able to estimate the biomass density of the culture at regular intervals. The choice of method for measuring biomass in the system was limited by the requirement for a continuous, rapid

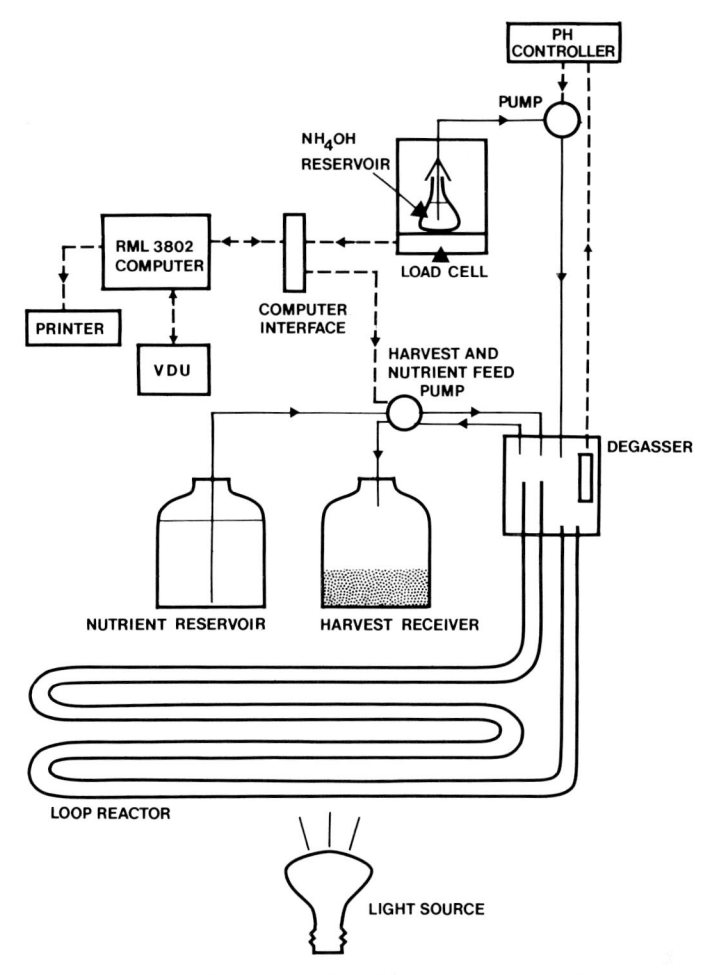

FIG. 1. Diagrammatic representation of the computer-controlled system.

estimate. Continuous turbidity measurements were found to be inadequate since the cells attached to the walls of the cuvette through which the light beam was passed. Therefore an indirect method based on the rate of nitrogen consumption by the algae was used. As shown in Fig. 2, under the conditions of operation, the amount of biomass formed is directly proportional to the amount of nitrogen utilised. This implies that cellular nitrogen remains a fixed proportion of the algal biomass and therefore the yield of biomass per unit of nitrogen source consumed (Y_N) is constant. As cells utilize NH_4OH as the sole nitrogen source, the pH value of the culture decreases. If more NH_4OH is pumped in to restore the pH to its original

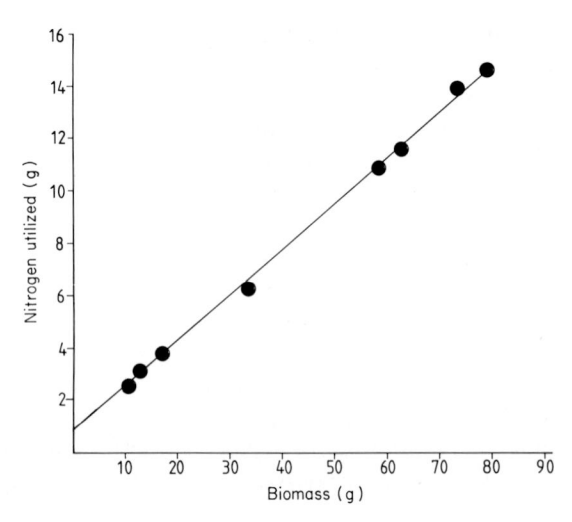

FIG. 2. Relationship between the amount of nitrogen utilised and the amount of biomass produced.

value and the amount is monitored, then it is possible to estimate the amount of biomass formed per unit time. Hence the specific growth rate of the culture may be determined.

In the system, a reservoir vessel containing NH_4OH is positioned on the pan of an electronic digital balance which is connected to a microcomputer via an interface. Ammonium hydroxide is pumped into the culture by a pH controller, the required pH value being set at 6.5. At half-hourly intervals the weight of reservoir + NH_4OH is recorded by the computer. The computer was programmed to estimate the rate of NH_4OH consumption and to calculate the current biomass density and specific growth rate of the culture using a value for Y_N previously determined. Information is then sent from the computer to set the nutrient/harvest pump at a speed just sufficient to keep the biomass density constant. Hence, in effect, the computer adjusts the dilution rate so that it equals the specific growth rate of the algae in the loop reactor.

Practical Considerations

Initially, several practical problems were encountered. It was found that if the feedline was physically attached to the NH_4OH reservoir, e.g. by means of a rubber bung, the recorded weight of the reservoir fluctuated, resulting in spurious estimates of the NH_4OH consumed by the algae. This was probably due to vibrations caused by the feedline when the NH_4OH was pumped out and also the varying effect of the feedline as the amount of NH_4OH in the reservoir decreased. The situation was remedied by passing a glass tube, to which the feedline was attached, down the neck of a narrow-necked reservoir vessel so that there was no contact between the two. The reservoir and the digital balance were then placed inside a Perspex containment cabinet to prevent the loss of NH_4OH to the atmosphere. Periodic checks of the concentration of NH_4OH in the reservoir vessel indicated that the containment cabinet prevented significant loss of NH_3.

To eliminate inaccuracies in weighing due to floor vibrations which were sometimes encountered, the NH_4OH reservoir, balance and containment cabinet were placed on a heavy concrete slab supported by rubber pedestals.

A safety device was incorporated into the system to alleviate the effect of computer failure. When this occurred and the nutrient/harvest pump was set at a high speed wash out could, and sometimes did, occur. Therefore, the main power supply to the nutrient/harvest pump was controlled by means of a relay. The relay is connected to a timer designed to switch off after each hour unless reset by a signal from the computer. The computer is programmed to send a signal to the timer at half-hourly intervals. Thus, in the event of computer failure, the timer would not reset and the relay would switch off the power to the nutrient/harvest pump.

Instrumentation

The language chosen for the bioreactor control programme was BASIC (RML BASIC version 5.0), although it was necessary to use Z80 assembler language for some of the routines connected with the transfer of data between the instruments and the computer. The programme employed, the interface boards and the instrumentation are described in detail by Walach *et al.* (1983). In addition to the variables necessary to estimate biomass density, temperature, light intensity, pH value, CO_2 partial pressure and gas flow rates were recorded by the computer. The results were automatically copied onto floppy disks thus providing a permanent record of each experiment and facilitating the analysis of the large amount of data generated.

Simulating the Day/Night Cycle

In view of the long-term aim of operating the photobioreactor using daylight, it was important to determine the behaviour of the system when it was subjected to diurnally varying illumination. A method for doing this was chosen which allowed the spectral properties of the experimental light source to be maintained while the intensity changed. The method consisted

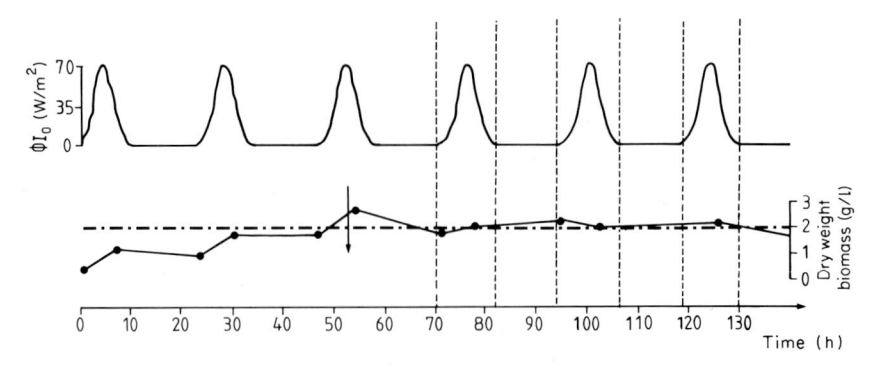

FIG. 3. An example of the control achieved under diurnal illumination. The arrow indicates the point at which the amount of NH$_4$OH equivalent to a biomass density of 2 g/litre had been pumped in, at which point the nutrient/harvest pump came under computer control. ϕI_0, radiation available for photosynthesis incident on the culture.

of interposing a motorised venetian blind between the light source and the loop reactor. The blind consisted of 17 vertical vanes mounted on a wooden frame and positioned in front of the reactor. The light intensity incident on the reactor was governed by the position of the vanes. The computer was connected to the motor which controlled the position of the vanes and a programme was written to set the position of the vanes and hence the intensity of light reaching the reactor. During experiments in which the light intensity was varied, the vanes were moved at half-hourly intervals over the course of a 12-h period in such a way that they were fully closed at the beginning, fully open after 6 h, and fully closed at the end of the time period. For the next 12 h they were kept closed. This cycle was used to simulate the day/night cycle of light intensity.

A profile of the light intensity reaching the reactor is shown in Fig. 3. Light intensity at the bioreactor surface was measured at several different positions so as to obtain a better estimate of the overall light intensity reaching it.

Experiments to Test the System

Two experiments were performed to test the efficiency of the computer at maintaining the biomass of the algal culture at a constant density. In the first experiment the culture was illuminated continuously and other environmental parameters were varied. In the second experiment the algae were subjected to a simulated diurnal variation in light intensity and other environmental parameters were constant.

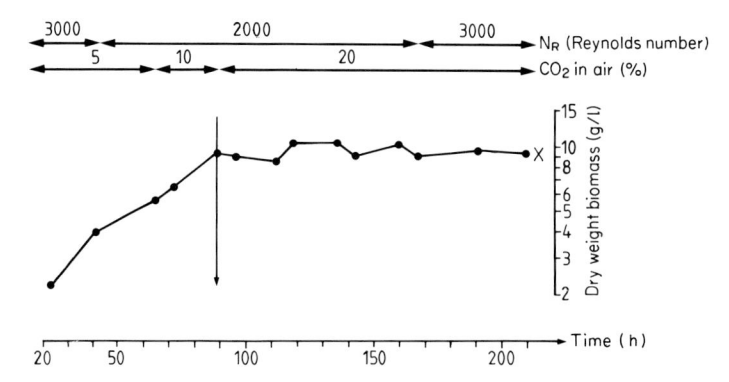

FIG. 4. An example of the control achieved under continuous illumination (40 W/m^2). The efficiency of control was tested by changing the CO_2 partial pressure and the turbulence of the system at the indicated times. The vertical arrow represents the point at which the amount of NH_4OH equivalent to a biomass density of 10 g/litre had been pumped in, at which point the nutrient/harvest pump came under computer control.

Continuous Illumination

In this experiment the computer was programmed to keep the biomass density of the culture at 10.0 g/litre, and after computer control had come into operation the effect of changing the CO_2 partial pressure and the turbulence of the system was tested. The results of the experiment are summarised in Fig. 4. After inoculating the bioreactor, the cells grew in batch culture until the amount of NH_4OH representing 10 g/biomass/litre had been pumped in. At this time (89 h) the computer activated the nutrient/harvest pump for the first time so that the dilution rate equalled the harvest rate and the current density of biomass was maintained. The efficiency of the control system was checked periodically by collecting samples for dry weight determinations. The mean dry weight of biomass from 89 to 220 h was 9.6 g/litre with a standard error of the mean of 0.2 g/litre. This variability is within the experimental error associated with dry weight estimations. Neither changing the degree of turbulence nor the CO_2 partial pressure by the amounts indicated in Fig. 4 appeared to affect the control system markedly.

Diurnal Illumination

The effect of diurnal illumination on the efficiency of the control system is shown in Fig. 3. Previous results indicated that if the inoculum culture was grown under constant illumination, a lag of 7–8 h was observed before nitrogen utilisation began. However, if the inoculum culture was first adapted to cycles of 12 h in the light followed by 12 h in the dark, then this lag period was reduced to ca. 2 h. The results shown in Fig. 3 were taken from an experiment in which such an adapted culture was used. The biomass density required was set at 2.0 g/litre and computer control started at 55 h after inoculation. From 55 to 121 h the mean dry weight biomass was 2.17 g/litre with a standard error of the mean of 0.12 g/litre. Since this variability is within experimental error associated with dry weight estimations, the results indicate that the computer control operated successfully.

This result is significant since in the natural environment, the rate of light absorption, and hence the algal specific growth rate, will reflect the fluctuating light intensities. In order to utilise as much of the available light as possible during periods when the light intensity is high, it is necessary to maintain large biomass concentrations. The computer-controlled system described achieved this result and is therefore one way in which the rate of light uptake by algal systems might be optimised.

Discussion

Computers are now commonly used for storing data supplied from fermentation systems, effluent purification processes and the natural environment. In many cases measuring devices have been linked with computers and data acquisition is an automatic procedure. However, feedback control systems based on some form of computer-calculated optimisation procedure still seem to be relatively rare. For example, at the 1982 international conference on *Computer Applications in Fermentation Systems* only one paper described a fully automatic computer-control system (Albrecht *et al.* 1982). One reason for the dearth of feedback control and optimisation systems associated with fermentation processes is that reliable sensors for the direct measurement of cell mass still do not exist (Cooney & Mou 1982). For this reason component balancing methods are used (Cooney *et al.* 1977; Zabriskie & Humphrey 1978a). Unfortunately, even this method is not always reliable. In our system, for example, we use the amount of nitrogen utilised by the algae to estimate the biomass density of the culture. The calculation involved assumes that the yield of biomass formed per unit of nitrogen consumed (Y_N) is constant. This assumption is only valid under a relatively narrow range of physiological conditions. For example, at low growth rates when nitrogen is limiting and CO_2 and light intensity are sufficient, the algae accumulate starch as a carbohydrate reserve material and the C:N ratio changes. This results in a change in Y_N and thus the calculation of biomass density based on nitrogen consumption alone is not valid. In the system described here we have the capacity to measure the carbon balance by monitoring the rates of flow and concentrations of CO_2 in and out of the bioreactor and therefore are able to compute biomass in terms of both nitrogen and carbon uptake. However, even the combination of these two estimates will not be completely reliable as the secretion of metabolic products into the culture medium is not taken into account. This phenomenon may be of significant proportions. Fogg (1975) reported that considerable amounts of glycollic acid are liberated by *Chlorella*. What is needed, therefore, is some more direct estimate of biomass density, preferably in association with component balancing methods. The most promising avenues for direct estimation of biomass seem to involve devices in which light is passed through a suspension of microorganisms and the amount of scatter or the culture fluorescence is recorded (Hatch *et al*. 1978; Zabriskie & Humphrey 1978b; Albrecht *et al*. 1982). It is to be hoped that routine and reliable methods for on-line estimation of microbial biomass density will soon be available.

References

ALBRECHT, CH., POHLAND, D., PRAUSE, M. & RINGPFEIL, M. 1982 Optimal control of SCP fermentation processes. In *Computer Applications in Fermentation Technology* pp. 191–197. London: Society of Chemical Industry.

COONEY, C. L. & MOU, D.-G. 1982 Application of computer monitoring and control to penicillin fermentation. In *Computer Applications in Fermentation Technology* pp. 217–225. London: Society of Chemical Industry.

COONEY, C. L., WANG, H. Y. & WANG, D. I. C. 1977 Computer-aided material balancing for prediction of fermentation parameters. *Biotechnology and Bioengineering* **19**, 55–60.

FOGG, G. E. 1975 *Algal Cultures and Phytoplankton Ecology*, 2nd ed. Wisconsin: University Press.

HATCH, R. T., WILDER, C. & CADMAN, T. W. 1978 Analysis and control of mixed cultures. *Second International Conference on Computer Applications in Fermentation Technology*. Philadelphia: University of Pennsylvania.

PIRT, S. J. 1981 Extension of the limits of photosynthesis by novel solar reactors. *Rivista di Biologia* **74**, 27–47.

PIRT, S. J., LEE, Y.-K., WALACH, M. R., WATTS-PIRT, M., BALYUZI, H. H. M. & BAZIN, M. J. 1983 A tubular bioreactor for photosynthetic production of biomass from CO_2: design and performance. *Journal of Chemical Technology and Biotechnology* **33B**, 35–58.

WALACH, M. R., BALYUZI, H. H. M., BAZIN, M. J., LEE, Y-K., & PIRT, S. J. 1983 Computer control of an algal bioreactor with simulated diurnal illumination. *Journal of Chemical Technology and Biotechnology* **33B**, 59–75.

ZABRISKIE, D. W. & HUMPHREY, A. E. 1978a Real-time estimation of aerobic batch fermentation biomass by component balancing. *American Institute of Chemical Engineers Journal* **24**, 138–146.

ZABRISKIE, D. W. & HUMPHREY, A. E. 1978b Estimation of fermentation biomass concentration by measuring culture fluorescence. *Applied and Environmental Microbiology* **35**, 337–343.

Viruses as Pathogens for the Control of Insects

Frances R. Hunter

Department of Microbiology, The University, Reading, UK

N. E. Crook

Glasshouse Crops Research Institute, Littlehampton, West Sussex, UK

P. F. Entwistle

Natural Environment Research Council, Institute of Virology, Oxford, UK

Viruses from seven families and a number of as yet unclassified groups are known to occur in insects (Table 1), in most instances causing lethal infections. Members of the Baculoviridae, Reoviridae (cytoplasmic polyhedrosis viruses), Parvoviridae and Picornaviridae have been shown to provide control of certain insect pests when applied to crops with conventional insecticidal spray equipment. Only one family, the Baculoviridae, is exclusive to invertebrates, occurring mainly in insects but also in some other arthropods. All other families contain members causing disease in humans or other vertebrates and as Harrap (1982) states, the molecular similarity of some of the insect viruses to vertebrate pathogens is close. For this reason, and because of their known potential, the development of insect viruses as biopesticides has tended to be concentrated on baculoviruses. This attitude was strongly endorsed by a joint FAO/WHO meeting on the use of insect viruses for pest control (World Health Organisation 1973).

At present, about 40 separate baculoviruses are being developed to control a wide variety of pests of field crops, grasslands, forests and stored products (Entwistle 1978). Fourteen examples are listed in Table 2 and areas of involvement by workers in the UK are indicated. Several baculoviruses have already been registered as pesticides and are available commercially; others are progressing rapidly towards this goal (Table 3). Baculovirus infections have been recorded in some mosquitoes but have not so far shown promise for control. A catalogue of viruses in insects, mites and ticks is given by Martignoni & Iwai (1981).

TABLE 1

Classification, nomenclature and basic properties of viruses pathogenic to insects and mites[*]

	Nucleic acid			
Virus group	Type[†]	No. of segments	Particle symmetry	Inclusion body
Family Baculoviridae				
Genus Baculovirus				
A: Nuclear polyhedrosis virus	ds DNA	1	Bacilliform	+
B: Granulosis virus			Bacilliform	+
C: Oryctes virus			Bacilliform	−
D: Polydisperse DNA genome	ds DNA	25		−
Family Reoviridae				
Genus Cytoplasmic polyhedrosis virus	ds DNA	10	Isometric	+
Family Poxviridae				
Subfamily Entomopoxvirinae Genera A, B and C	ds DNA	1	Ovoid or brick-shaped	+
Family Iridoviridae				
Genus Iridovirus	ds DNA	1	Isometric	−
Family Parvoviridae				
Genus Densovirus	ss DNA	1	Isometric	−
Family Picornaviridae				
Genus Enterovirus	ss RNA	1	Isometric	−
Unclassified small RNA viruses				
Divided genome group	ss RNA	2	Isometric	−
Nudaurelia β virus group	ss RNA	1	Isometric	−
Kelpfly virus group	ss RNA	1	Isometric	−
"Group 5" viruses	ss RNA	1	Isometric	−
"Minivirus" group	ss RNA	1	Isometric	−
Ovoid viruses	ss RNA	1	Ovoid	−
Family Rhabdoviridae				
Sigmavirus	Not known		Bullet-shaped	−
Unclassified viruses				
Drosophila X virus	ds RNA	2	Isometric	−

[*] Adapted from Payne & Kelly (1981).
[†] ds, Double stranded; ss, single stranded.

TABLE 2

Salient examples of current baculovirus control development programmes

Host		Virus
Common name	Formal name	
Agriculture		
Codling moth	*Cydia pomonella**	GV[†]
Cotton bollworm	*Heliothis spp.**	NPV
Cotton leafworm	*Spodoptera littoralis**	NPV
Potato tuber moth	*Phthorimaea opercullela*	GV
Cabbage looper	*Trichoplusia ni*	NPV
Cabbage moth	*Mamestra brassicae**	NPV
Rhinoceros beetle	*Oryctes rhinoceros**	Group C
Grass grub	*Wiseana* spp.	NPV
Forestry		
Gypsy moth	*Lymantria dispar*	NPV
Douglas fir tussock moth	*Orgyia pseudotsugata*	NPV
Spruce budworm	*Choristoneura fumiferana*	NPV
Pine sawfly	*Neodiprion sertifer**	NPV
Pine beauty moth	*Panolis flammea**	NPV
Stored products		
Indian meal moth	*Plodia interpunctella*	GV

* Programmes with UK involvement.
[†] GV, Granulosis virus; NPV, nuclear polyhedrosis virus.

TABLE 3

Commercial status of nuclear polyhydrosis viruses (NPV)

Host	Product name	Country
Registered		
Heliothis spp.	Elcar	US
Orgyia pseudotsugata	TM Biocontrol-1	US
Lymantria dispar	Gypchek	US
*Neodiprion sertifer**	Neochek-S, Virox	US/Canada, UK
Under safety test		
Neodiprion lecontei	—	Canada/US
Choristoneura fumiferana	—	Canada

* An NPV of this species is commercially available in Finland where it appears not to have been safety tested.
Note: In Japan the cytoplasmic polyhydrosis virus of *Dendrolimus spectabilis* has received some safety testing and is commercially available.

Characteristics of the Baculoviridae

Baculovirus Nomenclature

Subgroups A (nuclear polyhedrosis viruses, NPV) and B (granulosis viruses, GV) of the Baculoviridae are characterised by the presence of proteinaceous inclusion bodies within which either a single virus particle (GV) or many virus particles (NPV) are embedded. The matrix protein within which the virus particles are embedded is known as polyhedrin for NPV and granulin for GV. The polyhedral inclusion bodies (PIB) of NPV are up to 5 μm in diameter and may each contain several hundred virus particles (Figs 1 and 2a). Each virus particle may contain one (singly embedded, SNPV), or more than one (multiply embedded, MNPV), nucleocapsid. The inclusion bodies of GV are called capsules or granules and are ca. 400–500 nm × 200–300 nm. Each capsule normally contains a single virus particle (Figs. 1 and 2b). A much smaller subgroup (C) of baculoviruses includes *Oryctes* virus and possibly a few other viruses, all isolated from non-lepidopterous arthropods. Inclusion bodies are not produced but the virus particles resemble those of other baculoviruses.

Initially, the host range of individual baculoviruses was thought to be very restricted and this led to their being named after the insect from which they were originally isolated. Viruses which have been more extensively studied, however, have usually been found to infect at least two, and sometimes more, species within a genus. *Autographa californica* MNPV, for example, can infect insects from several genera.

Biochemical differences are frequently observed between isolates of virus from the same species and, in some instances, two isolates from a single species are not even closely related. Less reliance is being placed on morphological and biological criteria as the sole means of identifying viruses and more on biochemical characteristics.

Replication of Baculoviruses in the Insect

The susceptibility of larval Lepidoptera and sawflies to baculovirus infection declines with age, early instars (stages of development between moults) being most susceptible (Stairs 1965; Evans 1981). Decreases in susceptibility tend to be related to increase in weight such that the LD_{50}/mg body wt is often independent of host age. However, at some point, usually during the final feeding instar, susceptibility may depart markedly from this pattern to the extent that larvae become virtually resistant.

Infection occurs most commonly via the midgut following ingestion of inclusion bodies but can also be initiated by injection of virus into the

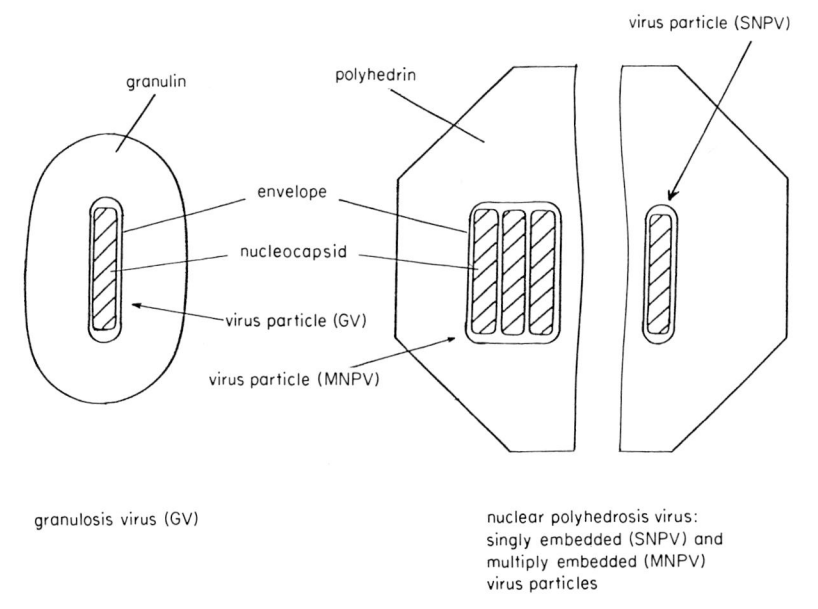

FIG. 1. Diagrammatic representation of the inclusion bodies of a granulosis virus and a nuclear polyhedrosis virus with virus particles shown in longitudinal section.

haemocoel, either by parasitoid wasps during oviposition, or by artificial means using non-occluded virus or dissolved inclusion bodies. When ingested, and under the alkaline conditions of lepidopterous and sawfly midgut, the inclusion bodies break down releasing virus particles. In Lepidoptera, the virus particles pass through the midgut cells, sometimes with a replicative phase, and enter the haemocoel. The major sites of replication are the nuclei of fat body, haemocytes and hypodermis. In sawflies, replication is restricted to midgut secretory cells. In the terminal phase of baculovirus infection, larval Lepidoptera usually become flaccid and the body wall is often very fragile. Provided cuticular pigmentation is not intense, infected larvae appear unnaturally pale. Sawfly larvae often show a pale band in the infected midgut region. Larvae of *Oryctes rhinoceros* (coconut rhinoceros beetle), infected by a baculovirus of group C, have turgid transluscent abdomens with, in the terminal phase, chalky white bodies appearing under the abdominal integument; the rectum may be prolapsed (Bedford 1981), *Oryctes* virus multiplies extensively in adult gut but does not appear to be transmitted from infected larvae through pupae to the adult stage (Zelazny 1976). Some NPV and GV infections may persist into the pupal stage of Lepidoptera; adults may have inapparent infections sometimes leading to overt disease in progeny. In sawflies the often prolonged prepupal resting stage is

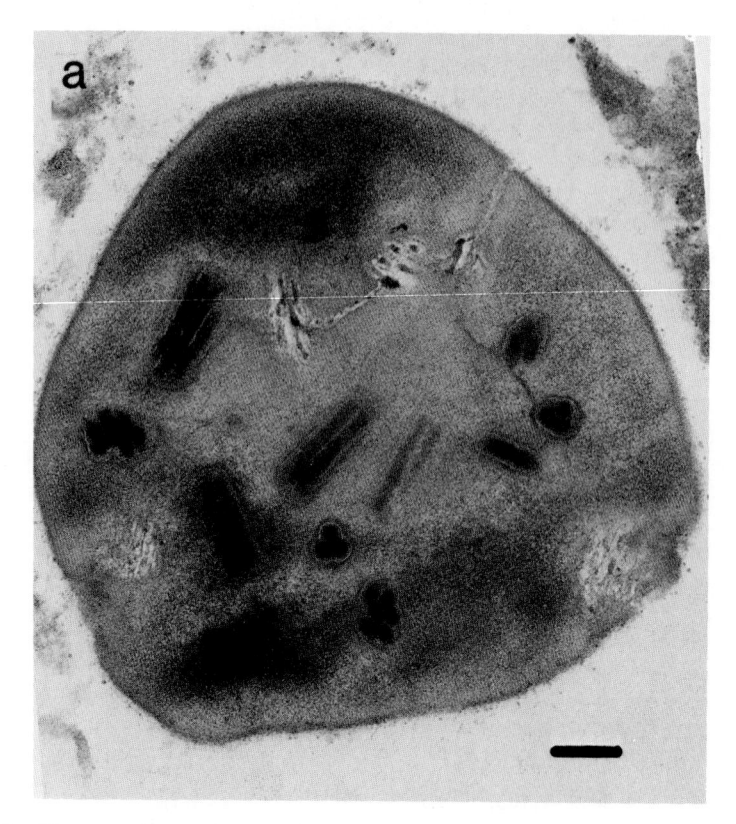

FIG. 2. Electron micrograph of (a) *Gonometa podocarpi* multiply embedded nuclear polyhedrosis virus (MNPV) and (b) *Laconobia oleracea* granulosis virus (GV). Bar = 200 nm in (a) and 100 nm in (b).

immune but replication has been observed in pupal and adult diprionid sawflies.

Major Ecological Aspects of Baculoviruses

Epizootics (disease affecting many animals of one kind in any region simultaneously) of baculovirus diseases are frequent in both Lepidoptera and sawflies often with very high larval mortalities resulting in strong population depression. Epizootic development usually requires several host generations and, apart from the obvious essential of high virus-infecting capacity (a conjunction of virus infectivity and productivity), the tendency of a baculovirus disease to become epizootic depends on the scale of virus dispersal (the spatial component) together with persistence of virus outside and within the host (the temporal component).

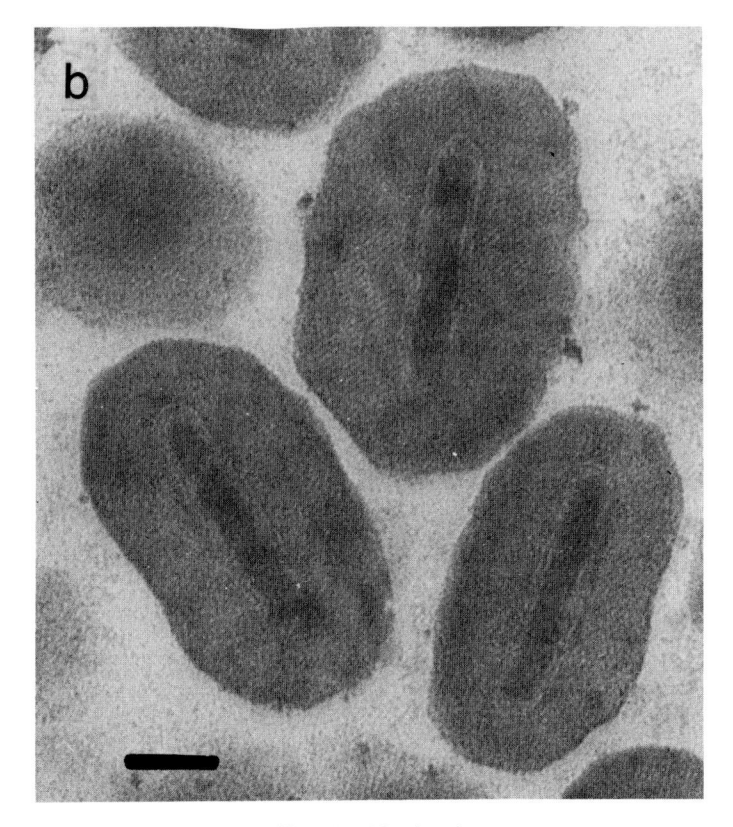

FIG. 2. (*Continued*)

Baculoviruses survive long periods in soil (Thompson *et al.* 1981). Methods of extraction and assessment of PIB from soil have been described by Evans *et al.* (1980). The recycling of virus from soil occurs by movement of animals from the soil to plants, by rain splash and by wind-blown soil particles and PIB onto leaves.

Residues of baculoviruses on plants seem especially important in the dynamics of individual epizootics, but reservoirs of baculovirus in soils probably have a more long-term importance, perhaps initiating fresh epizootics when insect populations resurge following a phase of low density (Entwistle 1976; Evans & Entwistle 1982).

Strain Differences and Resistance

Few studies have been reported of variation in baculoviruses either in terms of searches for naturally occurring variation in the field or through selective processes in the laboratory (e.g. Ossowski 1960; Shapiro & Ignoffo 1970;

Aizawa 1971). Only recently have attempts been made in the laboratory to clone NPV from naturally occurring isolates, e.g. from the wide host spectrum NPV of *Autographa californica* through plaque isolation [D. C. Kelly, Natural Environment Research Council (NERC) Institute of Virology, Oxford, personal communication].

Change in levels of host susceptibility to infection during the course of epizootics has been observed and may be a frequent phenomenon. The adequate demonstration of resistance requires the conduct of comparative log dose–probit mortality tests of susceptible populations and those suspected of being resistant. The interpretation of such tests and the pitfalls of single-dose comparisons, which are inadequate for resistance assessment, are discussed by Burges (1971). It is, of course, vital to determine if differences observed are, or are not, of a cyclical epizootic-associated nature.

Whilst the absence of clear examples of resistance to baculoviruses is encouraging, there has been insufficient experience of the long-term response of pest populations to repeated virus applications to be able to assess the possibility of control breakdown.

Laboratory Practice

In the UK a licence is required for importation of, and work with, non-indigenous plant pests. Facilities for working with such pests must be approved by the Ministry of Agriculture, Fisheries and Food (MAFF). Enquiries and application for a license should be made to the Plant Health Administrative Unit, MAFF, Eagle House, 90–96 Connor Street, London EC4N 6HT.

Uninfected stocks of insects for rearing purposes should be maintained at as great a distance as possible from those infected with virus. Ideally, they should be housed in separate buildings. The use of different-coloured laboratory coats for handling infected and uninfected insects helps to maintain discipline and prevent unintentional transfer of virus. Special precautions should be taken when handling larvae likely to cause allergic reactions.

Insects are most conveniently handled in plastic containers, preferably disposable. All unwanted cultures should be placed at $-20°C$ for at least 1 h to kill living insects. Containers which have been used with virus and are not to be re-cycled should be autoclaved at 121°C for 20 min before disposal. If they are to be re-used, their contents should be removed and autoclaved at 121°C for 20 min to inactivate virus. Containers should be soaked overnight in a 5% (v/v) aqueous solution of Chloros (industrial sodium hyporchlorite, ICI plc), rinsed in tap water, soaked in detergent solution, rinsed several times in tap water and twice in distilled water. A 10% (v/v) solution of formalin may be used instead of Chloros but is less easy to handle and the two disinfectants should not be used together as a carcinogenic product is formed. Paint brushes and instruments used for handling insects should be stored over formalin in a desiccator jar.

All work with infected insects and virus should be done on plastic trays, preferably within a Porton-type safety cabinet, to reduce the risk of cross-contamination. The trays can be decontaminated by soaking overnight in 2–5% (v/v) Chloros solution. Contaminated instruments are most easily sterilised in an instrument steriliser.

Rearing Insects in the Laboratory

Insects may be reared either on natural foods or on artificial, semi-synthetic diets. Under either conditions it is essential to avoid contamination with insect pathogens including unwanted viruses. Rearing rooms should be periodically sterilised overnight with formalin vapour, livestock having been removed first. Rooms must be thoroughly ventilated before being re-used. Except where pathogenic microorganisms are passed "vertically" (from one insect generation to the next) *within* the egg, it is usually possible to eliminate disease from cultures. Lepidopterous eggs may be surface sterilised by immersion in 10% (v/v) formalin for 25 min followed by rinsing in sterile water and drying in air on sterile absorbent tissue. Sawfly eggs are deposited in plant tissue, but for *Neodiprion sertifer* it has been shown that eggs on foliage will survive washing in 10% (v/v) formalin for up to 45 min. Nuclear polyhedrosis viruses on foliage may be inactivated by continuous agitation in warm 0.1-M Na_2CO_3 solution for 15 min followed by rinsing in water for 1 min (Kaupp 1981).

The fresh condition of cut foliage may be greatly prolonged if the ends of stems are surface sterilised by dipping in ethanol (and flaming, for woody tissue) and placing in containers of sterile water closed with sterile cotton wool plugs. This delays clogging of conductive tissue by bacteria. The procedure may be repeated by cutting off a few centimetres of stem and re-sterilising.

Microbial contamination of artificial diet must be avoided. Groups of young larvae can usually be fed on layers of diet poured into disposable plastic containers. As they grow, larvae of some species become cannibalistic, e.g. *Heliothis*, and must then be reared individually. Maintenance of larvae under conditions which do not lead to pupal diapause or selecting for non-diapausing strains eases problems of insect production. Singh (1977) has reviewed artificial diets and details are given in the Appendix of a variant of Hoffman's Tobacco Hornworm diet (modified by C. F. Rivers and R.

Warner of the NERC Institute of Virology, Oxford) suitable for rearing a wide range of Lepidoptera, a simplified diet for rearing the codling moth, *Cydia pomonella*, and one for rearing pests of stored products. Sawflies have so far been maintained only on natural foods.

Propagation and Purification of Virus

The production of baculoviruses varies greatly with larval age but apparently is consistent for larval unit weight; e.g. irrespective of instar, larvae of *Mammestra brassicae* yield ca. 8×10^6 PIB/mg dead body wt. Whilst large larvae yield numerically the largest numbers of PIB, they require relatively greater doses to induce infection. Thus the productivity ratio (PR = no. PIB to give 99.9% infection:no. of PIB yielded at death) for *Mammestra brassicae* declines with larval size from 8300 for first instar larvae to 1300 for fifth instar larvae. Taking the productivity of a fifth instar larva as unity, the number of first instar larvae required for equivalent virus production (EVP) in *Mammestra brassicae* is ca. 12,500 (there is an approximately log–linear relationship between EVP and instar) (Evans *et al.* 1981). Thus in practical production of baculoviruses, the advantage of the larger PR of young larvae must be set against the problems of handling greater numbers of small larvae. An additional consideration is that larval supply is usually finite; it is therefore common practice to infect larvae at the stage (often about late third instar) which will result in maximum productivity per individual at death (Table 4).

Infected larvae are commonly harvested shortly before the anticipated time of death. Although yields of virus from dying larvae may be less than from dead larvae (Ignoffo & Shapiro 1977), the latter are generally very fragile and may rupture with loss of virus during handling. In addition, the purification of inclusion bodies is easier and quality of the final product better if the larvae have not putrefied or melanised. Where the objective is to

TABLE 4

Yield of baculovirus inclusion bodies

Virus	Host	No. of species considered	Mean yield/larva
NPV*	Sawflies	3	1.3×10^8
NPV	Lepidoptera	5	5.8×10^9
GV	Lepidoptera	2	1.9×10^{11}

* NPV, Nuclear polyhedrosis virus; GV, granulosis virus.

Protocol	Relative velocities	
	Nuclear polyhedrosis virus	Granulosis virus
Weigh and macerate infected larvae in an equivalent volume SDS.* ↓		
Filter through muslin to remove gross debris. ↓		
Centrifuge to remove large contaminants. ↓	100 g, 0.5 min	400 g, 5 min
Resuspend pellet and repeat centrifugation. Combine supernatants and discard final pellet. ↓		
Centrifuge to remove lipid, soluble material and small contaminants. ↓	2500 g, 10 min	10,000 g, 30 min
Discard supernatant. Resuspend pellet in small volume SDS. ↓		
Centrifuge on rate-zonal gradients 30–80% (v/v) glycerol in SDS. ↓	1500 g, 8 min	12,000 g, 40 min
Collect bands containing virus, dilute at least 1/2 with SDS and centrifuge. ↓	2500 g, 10 min	10,000 g, 30 min
Discard supernatant and resuspend pellet in 0.1% SDS. Centrifuge on isopycnic gradients 45–60% (w/w) sucrose layers in SDS. ↓	50,000 g, 60 min	50,000 g, 60 min
Collect bands containing virus, dilute with SDS and centrifuge. ↓	2500 g, 10 min	10,000 g, 30 min
Discard supernatant, wash inclusion bodies twice by resuspending in sterile distilled water and centrifuging. Resuspend in distilled water. ↓	2500 g, 10 min	10,000 g, 60 min
Estimate protein concentration by the method of Lowry et al. (1951). Store virus at −20°C.		

* 0.1% (w/v) sodium dodecyl sulphate.

FIG. 3. Purification of baculovirus inclusion bodies.

make final preparations by freeze-drying and grinding, such early harvesting may be essential since a higher than stipulated count of bacteria would contravene the regulations of registering authorities.

Schemes for purifying inclusion bodies, virus particles and nucleocapsids are shown in Figs 3, 4 and 5, respectively. Purified inclusion bodies can be

---

Dissolve inclusion bodies at a concentration of 10 mg/ml in 0.05-M Na$_2$CO$_3$ at 37°C for 30 min.
↓
Centrifuge on 10–50% (w/w) sucrose gradient at 50,000 g for 45 min.
↓
Collect band (GV and SNPV) or bands (MNPV) corresponding to 1, 2, 3 or more nucleocapsids; centrifuge on 30–60% (w/w) sucrose gradient at 50,000 g for 3 h.
↓
Collect band(s), dilute in sterile distilled water and sediment virus particles by centrifugation at 50,000 g for 1 h.
↓
Resuspend in desired volume of sterile distilled water.

Note: Subgroup C baculoviruses are first treated to one cycle of low and high speed centrifugation.

FIG. 4. Preparation and purification of baculovirus particles. Note that subgroup C baculoviruses are first treated to one cycle of low- and high-speed centrifugation.

Incubate ca. 1 mg virus particles/ml with 0.05-M Tris–HCl buffer, pH 7.4 containing 1% (w/v) Nonidet P40 (BDH Chemicals Ltd) for 30 min at 30°C.
↓
Centrifuge on 10–45% (w/w) sucrose gradient in 0.05-M Tris–HCl buffer, pH 7.4, for 45 min at 40,000 g.
↓
Dilute band of nucleocapsids in sterile distilled water and sediment at 50,000 g for 60 min.
↓
Resuspend nucleocapsids in distilled water.

FIG. 5. Preparation of baculovirus nucleocapsids.

stored in suspension for several years at −20°C with little loss of infectivity. Neither virus particles nor nucleocapsids can be stored for long. They should be used as soon as possible after being purified.

Identification and Characterisation of Baculoviruses

Identification and characterisation of baculoviruses is essential to their development as pest control agents. Such studies are required by registering authorities, form the basis of diagnostic tests for virus and virus components in target and non-target cells and tissues and permit meaningful discussion in the scientific community.

The structural properties of insect viruses and methods for their identification have been reviewed in detail by Harrap & Payne (1979). The most important methods used for identification and characterisation of baculoviruses are restriction enzyme analysis of the DNA, electrophoresis of virus proteins in sodium dodecyl sulphate (SDS)–polyacrylamide gels and serological tests.

Restriction Endonuclease Analysis

The baculovirus genome is a supercoiled circular molecule of double-stranded DNA with a molecular weight of 60×10^6–100×10^6. Analysis of the DNA using restriction endonucleases is the most precise method for virus identification. Examination of the electrophoretic profile of DNA fragments produced by a single enzyme is usually sufficient to distinguish between even closely related strains of virus. It is not usually necessary to purify virus particles before extracting the DNA. A suitable protocol is as follows: 20-mg inclusion bodies are incubated in 1 ml 0.05-M Na_2CO_3 at 37°C for 30 min and then for a further 1 h at 37°C after addition of 100 µl 10% (w/v) SDS. The DNA is purified by extracting three times with an equal volume of phenol saturated with 10-mM Tris, 1-mM EDTA, pH 8.0, and once with an equal volume of chloroform:isobutanol (24:1). The DNA may either be precipitated by addition of 2 volumes of ethanol or dialysed against 10-mM Tris, 1-mM EDTA, pH 8.0.

The restriction endonuclease most commonly used for identification is *Eco*RI, which typically produces 10–20 fragments, mostly with a molecular weight of 1×10^6–20×10^6. *Hin*dIII and *Bam*HI have also been widely used and yield similar size distributions of fragments. Fragments are normally produced in equimolar proportions but submolar bands have been found for some viruses (Smith & Summers 1978). Further investigation has shown such viruses to be a mixture of variants which can be individually isolated by cloning. The submolar bands are then found to be molar bands from one of the lesser components in the mixture.

Electrophoresis in SDS–Polyacrylamide Gels

This is the standard technique for analysing virus structural proteins and the discontinuous Tris–HCl buffer system of Laemmli (1970) is most frequently used with gel concentrations of 10–12% (w/v) acrylamide. Electrophoresis of whole inclusion bodies produces only a single band (molecular weight ca. 3×10^4) since, unless the gel is heavily loaded, only the matrix protein will be detected. If the inclusion bodies are first dissolved in carbonate buffer, endogenous alkaline proteases digest the polyhedral protein resulting in smaller molecular weight bands corresponding to breakdown products. Between 15 and 25 polypeptide bands with molecular weight 12×10^3–100×10^3 are normally detected by electrophoresis of purified virus particles. Although there is much variation in the structural proteins of different viruses, a major band of low-molecular-weight basic protein is common to many of them and is associated with the nucleocapsid.

Serology

Serological methods are very useful for detecting and identifying baculoviruses especially when large numbers of samples must be tested or for work in field stations with limited facilities.

Since baculoviruses do not give rise to infection in vertebrates, they must be injected in relatively large amounts to produce an immune response. Care must be taken to remove contaminating antigens which, if present, will also give rise to antibodies. For this reason, virus immunogens must be very pure for production of antisera.

Radioimmunoassay (RIA), immunoradiometric assay (IRMA) and enzyme-linked immunosorbent assay (ELISA) are now most commonly used for studying baculoviruses. Immunodiffusion tests are less sensitive but are simple to perform and can provide much information concerning relationships between different viruses. Standardisation of test conditions and careful preparation of immunogen is necessary because the degree of cross-reaction between the matrix proteins (major polypeptide) varies considerably depending on whether antiserum is prepared using whole inclusion bodies, alkali-dissolved inclusion bodies or purified inclusion body protein produced by either acid precipitation or electrophoresis in preparative SDS–polyacrylamide gels. Radioimmunoassay is particularly useful for comparing cross-reactions between viruses and virus components (Summers & Hoops 1980). The extent to which different viruses compete with labelled homologous virus can be accurately quantified and the degree of cross-reaction assessed over several orders of magnitude.

Labelled antibody assays such as ELISA (see Hill, this volume) and IRMA are now widely used for detecting small amounts of virus (1–100 ng/ml). Various protocols for ELISA (Crook & Payne 1980) and IRMA (Crawford *et al.* 1978) have been developed for baculoviruses but the double antibody sandwich technique is most widely used. Microtitre plates are coated with antibody, reacted with antigen and finally incubated with labeled antibody. The amount of label bound to antigen is thus proportional to the amount of antibody in the sample. The ELISA, RIA and IRMA techniques are much more specific than many of the other serological methods employed so that cross-reactions between different granulins and polyhedrins can be reduced to almost background amounts.

Virus antigens can be compared in greatest detail by the sequential use of electrophoresis in polyacrylamide gel and serology (Smith & Summers 1981). Samples are electrophoresed through SDS–polyacrylamide gels and the resultant bands are blotted onto nitrocellulose sheets (Western blotting) and allowed to react with labelled antibody. After washing, the amount and position of label is compared with the profile obtained after normal staining.

Differences between bands revealed by reactions with homologous and heterologous antisera indicate the degree of relatedness between structural proteins of different viruses.

Counting Baculoviruses

Calibration of very pure preparations of baculovirus may be achieved by examination of total protein following carbonate dissolution of the inclusion bodies. However, for purposes of bioassay, and to allow measurement of doses for field control work, it is common practice to rely on direct counts of inclusion bodies using the optical microscope and one of the following three techniques. However, light microscope techniques are inaccurate for inclusion body numbers of less than ca. 1×10^6/ml.

Haemocytometer Counts

The PIB may be readily counted in blood cell counting chambers if preparations are free from contaminants of similar appearance and size. A suitable counting chamber has a depth of 0.1 mm and $1/400$ mm^2 grid. After settling for 10–15 min, PIB within each square and those touching the top and left hand square margins are counted. Adjustment of the PIB concentration to ca. 2–10/grid square minimises counting difficulties originating from Brownian motion. The contents of ca. 100 squares should be counted. The distribution of counts of non-aggregated PIB follows the Poisson distribution. The standard error (SE) of the mean is given by

$$SE = \sqrt{\bar{x}/N}$$

where $\bar{x}$ is mean number of PIB/count (i.e. total PIB/number of squares counted) and N is the number of squares counted [see Wigley (1980) for calculation of confidence limits].

Highly purified preparations of GV inclusion bodies (capsules) can also be counted in a haemocytometer chamber of either 0.01- or 0.02-mm depth by dark-field illumination.

Dry Films

This method depends upon counts of inclusion bodies in a measured volume of suspension spread evenly over a known area of microscope slide. Its particular advantages are that preparations are permanent and can serve as a record, and that when used with a specific stain may be made on any impure preparation in which inclusion bodies are not aggregated.

Inclusion body suspensions are fixed to alcohol-cleaned microscope slides by mixing with an equal volume of either 0.1% (w/v) gelatin, 1.0% (v/v) bovine serum or Mayer's albumin. Five microlitre volumes of the mixture are then spread over 14-mm-diameter circular areas, four of which can be fitted on one slide. To achieve this accurately a template is placed beneath the slide; spreading is by the spiral motion of a seeker from the centre of the circle outwards. Films are fixed in Carnoys fluid (1 part glacial acetic acid, 3 parts chloroform and 6 parts absolute alcohol) for 2 min, rinsed in ethanol, air dried, stained with naphthalene black 12B (1.5% (w/v) in 3% (v/v) glacial acetic acid) at 40°C for 1 min, rinsed in water and dried. The PIB are stained blue-black and GV capsules are stained less intensely blue-black.

Using an oil-immersion objective, counts are made along one radius of each of four replicate films, selecting a different radius each time. Ten counts (part or full field areas) are made along each radius: starting 0.25 mm from the circumference, 8 counts are made at successive 0.5-mm intervals and 2 counts are made at 1-mm intervals. Treatment of the data to give a count of inclusion bodies/millilitre is complex and involves, for instance, the weighting of values for each field counted according to the area of the annular stratum at that point (Wigley 1980). For routine use count data can be most rapidly evaluated by a computerized programme.

A "Sellotape" impression method has been described by Elleman *et al.* (1979) for assessing numbers of PIB on leaf surfaces. Leaves are pressed briefly onto double-sided Sellotape on microscope slides. The preparation is then stained with naphthalene black 12B and PIB counted at high magnification.

Proportional Counts

Measured volumes of PIB suspension are mixed with a suspension of known concentration of indicator particles of similar size, e.g. 2-μm-diameter latex spheres or spores of the giant puff ball *Calvatia gigantea*. Following appropriate staining on a microscope slide, counts of both types of particle are made in a number of part or full field areas and the concentration of the PIB suspension calculated from the PIB:indicator particle ratio. The statistical basis for the calculation is given by Wigley (1980). As with the dry film method, it is important to check that neither PIB nor indicator particles are lost during staining; this can usually be done by comparison with dry smears examined by phase contrast microscopy.

Electron Microscopy for GV Granules

Purified virus is mixed with an equal volume of latex particles (0.234-μm diameter) of known concentration (Payne *et al.* 1981). After addition of ovalbumin to a final concentration of 0.1% (v/v), the mixture is sprayed onto carbon grids and unstained, complete droplets are photographed in the electron microscope. The capsule concentration is calculated from the capsules:latex particles ratio.

Infectivity Assays

The response of whole insects to baculoviruses is usually measured by estimating the median lethal dose (LD_{50}) in terms of numbers of inclusion bodies, following feeding a series of doses to batches of test insects. The shape of the percentage mortality–dose curve is sigmoid, but when plotted as probit mortality against the logarithm of the dose, closely approximates to a straight line. Confidence limits are narrowest at the 50% mortality point which is thus the most sensitive measure of response. In LD_{50} tests, it is common practice to include ca. six dose levels of virus and to have ca. 50 test larvae at each dose. Decrease in variability can be achieved by selecting test larvae of similar genetic constitution, physiological condition and weight. Commonly the slope of the probit mortality–log dose line is rather small ($b <$ 2.0; Burges & Thomson 1971).

Prestarving test larvae, often considered a useful means of ensuring that a large proportion consumes the test dose, can alter the LD_{50} value compared with larvae fed normally. LD_{50} values may also be influenced by the nature of the food on which the dose is presented, e.g. artificial diet or natural plant food. Thus comparisons between bioassay results can easily be vitiated if test conditions have differed.

In LD_{50} tests, experimental larvae should be maintained in mutual isolation and fed individual measured doses of virus on amounts of food small enough to be consumed rapidly (12–24 h). This is most conveniently done by placing the inoculum on small pieces of artificial diet or leaf. The wells of microtitre plates make useful containers. After loading with diet, inoculum and test larvae, a piece of moist tissue is placed over the top of the plate to prevent diet desiccation and the whole sealed with plastic film (Clingfilm). After 24 h, those larvae which have eaten the dose of virus are transferred to individual small pots containing enough artificial diet for development; they are inspected periodically and dead larvae are frozen pending diagnosis for disease. Not all species will consume artificial diet and even with those which will, it may be an impractical means of dosing very young larvae. An alternative method is to sandwich a piece of leaf between two thick (0.5 cm) plastic plates identically perforated with holes of diameter appropriate to the insects being tested. By micropipette, the dose is placed on the leaf at the bottom of each well; the test larva is able to eat completely through the leaf to the opposite well. The whole is wrapped in damp tissue and Clingfilm

(Wigley 1976; Evans 1981). Larvae of diprionid sawflies may be fed individual doses as microdroplets dried on tips of coniferous needles; these may first be pushed into microtitre tray wells through small holes drilled in the base (Cunningham & Entwistle 1981). In some instances, a ball bearing may be suitable for closing holes (Griffiths & Smith 1977).

These techniques are used to determine responses in terms of absolute numbers of inclusion bodies and their conduct can be laborious. When a comparison between two or more virus samples on one host species only is needed, it is more convenient to use comparative non-absolute tests. These involve either spreading test inoculum evenly on the free surface of artificial diet in pots or incorporating it by thorough mixing into the diet before this cools to gelling point, taking care to avoid heat inactivation of virus. Virus may also be spread or sprayed, e.g. in an insecticidal spray tower, onto plant food.

A series of concentrations of each virus test sample is fed to batches of larvae for a standard length of time. Mortality data are subjected to probit analysis (Finney 1971) as for LD_{50} tests but the results are expressed as the lethal concentration to produce 50% mortality (LC_{50}). Such a technique is of limited value. For instance, it is unlikely to be useful for comparison of the same virus isolates on different host species or even different instars of the same species since in either case innately different feeding rates will result in the consumption of different quantities of inoculum.

Various aspects of the conduct of infectivity tests are considered by Burges & Thompson (1971), Burges (1971) and Evans (1981).

Safety Testing and Registration of Baculoviruses

Sophisticated testing of chemical pesticides and regulations governing their use and marketing have long been in operation in various parts of the world. However, the problem of what requirements should be satisfied by baculoviruses and other types of microbial agent intended for use as pesticides has not been considered until recent years. The development of thought, as expressed publicly through a series of international meetings, has been sensitively reviewed by Harrap (1982), who also gives details of the present attitudes and regulations of registering authorities.

The first set of guidelines to incorporate aspects of pathogenesis in safety testing emanated from the UK and was the result of a series of meetings in 1978 and 1979 of a panel for the Registration Criteria for Biological Agents Used as Pesticides under the Scientific Sub-committee on Pesticides of MAFF. Their findings, "Guidance on registration requirements of bacteria, protozoa, fungi and viruses used as pesticides" (Papworth 1980), have been incorporated in the UK Pesticides Safety Precautions Scheme (PSPS). This document represents as balanced and comprehensive an approach as can currently be achieved but in the light of future experiences must be expected to evolve. It requires that information be provided on the following main areas.

- Identity of and information on the formulated proprietary product,
- Identity of the active agent,
- Biological properties of the active agent,
- Manufacture and formulations,
- Application,
- Experimental data on efficiency,
- Experimental data on infectivity, allergenicity and toxicity including carcinogenicity and teratogenicity,
- Effects on humans,
- Residues,
- Information on environmental and wildlife hazards, and
- Labelling.

The nature of the safety tests required is detailed. A tier system is incorporated such that certain tests are required only if indicated by the results of 90-day animal studies. Further details of the UK, US and the currently anomalous EEC attitudes to baculovirus safety testing are given by Harrap (1982).

Utilisation of Baculoviruses for Pest Control

Three main methods have emerged by which baculoviruses may be utilised for pest control:

Management of Baculoviruses

This approach is best characterised by the example of *Wiseana* spp. in New Zealand, the larvae of which live in the soil and eat grass stems near ground level. Fresh pasture leys are especially susceptible to attack. Following observations that NPV is more frequently present in older pastures, systems of pasture management have been developed which maintain productivity by untraditional methods, maximise virus spread and minimise soil disturbance with concomitant loss of accessibility of NPV inoculum (Kalmakoff & Crawford 1982).

TABLE 5

Dose of virus necessary to control pests[*]

Virus	Pest	Crop situation	No. of cases	Mean total dose[†]/ha/ season	Effectiveness factor[‡]
NPV[§]	Lepidoptera	Cotton	6	1.8×10^{13}	—
	Lepidoptera	Other field crops	6	1.0×10^{13}	1.8
	Lepidoptera	Broad-leafed trees	4	5.0×10^{12}	3.6
	Lepidoptera	Conifers	4	4.5×10^{11}	40
	Sawflies	Conifers	4	5.0×10^{10}	360
GV	Lepidoptera	Field crops and trees	6	3.0×10^{13}	—

[*] From Entwistle (1978).
[†] Virus inclusion bodies (polyhedra or granules).
[‡] Compared with dose of NPV on cotton.
[§] NPV, Nuclear polyhedrosis virus; GV, granulosis virus.

Classical Biological Control

In this approach liberations of limited amounts of virus in pest scale populations result in adequate natural dispersal and suppression of insect numbers to an economically acceptable level.

Viruses as Sprayable Entities

For crop conditions unfavourable to epizootic development and especially where the natural growth of disease is too slow to prevent an unacceptable level of insect damage, baculoviruses can be sprayed like chemical pesticides. All the control examples in Table 2, except *Oryctes* and *Wiseana*, depend on spray applications. An approximation of the quantities of inclusion bodies/hectare necessary to achieve satisfactory control of a variety of pests on different crop types is given in Table 5.

Baculoviruses versus Chemical Pesticides

Several factors influence the question of acceptability of baculoviruses for pest control.

Success Rate per se *in Pest Control and Economic Competitiveness*

In successful classical biological control programmes there can be no doubt of the economic superiority of baculoviruses to the alternative of chemical suppression of pests. But because of very limited experience of commercial production of baculoviruses, and hence of their unit costs, it is at present less easy to pronounce on their economic status as sprayable entities in competi-

TABLE 6

*Relative efficiencies of baculoviruses and chemical pesticides in pest control**

No of pest crop situations tested	Best control		Equal control
	By virus	By chemical	
32	8	8	16

* From Entwistle (1978).

tion with chemicals. Results of experiments on the relative capacities of baculovirus and chemical sprays to control pests (Table 6) show that baculovirus applications were as, or more, effective. Different experiments are not always assessed in the same terms; therefore the table incorporates some results in terms of insects killed and others as increases in crop yield. Most of the results in Table 6 are based on the latter approach, which is generally the ultimate measure of effectiveness.

Environmental Acceptability Relative to Chemical Pesticides

The assessment of risk in the use of baculoviruses is at an early stage of development compared with chemical pesticides. Their position in the area of environmental impact seems very promising since no adverse effects in terms of unacceptable perturbations of non-target animals have been observed either in control programmes or in natural epizootics. The pathogenic nature of baculoviruses is increasingly being officially recognised in the structure of safety test procedures designed to inspect their direct effect on humans, domestic animals and other vertebrates. No adverse effects have been detected from any baculovirus but a generally approved view is that more detailed test data are required before any absolute assessment can be achieved.

Operational Advantages and Disadvantages

There is no doubt that, in general, baculoviruses have a distinct advantage over chemical pesticides because, as beneficial insects are unaffected, the contribution of parasites and predators to the control of insect pests is not impeded. Field trials with baculoviruses have often been on small plots within areas of general chemical use. Under such conditions the natural enemies of insect pests are in a general state of chemical suppression and therefore their capacity to supplement control cannot be assessed. This potential hidden benefit will only be demonstrable in large-scale, long-term experiments mimicking a putative operational control of insect pests.

Increases in insects not normally pests is a frequent phenomenon of chemical pesticide usage and has sometimes resulted in a "treadmill" situation where the more chemicals are applied the more are needed in both terms of concentration and number. Resistance problems, so far unknown with baculoviruses, exacerbate this effect.

Conclusion

At present, there are three main obstacles to the commercial use of baculoviruses. Firstly, there is a shortage of such viruses in pests other than Lepidoptera and sawflies. Secondly, where cosmetic injury to crops is economically significant, baculoviruses may not provide the necessary rapidity of control. Thirdly, because the spectrum of activity of baculoviruses is comparatively narrow, commercial outlets for individual viruses will be narrower than for individual chemicals. Competitiveness will depend on a satisfactory benefit–risk assessment in comparison with chemicals. At present this is likely to hinge purely on the economics of cost effectiveness since the risks of using baculoviruses have not yet emerged.

Appendix

Semi-Synthetic Diet for Rearing Lepidoptera [Hoffman et al. 1966 Modified by C. F. Rivers & R. Warner]

The following quantities make 3.75 litres of diet which is the capacity of a domestic food mixer.

Vitamin and Antibiotic Mixture

Mix together nicotinic acid, 5.0 g; calcium pantothenate, 5.0 g; riboflavine (B_2), 2.5 g; aneurine hydrochloride (B_1), 1.25 g; pyridoxine hydrochloride (B_6), 1.25 g; folic acid, 1.25 g; D-biotin, 0.10 g; cyanocobalamine (B_{12}), 0.01 g. To 1.0 g vitamin mixture add streptomycin, 1.0 g; aureomycin (veterinary soluble powder, 25 g/lb) 18.0 g; ascorbic acid (vitamin C), 40.0 g. Mix well and store at −20°C.

Dry Ingredients

Agar, 75.0 g; casein, 132.0 g; wheat germ ("Bemax"), 288.0 g; Wessons salts (or alternative, cf. below), 37.5 g; dried brewers yeast, 57.0 g; sugar, 117.0 g; cholesterol, 3.75 g; methyl-4-hydroxy benzoate, 3.75 g; sorbic acid, 6.99 g; dried cabbage leaf,* 75.0 g.

*Ingredients only added in preparing diet for species which will not feed readily on the basic diet.

Alternative to Wessons Salts (Beckman et al. 1953)

$CaCO_3$, 1200.0 g; K_2HPO_4, 1290.0 g; $CaHPO_4 \cdot 2H_2O$, 300.0 g; $MgSO_4 \cdot 7H_2O$, 408.0 g; NaCl, 670.0 g; $FeC_6H_5O_7 \cdot 6H_2O$, 110.0 g; KI, 3.2 g; $MnSO_4 \cdot 4H_2O$, 20.0 g; $ZnCl_2$, 1.0 g; $CuSO_4 \cdot 5H_2O$, 1.2 g.

Method

Mix dry ingredients together thoroughly in the bowl of a food mixer, if available. While mixing add water, 3450.0 ml; linseed oil, 7.5 ml; Sinigrin† (1%, w/v soln), 15.0 ml. Autoclave at 121°C for 20 min. Pour into the mixer bowl and whisk the mixture until it cools to 60–70°C. Add vitamin and antibiotic mixture, 22.5 g; choline chloride, 3.75 g; Carbendazim or Benlate,‡ 1.5 g.

Simplified Diet for Rearing Cydia pomonella *(Guennelon et al. 1981)*

Ingredients

Water, 1000.0 ml; agar, 26.6 g; maize meal, 190.0 g; dried yeast powder, 50.0 g; wheat germ, 46.0 g; ascorbic acid, 7.0 g; benzoic acid, 3.0 g; formaldehyde (30%), 1.6 ml; methyl-4-hydroxy benzoate, 2.3 g.

Method

Dissolve agar in 600 ml H_2O by boiling or steaming. Mix dry ingredients in blender, add remaining 400 ml H_2O and formaldehyde and mix thoroughly. Add agar after cooling to 70°C, stir and dispense.

Artificial Diet for Rearing Pests of Stored Products (Hunter & Boraston 1979)

Ingredients

Farex, 400 g; honey, 200 g (140 ml); glycerol, 200 g (180 ml); dried yeast powder, 100 g.

Method

Mix yeast and Farex in blender, add glycerol–honey mixture, mix well and dispense as required.

References

AIZAWA, K. 1971 Strain improvement and preservation of virulence of pathogens. In *Microbial Control of Insects and Mites* eds. Burges, H. D. & Hussey, N. W., pp. 655–672. London: Academic Press.

BECKMAN, H. F., BRUCKART, S. M. & REISER, R. 1953 Laboratory culture of the pink bollworm on chemically defined media. *Journal of Economic Entomology* **46**, 627–630.

†Sinigrin may be substituted for dried cabbage leaf.
‡May be added if the insect to be reared is known to have a microsporidian infection.

BEDFORD, G. D. 1981 Control of the Rhinoceros beetle by Baculovirus. In *Microbial Control of Pests and Plant Diseases 1970–1980* ed. Burges, H. D., pp. 409–426. London: Academic Press.

BURGES, H. D. 1971 Possibilities of pest resistance to microbial control agents. In *Microbial Control of Insects and Mites* eds. Burges, H. D. & Hussey, N. W., pp. 445–457. London: Academic Press.

BURGES, H. D. & THOMSON, E. M. 1971 Standardization and assay of microbial insecticides. In *Microbial Control of Insects and Mites* ed. Burges, H. D. & Hussey, N. W., pp. 591–622. London: Academic Press.

CRAWFORD, A. M., FAULKNER, P. & KALMAKOFF, J. 1978 Comparison of solid-phase radioimmunoassays for baculoviruses. *Applied and Environmental Microbiology* **36**, 18–24.

CROOK, N. E. & PAYNE, C. C. 1980 Comparison of three methods of ELISA for baculoviruses. *Journal of General Virology* **46**, 29–37.

CUNNINGHAM, J. C. & ENTWISTLE, P. F. 1981 Control of sawflies by baculovirus In *Microbial Control of Pests and Plant Diseases 1970–1980* ed. Burges, H. D., pp. 379–407. London: Academic Press.

ELLEMAN, C. J., ENTWISTLE, P. F. & HOYLE, S. R. 1979 Application of the impression film technique to counting inclusion bodies of nuclear polyhedrosis viruses on plant surfaces. *Journal of Invertebrate Pathology* **36**, 129–132.

ENTWISTLE, P. F. 1976 The development of an epizootic of a nuclear polyhedrosis virus disease in European spruce sawfly, *Gilpinia hercyniae*. *Proceedings of the First International Colloquium of Invertebrate Pathology, Kingston, Canada* pp. 184–188.

ENTWISTLE, P. F. 1978 Microbial control of insects and other pests. *British Association for the Advancement of Science, Report of the Annual Meeting, Bath University, UK, Section M (Agriculture) Crop Protection* pp. 71–96.

EVANS, H. F. 1981 Quantitative assessment of the relationships between dosage and response of the nuclear polyhedrosis virus of *Mamestra brassicae*. *Journal of Invertebrate Pathology* **37**, 101–109.

EVANS, H. F. & ENTWISTLE, P. F. 1982 Epizootiology of the nuclear polyhedrosis virus of European spruce sawfly with emphasis on persistence of virus outside the host. In *Microbial Pesticides* ed. Kurstak, K., pp. 449–461. New York: Dekker.

EVANS, H. F., BISHOP, J. M. & PAGE, E. A. 1980 Methods for the quantitative assessment of nuclear polyhedrosis virus in soil. *Journal of Invertebrate Pathology* **35**, 1–8.

EVANS, H. F., LOMER, C. J. & KELLY, D. C. 1981 Growth of nuclear polyhedrosis virus in larvae of the cabbage moth, *Mamestra brassicae* L. *Archives of Virology* **70**, 207–214.

FINNEY, D. J. *Probit Analysis*, 3rd edn. Cambridge: University Press.

GRIFFITHS, I. P. & SMITH, A. M. 1977 A convenient method for rearing lepidopteran larvae in isolation. *Journal of the Australian Entomological Society* **16**, 366.

GUENNELON, G., AUDEMARD, H., FREMOND, J. C. & EL IDRISSI AMMARI, M. A. 1981 Progrès réalisés dans l'élevage permanent du carpocapse (*Laspeyresia pomonella* L.) sur milieu artificiel. *Agronomie* **1**, 59–64.

HARRAP, K. A. 1982 Assessment of the human and ecological hazards of microbial pesticides. *Parasitology* **84**, 269–296.

HARRAP, K. A. & PAYNE, C. C. 1979 The structural properties and identification of insect viruses. *Advances in Virus Research* **25**, 273–355.

HOFFMAN, J. D., LAWSON, F. R. & YAMAMOTO, R. 1966 Tobacco Hornworms. In *Insect colonization and Mass Rearing* ed. Smith, C. N., pp. 479–486. London: Academic Press.

HUNTER, F. R. & BORASTON, R. C. 1979 Application of the Laurell immunoelectrophoresis technique to the study of serological relationships between granulosis viruses. *Journal of Invertebrate Pathology* **34**, 248–256.

IGNOFFO, C. M. & SHAPIRO, M. 1977 Characteristics of baculovirus preparations processed from living and dead larvae. *Journal of Economic Entomology* **71**, 186–188.

KALMAKOFF, J. & CRAWFORD, A. M. 1982 Enzootic virus control of *Wiseana* spp. in the pasture environment. In *Viral Pesticides* ed. Kurstak, E., pp. 435–448. New York: Dekker.

KAUPP, W. J. 1981 Studies on the ecology of the nuclear polyhedrosis virus of the European pine sawfly, *Neodiprion sertifer* Geoff. Ph.D. Thesis, University of Oxford, UK.

LAEMMLI, U. K. 1970 Cleavage of structural proteins during the assembly of the head of bacteriophage T4. *Nature, London* **227**, 680–685.

LOWRY, O. H., ROSENBROUGH, N. J., FARR, A. L. & RANDALL, R. J. 1951 Protein measurement with folin phenol reagent. *Journal of Biological Chemistry* **193**, 265–275.

MARTIGNONI, M. E. & IWAI, P. J. 1981 A catalogue of viral diseases of insects, mites and ticks. In *Microbial Control of Pests and Plant Diseases 1970–1980* ed. Burges, H. D., pp. 897–911. London: Academic Press.

OSSOWSKI, L. L. J. 1960 Variations in virulence of a wattle bagworm virus. *Journal of Insect Pathology* **2**, 35–43.

PAPWORTH, D. S. 1980 Registration requirements in the UK for bacteria, fungi and viruses used as pesticides. *Ecological Bulletins* No. 31, pp. 135–143.

PAYNE, C. C. & KELLY, D. C. 1981 Identification of insect and mite viruses. In *Microbial Control of Pests and Plant Diseases 1970–1980* ed. Burges, H. D., pp. 61–91. London: Academic Press.

PAYNE, C. C., TATCHELL, G. M. & WILLIAMS, C. F. 1981 The comparative susceptibilities of *Pieris brassicae* and *P. rapae* to a granulosis virus from *P. brassicae*. *Journal of Invertebrate Pathology* **38**, 273–280.

SHAPIRO, M. & IGNOFFO, C. M. 1970 Nucleopolyhedrosis of *Heliothis:* activity of isolates from *Heliothis zea*. *Journal of Invertebrate Pathology* **16**, 107–111.

SINGH, P. 1977 *Artificial Diets for Insects, Mites and Spiders*, i–xi + 594 pp. New York: Plenum Press.

SMITH, G. E. & SUMMERS, M. D. 1978 Analysis of baculovirus genomes with restriction endonucleases. *Virology* **89**, 517–527.

SMITH, G. E. & SUMMERS, M. D. 1981 Application of a novel radioimmunoassay to identify baculovirus structural proteins that share interspecies antigenic determinants. *Journal of Virology* **39**, 125–137.

STAIRS, G. R. 1965 Quantitative differences in susceptibility to nuclear polyhedrosis virus among larval instars of the forest tent caterpillar, *Malacosoma disstria* (Hubner). *Journal of Invertebrate Pathology* **7**, 427–429.

SUMMERS, M. D. & HOOPS, P. 1980 Radioimmunoassay analysis of baculovirus granulins and polyhedrins. *Virology* **103**, 89–98.

THOMPSON, C. G., SCOTT, D. W., & WICKMAN, B. E. 1981 Long term persistence of the nuclear polyhedrosis virus of the Douglas fir tussock moth, *Orgyia pseudotsugata* (Lepidoptera: Lymantriidae), in forest soil. *Environmental Entomology* **10**, 254–255.

WIGLEY, P. J. 1976 The epizootiology of a nuclear polyhedrosis virus disease of the winter moth, *Operophtera brumata* L. at Wistman's wood, Dartmoor. Ph.D. Thesis, University of Oxford, England.

WIGLEY, P. J. 1980 Practical: counting micro-organisms. *New Zealand Department of Science and Industrial Research, Bulletin* No. 228, 29–35.

WORLD HEALTH ORGANIZATION 1973 The use of viruses for the control of insect pests and disease vectors. Report of a joint FAO/WHO Meeting on insect viruses. *World Health Organization Technical Report Series* No. 531.

ZELAZNY, B. 1976 Transmission of a baculovirus in populations of *Oryctes rhinoceros*. *Journal of Invertebrate Pathology* **27**, 221–227.

The ELISA (Enzyme-Linked Immunosorbent Assay) Technique for the Detection of Plant Viruses

STEPHEN HILL

Virology Branch, Agricultural Development and Advisory Service, Ministry of Agriculture, Fisheries and Food, Cambridge, UK

Serological techniques have been in common use in the study of plant viruses for many years. Many plant viruses are strongly immunogenic and can be purified with comparative ease, and antisera of high titre are available. Many of the techniques of value to the researcher are inapplicable in the advisory or extension laboratory, being either too complex, too lengthy or requiring high-cost equipment or research expertise. The serological tests which have been used in routine testing for plant viruses have been the flocculation tests, which depend on the visual estimation of aggregates formed in liquid media, and the gel diffusion tests, in which precipitation lines form in agar gel media.

Of the flocculation tests, the most basic, chloroplast agglutination, is still used as a field test for detection of some potato viruses. A more sensitive modification of this technique which has found wide use is the latex test in which antibodies, adsorbed to the surface of the inert carrier, polystyrene latex particles, flocculate to give readily visible precipitates in the presence of virus. Whilst of great value as a field test, with sensitised latex suspensions being supplied from a central laboratory, the latex test may be less reliable where viruses reach a large concentration in their host and excess virus inhibits flocculation. Additionally, for some virus–host combinations flocculation may be inhibited by substances in the host sap requiring the use of additives, reducing or chelating agents.

Gel diffusion tests have most frequently been used in advisory testing for specific confirmation of plant virus identity. The main problem with this technique is that whilst it works well for spherical particles, most elongated

MICROBIOLOGICAL METHODS
FOR ENVIRONMENTAL BIOTECHNOLOGY
ISBN 0-12-295040-2

*Copyright © 1984 by the Society for Applied Bacteriology
All rights of reproduction in any form reserved*

virus particles do not diffuse readily through the pores of the agar medium. Techniques to break elongated particles into fragments to allow them to diffuse are available but provide an added complication to an already time-consuming test. Other problems with gel diffusion tests are inhibition of reactions by plant sap components and rather excessive use of antiserum, but perhaps the major source of error for the irregular use of this technique is the failure to define correctly the conditions needed to obtain the precipitation reaction. In particular, the balance of antigen and antibody must be precise.

Not only are many of the disadvantages encountered by the field laboratory user of the serological test involving precipitation or flocculation reactions overcome by the enzyme-linked immunosorbent assay technique (ELISA), but it offers many additional advantages in such situations. The most frequent requirement of an advisory or extension plant virologist is the routine detection of plant virus in large numbers of samples in surveys or for certification schemes. Until the development of ELISA, such routine processing of samples was impossible except by the employment of many staff, many test plants for sap transmission testing or interminable repetitive and probably insensitive precipitation or flocculation tests. The ELISA test as described by Voller *et al*. (1976) and Clark & Adams (1977) has the following advantages: (1) extreme sensitivity; (2) applicability to large numbers of samples; (3) economy in use of high-cost antisera; (4) semi-automatable; (5) quantitative; (6) independent of virus morphology; (7) independent of virus concentration.

The use of the ELISA technique for plant virus detection has been reviewed extensively (Bar-Joseph & Garnsey 1981; Clark 1981; Torrance & Jones 1981).

The Double Antibody Sandwich (DAS) ELISA

The double antibody sandwich (DAS) ELISA (Fig. 1) is the technique most frequently employed for the detection of plant viruses, and the enzyme usually used to label the specific antibody is alkaline phosphatase.

Technique

The following is a full account of the technique; a summary of the procedure is given in Appendix 1.

Preparation of Gamma-Globulin

Plant virus antibodies are found in the gamma-globulin fraction of the whole serum. Most workers use fractionated and partially purified gamma-

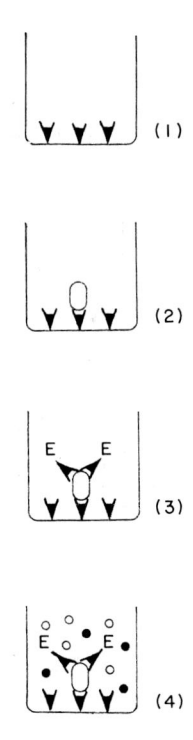

FIG. 1. Double antibody sandwich (DAS) ELISA protocol. (1) Specific antibody raised in rabbit adsorbed to plate (Y). Incubate 3 h at 37°C. Wash. (2) Add test extract containing virus (0). Incubate overnight at 6°C. Wash. (3) Add enzyme-labelled specific antibody conjugate. (E▲). Incubate 3 h at 37°C. Wash. (4) Add enzyme substrate (○). Evaluate colour intensity and virus concentration after 30–60 min incubation at room temperature. (●), Enzyme product.

globulin but others use less well-purified preparations, e.g. ammonium sulphate-precipitated gamma-globulins without further purification. The procedure adopted by Clark & Adams (1977) is used.

(1) To 1 ml antiserum add 9 ml distilled water.
(2) Add 10 ml saturated ammonium sulphate.
(3) Leave to precipitate for 30–60 min at room temperature.
(4) Centrifuge to collect precipitate.
(5) Dissolve precipitate in 2 ml half-strength phosphate-buffered saline (PBS; see Appendix 2).
(6) Dialyse three times (afternoon, overnight and morning) against 500 ml half-strength PBS.
(7) Filter through 3–5 ml DE 22 cellulose pre-equilibrated in half-strength PBS.
(8) Wash gamma-globulin through cellulose with half-strength PBS.

(9) Monitor effluent at 280 nm and collect first protein fraction.
(10) Measure OD_{280} and adjust strength of gamma-globulin to read ca. 1.4 OD (ca. 1 mg/ml).
(11) Store in silicone-treated glass tubes at $-18°C$ (but avoid freezing and thawing repeatedly).

Bar-Joseph & Garnsey (1981) have also purified gamma-globulins by adsorbtion by Protein A–Sepharose CL-4B (Pharmacia, Uppsala) followed by elution with 0.1-N glycine–HCl, pH 3.0. Antisera of high titre are preferable but not essential. Antibodies to host components may be removed by addition of host protein before fractionation, or alternatively by adsorption from conjugates immediately before use.

Preparation of Enzyme Conjugate

The one-step glutaraldehyde method is widely used for preparation of satisfactory antibody–alkaline phosphatase conjugates.

(1) Centrifuge 1 ml (equivalent to 5 mg) enzyme precipitate. Discard supernatant liquid.
(2) Dissolve precipitate directly in 2 ml (= 2 mg) purified gamma-globulin.
(3) Dialyse three times against 500 ml PBS.
(4) Add fresh glutaraldehyde solution to give 0.06% (w/v) final concentration; mix well.
(5) Leave for 4 h at room temperature during which time a yellow-brown colour should develop.
(6) Dialyse three times (afternoon, overnight, morning) against 500 ml PBS to remove glutaraldehyde.
(7) Add bovine serum albumin to give ca. 5 mg/ml and store at 4°C in the refrigerator.

Plate Coating

Reference to Fig. 1 will illustrate the importance of plate coating in the ELISA process. The sensitisation of the solid phase, whether polystyrene microtitre plate or polystyrene tubes, beads or stirring sticks, involves the adsorption of gamma-globulin proteins in an irreversible hydrophobic interaction. Non-ionic detergents prevent but do not reverse this interaction and are added in later steps to prevent non-specific binding. Passive sensitisation of the solid phase is achieved by exposure to a solution of partially purified gamma-globulin in sodium carbonate buffer, pH 9.6 (see Appendix), and allowing antibodies to adsorb to the solid phase. The optimal concentration of the gamma-globulin and time and temperature of incubation must be determined experimentally, but for many plant viruses gamma-globulin suspension of 1 mg/ml may be further diluted to $1/500$ or $1/1000$.

Concentrations of greater than 10 μg/ml are reported to reduce the strength of the virus-specific reaction and increase intensity of non-specific reaction. The incubation time most often found to be efficient is 3 h at a temperature of 37°C. During all incubation stages, plates should be covered to prevent evaporation of reagents; Cling-film provides a satisfactory inexpensive cover for this purpose. The microtitre plate provides the most convenient solid phase for ELISA; the most common configuration of 8 × 12 wells is suitable for insertion into most of the mechanical equipment for washing, plate reading, etc. However, other systems for specific purposes are available. The sensitivity of reaction which can be achieved varies from plates of one type of manufacture to another, and in some cases there has been differential and often non-specific variation within plates which probably relate to the gamma-globulin coating process. Coated plates, if carefully covered, may be stored under deep freeze conditions, retaining their sensitivity for many months.

Plate Washing

Thorough washing between component stages of the ELISA process is essential to prevent carry-over of reactants that are not part of the solid-phase double antibody sandwich complex. Usually after the coating, test sample and conjugate incubation processes, plates are washed at least three times with PBS Tween buffer (see Appendix 2) and often wash liquid is left in the plates for several minutes at each stage. Automatic plate-washing machines are available to provide standardisation for this process. With some host–virus systems, distilled or tap water has been used for washing without adverse results.

Extraction of the Test Sample

Sample preparation technique should be modified according to the concentration and stability of the virus and the presence of inhibiting host sap components. Most ELISA proponents have prepared the sample in a PBS Tween buffer with the addition of 2% (w/v) polyvinylpyrrolidone (PVP) of molecular weight 25,000 or 44,000 and 0.2% w/v ovalbumin. A buffer containing 2-mercaptoethanol was reported by McLaughlin *et al.* (1979) to give good results with purified virus preparations in ELISA. Alternatively, diethyldithiocarbamate as an additive also proved valuable, especially with sap suspensions. Polyvinylpyrrolidone decreases non-specific "background" reactions but may also reduce specific reactions. Similarly, the addition of dithiothreitol may decrease both specific and non-specific reactions. Sample extract dilution in buffer must be determined experimentally. With a large concentration of virus it is possible to use dilute extracts ($1/100$) to reduce or eliminate non-specific reactions due to host cell components. Long incubation of the sample extract in the plate has been found most effective, and incubation overnight at temperatures of 6°C has been found convenient for routine tests.

Addition of Conjugate

After thorough washing, conjugated gamma-globulin suitably diluted in freshly prepared PBS Tween + 2% (w/v) PVP + 0.2% (w/v) ovalbumin as pre-determined experimentally is added. Increasing conjugate dilution results in corresponding reductions in the ELISA reaction, but his can be partly compensated for by increasing incubation time. Conjugate incubation is usually at 30–37°C for 3 h, but regimes of 6°C overnight have also given good results.

Substrate Addition

Substrate freshly prepared at the required concentration and free from colour is added after final washing of the conjugate. For alkaline phosphatase conjugates, the substrate is p-nitrophenyl phosphate, which is available in tablet form and most often used at a concentration of 0.6 mg/ml in 10% (v/v) diethanolamine, pH 9.8 (see Appendix 2). Sufficient colour change has usually occurred after 30–60 min when the extent of the reaction can be evaluated. Alkaline phosphatase activity can be stopped by addition of 3-M NaOH and such 'stopped' plates may be stored (covered with Cling-film) at 4°C for several hours. It may be adequate to read plates directly by eye, which has a sensitivity of 0.15 OD_{405}. However, for more precise evaluation the absorbance of each well component should be determined spectrophotometrically at 405 nm in the alkaline phosphatase system.

Equipment for ELISA

Provided that the ELISA reaction can be read directly on plates and prepared conjugates are available commercially, the ELISA test can be done with the most rudimentary equipment. However, in order to take full advantage of the large sample number capability of the ELISA system, component processes may be mechanised. For loading of coating globulin, conjugates and substrates, multi-channel pipettes with disposable pipette tips adapted to fill either 8 or 12 wells simultaneously are available. A variety of plate-washing machines and plate readers of various degrees of sophistication can be purchased, and a completely automated system is available which dispenses and extracts component mixtures, washes and read plates.

The preparation of samples remains the most time-consuming part of the process. The means of preparing sample extracts must be varied according to the type of sample and must take account of the physical state of the material, the stability and concentration of the virus it may contain and the effects of host plants cell constituents on the ELISA reaction. For succulent, sappy leaves with stable viruses which are not easily mechanically transmissible, a roller press of the kind designed for use with flocculation tests is ideal. Where leaves are less sappy, an automatically operated pipette can be positioned to deliver buffer onto the rollers. Where viruses are mechanically transmissible, small quantities of leaf material may be crushed inside small polythene bags. From raw fibrous material, e.g. graminaceous hosts, freeze or air drying of samples followed by milling has given good results with the advantage that prepared samples can be subdivided and stored almost indefinitely. For extraction of sap from potato tubers, a dentist's drill provided with a sucking and dispensing diluter allows extraction, dilution and transfer to the coated ELISA plate in one operation.

Developing ELISA for Specific Host–Virus Combinations

Whilst the ELISA procedure is relatively simple, for each new virus–host combination a certain amount of experimentation and practice is required. An appreciation of the limitations of the antiserum available and knowledge of the likely virus concentration and distribution in the plant both in spatial and temporal terms are essential. The optimal concentrations for use of coating gamma-globulin, conjugate and sample extract must be determined experimentally. These may need to be repeated with each newly prepared specific component.

Cross-contamination of wells at the sample incubation and washing stages can cause spurious reactions; the adoption of standard routine loading and washing procedures helps to interpret such errors. Extreme cleanliness in handling all ELISA components must be exercised, preferably with different equipment being used for conjugate and substrate processes. The re-use of disposable equipment is possible but should be carefully monitored.

Interpretation of Results

Quantitative comparisons of ELISA values in different microtitre plates is not advisable due to possible plate-to-plate variation in sensitivity. Within each plate therefore, appropriate controls should be included for reference. Where non-specific reactions are small and specific reactions are large, plate reading is straightforward. Difficulty in interpretation arises when the range of non-specific and specific reaction values overlap. In such instances it may become necessary to include a large number of known healthy control samples and to determine statistically a threshold level for infection. Several authors have used the mean value for healthy controls plus three times their standard deviation ($\bar{x} + 3$ SD) to establish thresholds.

TABLE 1

Some concentrations of different plant viruses detectable by ELISA

	Concentration	
Virus	ng/ml	Dilution of extract
Apple chlorotic leaf spot	1–5	10^{-4}
Apple mosaic		10^{-4}
Barley yellow dwarf	30	
Cauliflower mosaic	25	
Citrus tristeza		10^{-3}
Clover yellow mosaic		10^{-6}
Cucumber mosaic	0.01	
Lettuce mosaic	9	
Plum pox	1	10^{-5}
Tobacco mosaic	50	
Tobacco rattle	30	
Tobacco ringspot	15	

Sensitivity and Specificity of ELISA

The approximate concentration of some plant viruses detected by ELISA shown in Table 1 demonstrates that ELISA provides a sensitive assay, although complete comparisons with the same virus host systems are lacking. Table 2 shows that ELISA compares well with other serologically based diagnostic tests.

The sensitivity of ELISA allows batch testing of samples in the testing of seed and herbaceous material for certification purposes: In the case of seeds a lengthy alternative procedure is avoided. Whilst there has been some debate about the specificity of ELISA, it is generally agreed that the test is

TABLE 2

*Sensitivity of some plant virus detection methods**

Method	Approximate concentration detectable (μg/ml)
ELISA	<0.01
Immunodiffusion	1.00
Microprecipitin	0.5
Latex flocculation	0.01
Electron microscopy	0.1
Immunosorbent electromicroscopy	<0.01

* Matthews (1970); Derrick (1973); Clark & Adams (1977).

highly specific and detection of a heterologous virus strain can be affected by the degree of serological relationship between it and the strain to which the antiserum used was prepared. The reason for the specificity is not known, but it has been suggested that in the conjugation process there is some impairment of serological activity, which is more evident in tests on heterologous than homologous strains. Such specificity can be disadvantageous in scheme or survey situations, where strain identity is not important, and it is essential to detect all virus. The specificity may be overcome by mixing antisera to known strains, but there still remains the risk of missing undiscovered strains. Indirect ELISA methods which will be described next are not as strain specific.

Modifications of the ELISA Procedure

A number of modifications of the double antibody sandwich ELISA technique have been described and used for detection of plant viruses, each proponent claiming some advantage for their technique. Indirect ELISA systems follow the DAS procedure except that the second layer of specific antibody is not enzyme labelled. Instead, the label is introduced as a conjugated immunoglobulin (Ig) in an additional step (Fig. 2). It is claimed that this allows the full binding property of the specific gamma-globulin to be used, giving greater sensitivity and overcoming the extreme specificity of DAS ELISA (Bar-Joseph & Malkinson 1980).

A different modification (Torrance 1980) uses C1q (a component of complement obtained from bovine serum) to trap virus antibody aggregates. Plates are first coated with C1q after which a mixture of infected plant sap and virus-specific Ig (raised in rabbit) is incubated overnight. Trapped virus–antibody aggregates are detected by the subsequent addition of enzyme-labelled anti-rabbit Ig followed by substrate. The C1q assay (Fig. 3) offers not only the advantages of indirect techniques but also uses non-specific coating and enzyme-labelled components. However, the technique is adversely affected by concentrated sap of certain plant species, e.g. potato, and where this occurs sap must be diluted. Therefore C1q assay may not be suitable for routine application where such non-specific sap reactions preclude dilution in grouped samples.

Another recent modification (Barbara & Clark 1982) combines the advantages of an indirect assay with those of DAS. Antigen is trapped on a solid phase as in DAS but using instead the $F(ab')_2$ part of the IgG molecule, which is prepared by incubation with pepsin. Trapped virus is detected using unlabelled Ig and this in turn is detected using an immunoglobulin-based enzyme conjugate specific for the Fc portion of the IgG. Pepsin digestion of the Fc portion of the trapping antibody permits the use of a

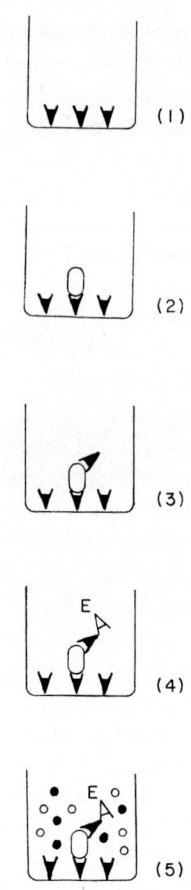

FIG. 2. Indirect ELISA protocol. (1), (2), Same as Fig. 1. (3) Add specific antibody (▲). Incubate 3 h at 37°C. Wash. (4) Add enzyme-labelled anti-rabbit–antibody conjugate (EA). Incubate 3 h at 37°C. Wash. (5) Add enzyme substrate (○). Evaluate as for DAS ELISA (Fig. 1). (●), Enzyme product.

general purpose enzyme conjugate to discriminate between trapping and detection antibody. Disadvantages of the method are that the specificity of the procedure is dependent on the concentration of the second antibody, larger concentrations giving decreased specificity. However, smaller background reactions are obtained and the procedure may be useful for investigation where the effort or expense of preparing individual virus specific conjugates is not justified.

The use of fluorogenic substrate (methyl umbelliferim substrate) has recently been compared with alkaline phosphatase for the detection of plant viruses (L. Torrance & R. Jones, unpublished observation). The fluorogenic

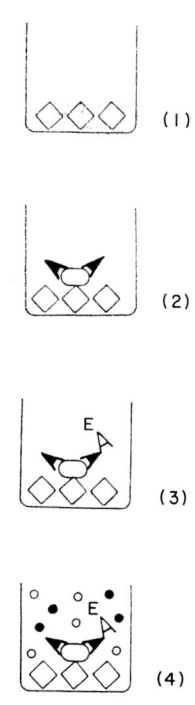

FIG. 3. C1q–ELISA protocol. (1) C1q adsorbed to plate (◊). Incubate 3 h at 34°C. Wash. (2) Add test sample and specific antibody raised in rabbit at dilutions to give optimal reaction in well (⌣). Incubate overnight at 4°C. Wash. (3) Add enzyme–labelled anti-rabbit–antibody conjugate (EA). Incubate 2 h at 34°C. Wash. (4) Add enzyme substrate (○). Evaluate as for DAS ELISA (Fig. 1). (●), Enzyme product.

substrate allowed increases in sensitivity of 2- to 16-fold with several viruses in leaf extracts and 2- to 4-fold in tuber extracts for potato virus and allowed more efficient detection of circulative virus in individual aphids and seed-borne virus in true potato seed.

Practical Applications of ELISA

Routine Detection of Plum Pox Virus

Double antibody sandwich ELISA has been in use for the routine detection of plum pox virus (PPV) in UK certification schemes for a number of years. Up to 8000 samples annually of plum leaves from English mother trees and rootstock hedges have been tested at the Agricultural Development and Advisory Service (ADAS) Harpenden Laboratory (Pemberton 1979). Leaf samples can be grouped in batches for initial testing and the sensitivity of ELISA allows the detection of one infected leaf in eight healthy ones. For such situations further testing determines the origin of infection. Before the introduction of ELISA for PPV detection these tests were done by chip grafting to peach seedlings with very many fewer tests being possible each year. Plum pox virus was only relatively recently introduced into the UK and nursery production of prunus material remains subject to statutory control to minimise further spread of the virus.

Detection of Virus in Potato Tubers

Under research laboratory conditions ELISA is sufficiently sensitive to detect many of the viruses which are transmitted through potato tubers. Growing crop inspections may detect tuber-borne virus diseases which show a severe infection in the emerging plant, but may overlook aphid-borne infection during the growing season. Traditional tuber indexing to demonstrate such late season spread of virus involves the growing-on of eyes cut from a random sample of seed-sized tubers either in glasshouse conditions in Europe or outdoors in more favourable winter climates. Such tests are time consuming, labour intensive and may be subjective. The use of a tuber ELISA test as an alternative method of detecting virus has been described in Europe and under research laboratory conditions in the UK (Tamada & Harrison 1980). However, in an investigation by the present author, constraints introduced in a practical ELISA test, i.e. the use of a drill dilutor extractor desctibed under *Equipment for ELISA*, page 354, together with the variable virus concentration and distribution in field-grown tubers of variable physiology, render tests unreliable in the present stage of development. Further work using modified ELISA methods may offer hope for greater success.

Routine Detection of Other Viruses

Poinsettia plants are propagated from cuttings bought annually from the US. Poinsettia mosaic virus induces a mosaic and may be responsible for leaf deformation which causes quality down-grading or rejection. Virus effects may be enhanced by lower growth temperatures so that UK production may suffer more than that in the US. Detection of poinsettia mosaic virus by ELISA has allowed the indexing of large quantities of plants on receipt to see if any are free of virus. The technique will be used to index material after heat treatment and micropropagation and during evaluation experiments. If virus-free material is of better quality, this would bring cost savings by allowing cultivation at lower temperatures. The author has used ELISA successfully to detect ryegrass mosaic virus in field samples. Several different isolates of the virus were readily detected using an antiserum prepared to a mild strain of the disease. As expected recent infections gave less definite ELISA reactions but in direct comparisons the technique was more sensitive than "quick-dip" electron microscopy (Chester *et al.*, 1983).

An added complication in this exercise was the presence of antibodies to ryegrass seed-borne virus, so that in initial tests, what were at first mistaken for variable non-specific background reactions related to different cultivars were eventually found to be different levels of the seed-borne virus.

Conclusion

The DAS ELISA provides a versatile, quick and relatively economical technique for plant virus diagnosis. The volume of literature describing uses of the technique is increasing rapidly with reports of viruses being detected in many plant parts, flowers, leaves, tubers, seeds, etc. The great value of the technique in survey and scheme diagnosis makes it attractive to advisory and statutory extension workers. The potential for testing large quantities of material and its relative simplicity also recommend it. The availability of equipment for automated component processes and reagent kits for an increasing number of plant viruses show that ELISA has established a firm place with the plant virus diagnostician.

Appendix 1

Summary of Procedure for Double Antibody Sandwich ELISA

(1) Add 200 µl purified gamma-globulin in coating buffer to each well of the microtitre plate. Incubate 2–4 h at 37°C.
(2) Wash by flooding wells with PBS–Tween. Leave at least 3 min. Repeat wash three times. Empty plate.
(3) Add 200-µl aliquots of the test sample to duplicate wells. Leave at 6°C overnight or at 37°C for 4–6 h.
(4) Wash plate three times as in (2) above.
(5) Add 200-µl aliquots of enzyme-labelled gamma-globulin to each well. Incubate at 37°C for 3–6 h.
(6) Wash plate three times as in (2) above.
(7) Add 300-µl aliquots of freshly prepared substrate to each well. Incubate at room temperature for 1 h, or as long as necessary to observe reaction.

(8) Stop reaction by adding 50 μl 3-*M* NaOH to each well.
(9) Assess results by (a) direct visual observations; (b) measurement of absorbance at 405 nm.

Appendix 2

Buffers Required for ELISA

Phosphate-buffered saline (PBS) (pH 7.4): NaCl, 8.0 g; KH_2PO_4, 0.2 g; $Na_2HPO_4 \cdot 12H_2O$, 2.9 g; KCl, 0.2 g. Make up to 1 litre with distilled water. Whole buffer best stored at 10× concentration.
PBS–Tween buffer: PBS + Tween 20, 0.5 ml/litre.
Coating buffer (pH 9.6): Na_2CO_3, 1.59 g; $NaHCO_3$, 2.93 g; NaN_3, 0.2 g in 1 litre H_2O.
Substrate buffer (pH 9.8): diethanolamine, 97 ml; H_2O, 800 ml; NaN_3, 0.2 g. Make up to 1 litre. Adjust pH with HCl.

References

BARBARA, D. J. & CLARK, M. F. 1982 A simple indirect ELISA using F(ab')₂ fragments of immunoglobulin. *Journal of General Virology* **58**, 315–322.
BAR-JOSEPH, M. & GARNSEY, S. M. 1981 Enzyme-linked immunosorbent assay (ELISA): principles and applications for diagnosis of plant viruses. In *Plant Diseases and Vectors—Ecology and Epidemiology* eds. Maramirosch, K. & Harris, K. F., pp. 35–60. London: Academic Press.
BAR-JOSEPH, M. & MALINKSON, M. 1980 Hen egg yolk as a source of antiviral antibodies in the enzyme-linked immunosorbent assay (ELISA): a comparison of two plant viruses. *Journal of Virological Methods* **1**, 179–183.
CHESTER, I. B., HILL, S. A. & WRIGHT, D. M. 1983 Serological detection of ryegrass mosaic virus and ryegrass seed-borne virus. *Annals of Applied Biology* **102**, 325–329.
CLARK, M. F. 1981 Immunosorbent assays in plant pathology. *Annual Review of Phytopathology* **19**, 83–106.
CLARK, M. F. & ADAMS, A. N. 1977 Characteristics of the microplate method of enzyme-linked immunosorbent assay for the detection of plant viruses. *Journal of General Virology* **34**, 475–483.
DERRICK, K. S. 1973 Quantitative assay for plant viruses using serologically specific electron microscopy. *Virology* **56**, 652–653.
McLAUGHLIN, M. R., BARNETT, O. W. & GIBSON, P. B. 1979 The influence of plant sap and antigen buffer additives on the enzyme-immunoassay of two plant viruses. In *Abstracts Ninth International Congress Plant Protection, Washington, D.C.* No. 246.
MATTHEWS, R. E. F. 1970 *Plant Virology.* New York & London: Academic Press.
PEMBERTON, A. W. 1979 Campaigns against plant virus diseases in England and Wales. In *Plant Health: the Scientific Basis for Administrative Control of Plant Diseases and Pests* eds. Ebbels, D. L. & King, J. E. Oxford: Blackwell.

TAMADA, T. & HARRISON, B. D. 1980 Application of enzyme-linked immunosorbent assay to the detection of potato leafroll virus in potato tubers. *Annals of Applied Biology* **96**, 67–78.

TORRANCE, L. 1980 Use of bovine C1q to detect plant viruses in an enzyme-linked immunosorbent-type assay. *Journal of General Virology* **51**, 229–232.

TORRANCE, L. & JONES, R. A. C. 1981 Serological methods in testing for plant viruses. *Plant Pathology* **30**, 1–24.

VOLLER, A., BARTLETT, A., BIDWELL, D. E., CLARK, M. F. & ADAMS, A. N. 1976 The detection of viruses by enzyme-linked immunosorbent assay (ELISA). *Journal of General Virology* **33**, 165–167.

Detection of Enteroviruses in Water by Suspended-Cell Cultures

J. S. SLADE, R. G. CHISHOLM AND N. R. HARRIS

Thames Water Authority, New River Head Laboratories, London, UK

Human enteric viruses are a normal constituent of sewage and may be present in very large numbers. Most sewage treatment processes reduce the numbers considerably, but with few exceptions none remove them entirely. Therefore virus is released into the environment with most sewage effluents and sludges, and as a result, practically all of the bodies of water associated with humans are contaminated to a greater or lesser extent. Human enteric viruses constitute a potential hazard as a variety of natural and human agencies may return them to the community, thereby establishing a cycle of infection.

A number of special techniques have been developed in order to evaluate this risk, and many are markedly different from those used in other branches of virology. Environmental samples often contain very small numbers of virus and may contain a wide range of different types. Furthermore, the very small minimum infectious dose, possibly as little as one virus particle, dictates that in certain sensitive situations, e.g. potable supplies, they have to be reduced to very small levels. Accordingly, sensitive methods of detection have had to be developed.

A variety of techniques is now available for concentrating viruses from large volumes of water (Gerba *et al.* 1978; Sobsey & Glass 1980), but the problem of virus detection has been only partly resolved. Cultural methods are the most sensitive for those viruses for which suitable cell systems exist. Among the most widely favoured are derivatives of the monolayer plaque method of Dulbecco (1952) which provide both a rapid indication of virus numbers and a simple means of isolation for subsequent identification.

The cell suspension plaque method was developed by Cooper (1955) in an

MICROBIOLOGICAL METHODS
FOR ENVIRONMENTAL BIOTECHNOLOGY
ISBN 0-12-295040-2

Copyright © 1984 by the Society for Applied Bacteriology
All rights of reproduction in any form reserved

attempt to adapt the versatile bacteriophage techniques to other groups of viruses. It was refined further (Cooper 1961) to give a four- to sixfold increase in plating efficiency. Bradish & Allner (1967) adapted the method to assay Langat and Louping-ill viruses in chick embryo cells, but it does not appear to have been widely used elsewhere. This is possibly due to the relatively large number of cells required. The method was evaluated by the Metropolitan Water Board (now part of the Thames Water Authority) as it was considered that in the field of water supply, the extra sensitivity may well justify the increased cost. The technique proved successful and has been employed since 1974.

In this chapter the adaptation of the cell suspension assay for use with monkey kidney cell lines is described and compared with the monolayer plaque method using both laboratory-cultured virus strains and environmental samples.

Procedures for Suspended-Cell Plaque Assays

Sample Preparation

A series of dip samples was collected from a variety of sources. Treated sewage effluent (100 ml) was obtained from Caddington Sewage Treatment Works near Luton, Bedfordshire. Samples (5 litres) were collected from the lower River Thames at Laleham Surrey at 14-day intervals, a small number from the River Wey at Shalford Surrey and the River Wear in Sunderland. Partially treated water samples (15 and 200 litres) were collected from the Coppermills Water Treatment Works of the Thames Water Authority at weekly intervals.

Viruses in the samples were concentrated by a two-stage method using membrane adsorption and protein precipitation. In the first stage, the sample was adjusted to pH 3.5 with HCl and the water passed through a cellulose nitrate membrane (Sartorius 142-mm diameter, 0.45-μm pore size) protected by a glass fibre prefilter (Whatman 12.5-cm diameter, grade GF/F). Viruses adsorbed to the filters were eluted with 70 ml beef extract (Oxoid Lab-Lemco, 3%, w/v, pH 9). For the 15- and 200-litre samples a membrane of 293-mm diameter was used and elution was with 400 ml beef extract. In the second stage, the beef extract was acidified with HCl to pH 3.5, and the resulting precipitate was sedimented by centrifugation at 2800 g for 20 min. The precipitate was dissolved in either 5 or 10 ml Na_2HPO_4 solution (AR grade 2%, w/v) and stored at $-30°C$.

All filtrations were done with Sartorius stainless steel pressure vessels and filters. The 200-litre samples were concentrated as far as the adsorption stage on site using a virus concentrator similar to that of Hill *et al.* (1974); all other stages in the process were done in the laboratory.

Virus Strains

Attenuated poliovirus type 1 LSc 2ab strain was used for the initial experimental work. Stock virus was prepared by inoculating confluent cultures of Vero cells with 10^7 plaque-forming units (pfu) virus/20-oz bottle. After 3 days of incubation the culture was subjected to three cycles of freezing and thawing and clarified by centrifugation. The number and size of viral aggregates were reduced by passing the suspension through a series of membranes (Millipore GS, VC and VM grades). Aliquots of the virus (1 ml) were stored at $-70°C$. The other types of viruses tested were only processed as far as the clarification stage. Virus strains isolated from contaminated waters in the Thames Water Region were given the prefix EV. The reovirus and echo virus type 4 strains were kindly supplied by Dr Craske of the Central Public Health Laboratory, Colindale.

Cell Culture

Cell Lines

Two monkey kidney cell lines were used, Vero and BGM derived from the African green monkey (Macfarlane & Sommerville 1969; Davis & Phillpots 1974; Barron *et al.* 1970; Dahling *et al.* 1974). These lines received similar treatment and were maintained on a 7-day cycle in both roller and stationary cultures. Roller cultures were grown in glass bottles ($\frac{1}{2}$ gallon, Flow Laboratories Ltd) inoculated with 2.5×10^7 cells in 40 ml growth medium and rotated at 2 rev/min at 37°C. The growth medium was replaced by maintenance medium on the fourth day of incubation. Cells to be used in the suspension plaque assay were harvested on either the fifth or sixth day yielding an average of 1.5×10^8 cells/bottle. Stationary cultures for the monolayer assay were grown in 4-oz bottles, inoculated with 8×10^5 cells in 10 ml growth medium, which was replaced by maintenance medium when the cell sheet was confluent, usually after 2 days, and used on the fourth day.

Vero Cells. Growth medium consisted of Medium 199 with Earle's salts (Wellcome Reagents Ltd); foetal calf serum (Flow Laboratories Ltd 5%, v/v); $NaHCO_3$–CO_2 buffer (0.11%, v/v) added as 2.5 ml solution, 4.4% (w/v)/100 ml medium; penicillin, 200 u/ml; streptomycin 100 µg/ml. Maintenance medium was of similar composition except that the serum concentration was reduced to 2.5% (v/v) and the $NaHCO_3$ buffer was increased to 0.22% (v/v).

BGM Cells. Growth medium contained equal volumes of Leibovitz L15 and Eagle's Minimum Essential Medium with Hank's salts (Gibco Europe Ltd);

foetal calf serum, 10% (v/v); NaHCO$_3$–CO$_2$ buffer, 0.11% (v/v); penicillin, 200 u/ml: streptomycin, 100 µg/ml. Maintenance medium consisted of Medium 199 with Hank's salts; foetal calf serum, 5% (v/v); NaHCO$_3$ buffer, 0.22% (v/v); penicillin, 200 u/ml; streptomycin, 100 µg/ml.

Overlay Medium

Overlay media for both cell lines consisted of the appropriate maintenance medium with the addition of agar (Difco, purified, 1.5%, w/v) and intravital neutral red stain (G. T. Gurr, 0.0015%, w/v). The concentration of neutral red was 0.00075% (w/v) for the cell suspension overlay. When assaying samples from contaminated sources, neomycin sulphate (70 µg/ml), polymixin B sulphate (100 u/ml) and fungizone (3.5 µg/ml) were included.

The overlay medium was prepared in two portions, one containing agar and the other containing the heat-sensitive ingredients. For example, for 100 ml of medium, 1.5 g agar was dissolved in 50 ml distilled water by autoclaving at 121°C for 15 min and cooled to 45°C. The other 50 ml of medium consisted of 10 ml concentrated (10 times) Medium 199, serum and NaHCO$_3$ buffer, made up to volume with sterile distilled water which was warmed to 43°C. The two portions were then mixed and allowed to equilibrate at 43°C. Neutral red stain and antibiotics were added just before use.

Plaque Assays

Monolayer Plaque Assay

Bottles containing confluent cell sheets were drained and inoculated with either 0.5 or 1.0 ml virus suspension. After 1 h at room temperature, 10 ml agar overlay medium was added, rocked over the cell layer and allowed to solidify under a black cloth. When set the cultures were incubated with the cell sheets and agar overlay medium uppermost in the dark at 37°C. Plaques were counted and marked as they appeared after 3–7 days. A selection of plaques was picked and subcultured into tube cultures for confirmation of the presence of virus and identification.

Cell Suspension Plaque Assay

Either 5- or 6-day-old roller culture bottles were drained and the cell layer washed with phosphate-buffered saline (20 ml). Ten millilitres of a solution of trypsin (0.1%, w/v) and EDTA (0.02% w/v) were added and the bottles rolled for a further 10–20 min. The loosened cells were washed off the glass surface, dispersed in maintenance medium and counted. They were then centrifuged at 150 g for 8 min and resuspended in maintenance medium to give a count of 1×10^7/ml. Two millilitres of this suspension were added to 1 ml virus concentrate in a 90-mm-diameter vented plastic petri dish (tissue culture grade plastic was not necessary). To this mixture was added 10 ml overlay medium at 43°C. The suspension was mixed well, allowed to set and incubated at 37°C in the dark in an atmosphere of 5% (v/v) CO_2 in air. The atmosphere was provided by a CO_2–air mixture in a gas-tight desiccator cabinet from which the air was flushed by the gas mixture. Plaques were marked as they appeared after 2–7 days. The presence of virus was confirmed by sub-culturing into tube cultures; identification was made by means of specific antisera.

Optimisation of the Cell Suspension Assay

Investigations were conducted initially on Vero cells using attenuated poliovirus type 1 and later extended to BGM cells. The results from the two cell lines were found to be similar. In the procedure finally adopted for the cell suspension assay, 20×10^6 cells in 2 ml maintenance medium and up to 1 ml virus concentrate were placed in a 90-mm-diameter plastic petri dish and mixed with 10–12 ml overlay medium. A consideration of some of the factors that were investigated during the preliminary work that led to the adoption of this procedure will indicate the aspects to which particular attention should be given.

The use of an uninoculated basal layer of overlay medium and the type of container, i.e. glass, tissue culture grade plastic or normal bacteriological grade plastic petri dishes, were found not to be important. The volume of medium used was not critical but 10–15 ml appeared to be optimal.

Cell density was an important factor. Concentrations ranging from 5×10^6 to 40×10^6 cells/dish were examined. Cell survival was poor with $<10 \times 10^6$ cells/dish. As cell concentrations were increased, plaques became smaller, more clearly defined and more numerous. The number of plaques increased ca. twofold when the cell density was increased from 10×10^6 to 40×10^6/dish. The concentration of 20×10^6 cells/dish finally chosen was a compromise between the number of cells required to give maximum plaque yield and the number of cells consumed.

The use of 1.5 ml and more virus concentrate/dish was occasionally found to give rise to toxic effects. The toxicity varied with different batches of beef extract and, less predictably, with the volume and characteristics of the water sampled.

Control of pH value of the medium was found to be important. Increases in acidity caused by CO_2 accumulation and increases in alkalinity due to CO_2 loss adversely affected cell survival. The use of vented petri dishes and the precaution of only half filling the gas-tight containers with dishes during incubation avoided the problem of CO_2 accumulation. Loss of CO_2 was controlled by minimising the time of exposure to air during plaque counting.

The possibility of controlling pH value by using HEPES buffer (20 mM) in place of the NaHCO$_3$ buffer was examined briefly but gave variable success and was not pursued further. Exposure of the dishes to light was minimised to avoid the adverse photosensitising effect on the neutral red dye.

Comparison of the Cell Suspension and Monolayer Assays

Cultured Virus Strains

When poliovirus type 1 was titrated, the suspension assay produced a markedly higher average titre (4- to 5-fold greater) than the monolayer assay for Vero and BGM cell lines (Tables 1 and 2). With the other virus types examined with BGM cells (Table 2) there were two distinct responses, i.e. Coxsackie and polio type 3 viruses showed a higher titre with the suspension assay (2.7- to 9.7-fold), whereas the echo and reovirus strains showed a lesser titre.

It was also noticed that the cell suspension plates produced results more rapidly. Typically some 95% of the plaques were visible in 3 days after overlaying and very few more appeared after 5 days. In the monolayer bottles however, ca. 85% of the plaques appeared after 3 days and new plaques continued to appear for up to 7 days.

Environmental Samples

Much larger numbers of virus plaques were also isolated from environmental samples with the cell suspension assay for the two cell lines (Tables 3 and 4), although the increase over the monolayer method appears to be much more variable than was shown in Tables 1 and 2 for cultured virus strains, i.e. increases of 1.5- to 25-fold with Vero cells (Table 3) and 1.0- to fold with

TABLE 1

Comparison of titres of cultured enterovirus suspensions obtained by the monolayer and cell suspension methods using Vero cells

Virus Type	Strain	Monolayer (M)	Suspension (S)	S/M
Polio 1	LSc	3.7 × 10^7	20.0 × 10^7	5.4
Polio 1	LSc	4.3 × 10^7	22.5 × 10^7	5.2
			Mean	5.3

TABLE 2

Comparison of titres of cultured enterovirus suspensions obtained by the monolayer and cell suspension methods using BGM cells

Virus		Titre (pfu/ml)		
Type	Strain	Monolayer (M)	Suspension (S)	S/M
Polio 1	LSc	5.1×10^8	23.7×10^8	4.6
Polio 1	LSc	2.1×10^7	9.0×10^7	4.3
Polio 3	EV 304	2.2×10^9	21.3×10^9	9.7
Coxsackie B1	EV 322	8.2×10^9	22.3×10^9	2.7
Coxsackie B4	EV 302	1.1×10^9	7.9×10^9	7.2
Coxsackie B5	EV 342	4.9×10^9	23.0×10^9	4.7
Coxsackie A9	EV 225	1.1×10^7	6.3×10^7	5.7
			Mean	5.6
Echo 1	EV 298	13.8×10^7	8.3×10^7	0.60
Echo 4	Du Toit	54.8×10^7	6.7×10^7	0.12
Reo 1	Wallis 716	2.0×10^6	$<0.3 \times 10^6$	<0.15
			Mean	<0.29

BGM cells (Table 4). Possibly of much greater significance, however, is the isolation of viruses from the environment by the suspension assay which were not detected by the monolayer method.

The results of the application of the cell suspension and monolayer methods to a wide variety of samples over a period of several years are shown in Table 5. It is apparent that only a limited range of virus types have been isolated, i.e. polio and Coxsackie B, although other virus types must have been present on occasions.

TABLE 3

Comparison of titres of virus from environmental samples obtained by the monolayer and cell suspension methods using Vero cells

Sample		Titre (pfu/volume sampled)		
Site	Volume (litres)	Monolayer (M)	Suspension (S)	S/M
River Thames, Laleham	2.5	2	3	1.5
	2.5	1	25	25.0
	2.5	0	21	>21.0
	2.5	1	8	8.0
	2.5	0	8	>8.0
River Wey, Shalford	1.0	24	188	7.8
			Mean	>11.9

Table 4

Comparison of titres of virus from environmental samples obtained by the monolayer and cell suspension methods using BGM cells

Sample Site	Volume (litres)	Monolayer (M)	Suspension (S)	S/M
Treated sewage effluent				
Caddington	0.1	60	347	5.8
River water				
River Thames, Laleham	2.5	14	22	1.6
	2.5	11	19	1.7
	2.5	19	63	3.3
	2.5	20	51	2.6
	2.5	20	104	5.2
River Wear	2.5	80	301	3.8
	2.5	230	526	2.3
River Wey, Shalford	2.5	43	220	5.1
Partially treated drinking water				
Coppermills Water Works	7.5	2	7	3.5
After rapid sand filtration	7.5	6	44	7.3
	7.5	0	16	>16.0
	7.5	1	1	1.0
	7.5	0	1	> 1.0
	7.5	0	8	> 8.0
	7.5	0	15	>15.0
	7.5	3	3	1.0
	7.5	0	1	> 1.0
	7.5	0	1	> 1.0
After slow sand filtration	100	3	17	5.7
	100	0	4	> 4.0
			Mean	5.1

Table 5

The relative proportions of virus types isolated on Vero and BGM cells by the monolayer and cell suspension plaque methods

Cell line	Assay method	No. of isolates	Polio 1	Polio 2	Polio 3	Coxsackie B 1	Coxsackie B 2	Coxsackie B 3	Coxsackie B 4	Coxsackie B 5	Coxsackie B 6
Vero	Monolayer	241	1	21	2	1	0	17	8	50	0
Vero	Suspension	508	2	25	2	0	2	2	30	37	0
BGM	Suspension	5542	4	28	5	4	8	9	19	22	1

Discussion

Clearly the cell suspension assay produces greater counts of polio and Coxsackie B viruses than the monolayer assay. What is more important, however, is the demonstration by the cell suspension method of the presence of viruses in samples which were not detected by the monolayer method. This better sensitivity could be of great importance in situations where low levels of virus contamination are considered to be significant.

The range of virus types isolated with the methods used here is rather limited, but the data in Table 5 suggest that this limitation is mainly due to either the concentration process, or, more probably, the cell lines, or both. Although the introduction of the cell suspension assay did not substantially alter the range of virus types isolated, it should be noted that the results are not fully comparable. This is because the range of samples tested by the two methods was similar but they were collected over different periods of time, when different virus types may have been excreted by the local population; this is a factor which is always changing.

The increased speed with which results become available with the cell suspension method is an advantage but it is not a sufficient one to make a major change in laboratory practice. Possibly more important is the ready availability of cells for the suspension method. Provided that cells are available in reasonable condition, a suitable suspension can be prepared and used at short notice, whereas in the monolayer method suitable bottles of cells take several days to prepare. A method of storing suspended cells for periods up to 4 days was reported by Cooper (1961) but this was not investigated.

A little extra care needs to be taken when using suspended cells as in this state they appear to be more sensitive to a variety of factors than when in monolayers, i.e. toxicity. Any difficulty that may be experienced in reliably detecting plaques in the cell suspension assay may be easily overcome with a little practice.

The main disadvantage of the cell suspension assay is the large number of cells consumed, possibly as many as 100×10^6/sample. This requires culture on a larger scale than is normal in most laboratories that use the monolayer method. A further possible difficulty is the provision of a CO_2–air gas mixture, although facilities such as CO_2-gassed incubators are becoming more widely available.

The reason why cells in suspension exhibit greater sensitivity to virus than when in monolayer is not known. It is probably related to the chelating agents and digestive enzymes used to prepare the cell suspension. The resuspension process will have a great effect on the surface of the cells and this may affect virus attachment or penetration.

The cell suspension plaque assay method appears to have potential. The use of other cell lines should widen the range of viruses isolated and better indicator dyes may improve plaque visibility. If cell survival could be prolonged, this should permit greater recovery of slowly growing viruses. Alternative media formulations could facilitate pH control and possibly remove the need for a CO_2–air atmosphere.

Acknowledgement

This chapter is published with the permission of Dr. M. C. Dart, Director of Scientific Services, Thames Water Authority. Views expressed do not necessarily correspond to those of the authority.

References

BARRON, A. L., OLSHEVSKY, C. & COHEN, M. M. 1970 Characteristics of the BGM line of cells from African Green Monkey Kidney. *Archiv fuer die Gesamte Virusforshung* **32**, 289–392.

BRADISH, C. J. & ALLNER, K. 1967 The assay of Langat and Louping-ill viruses and their specific antiserum in agar-suspensions of chick-embryo fibroblasts. *Proceedings of the Conference on Rapid Identification of Biological Agents* pp. 234–244.

COOPER, P. D. 1955 A method for producing plaques in agar suspensions of animal cells. *Virology* **1**, 397–401.

COOPER, P. D. 1961 An improved agar cell-suspension plaque assay for poliovirus: some factors affecting efficiency of plating. *Virology* **13**, 153–157.

DAHLING, D. R., BERG, G. & BERMAN, D. 1974 BGM, a continuous cell line more sensitive than primary Rhesus and African Green kidney cells for the recovery of viruses from water. *Health Laboratory Sciences* **11**, 275–282.

DAVIS, P. M. & PHILLPOTTS, R. J. 1974 Susceptibility of the the VERO line of African green monkey kidney cells to human enteroviruses. *Journal of Hygiene* **72**, 23–30.

DULBECCO, R. 1952 Production of plaques in monolayer tissue cultures by single particles of an animal virus. *Proceedings of the National Academy of Sciences of the United States of America* **38**, 747–752.

GERBA, C. P., FARRAH, S. R., GOYAL, S. M., WALLIS, C., & MELNICK, J. L. 1978 Concentration of enterovirus from large volumes of tap water, treated sewage, and sea water. *Applied and Environmental Microbiology* **35**, 540–548.

HILL, W. F., AKIN, E. W., BENTON, W. H., MAYHEW, C. J. & JAKUBOWSKI, W. 1974 Apparatus for conditioning unlimited quantities of finished waters for enteric virus detection. *Applied Microbiology* **27**, 1177–1178.

MACFARLANE, D. E. & SOMMERVILLE, R. G. 1969 VERO cells (*Cercopithecus aethiops* kidney)—growth characteristics and viral susceptibility for use in diagnostic virology. *Archiv fuer die Gesamte Virusforschung* **27**, 379–385.

SOBSEY, M. D. & GLASS, J. S. 1980 Improved electropositive filters for concentrating viruses from large volumes of water. In *Viruses and Wastewater Treatment* eds. Goddard, M. & Butler, M., pp. 239–245. Oxford: Pergamon Press.

Mutagenicity Testing of Drinking Water Using Freeze-Dried Extracts

R. FORSTER*

Water Research Centre, Medmenham Laboratory, Medmenham, Marlow, Buckinghamshire, UK

Drinking waters are known to be contaminated with a wide range of organic chemicals present at very small concentrations (Fielding & Packham 1977). These chemicals are of natural and anthropogenic origin and frequently include chlorinated organic compounds formed during chlorination for disinfection. There has naturally been some concern about the possible long-term health effects of the ingestion of these micropollutants, kindled by epidemiological studies in the United States which appeared to show links between different drinking water supplies and the incidence of some cancers (Page *et al.* 1976), and further fired by reports of mutagenic activity detected in extracts of drinking waters (Loper 1980). This chapter deals with a part of a wider study of the possible health implications of these micropollutants made by workers at the Water Research Centre, Medmenham Laboratory. An examination of methods for concentration and extraction of samples and the application of the chosen procedure to testing the drinking water supplies for mutagenic activity are reported.

Sample Collection and Analysis

Drinking water samples were collected at the water treatment works immediately before distribution; the samples were processed either immediately on return to the laboratory or after storage at low temperature to avoid

---

*Present address: Life Science Research, Roma Toxicology Centre, Via Tito Speri 14, 00040 Pomezia, Rome, Italy.

MICROBIOLOGICAL METHODS
FOR ENVIRONMENTAL BIOTECHNOLOGY
ISBN 0-12-295040-2

Copyright © 1984 by the Society for Applied Bacteriology
All rights of reproduction in any form reserved

degradation. Procedural blanks (see p. 382) consisted of extracts prepared from double-glass-distilled water, to which 0.3–0.5 g/litre KCl had been added to produce a realistic amount of solid residue for extraction. Samples were analysed for total organic carbon (TOC) and residual disinfectant; the mutagenicity of water extracts is believed to be related to the organic content of the water and chlorination practice (Loper 1980). The level of inorganic nitrite was also measured since nitrite is mutagenic to bacteria (Zimmermann 1977). A sample of 15 litres was found to be adequate to allow a full set of mutagenicity tests in duplicate.

Preparation of Samples for Mutagenicity Testing

Concentration of Samples by Freeze-Drying

A number of methods are available for concentrating the organic material from water, such as solvent extraction, adsorptive methods (using XAD resins, alumina, activated carbon or polyurethane foam), freeze concentration, reverse osmosis or rotary evaporation (reviewed by Jolley 1981). Freeze-drying offers a very gentle method of concentration which avoids high temperatures or extremes of pH value and should, therefore, minimise the chances of altering the chemical composition of the organic chemicals. In addition, apart from the loss of some of the more volatile materials, freeze-drying concentrates the entire spectrum of organic chemicals in a sample. In this respect it differs from methods such as XAD adsorption, which are effective in concentrating mutagenic activity but adsorb only a small proportion of the organic matter in water.

In preliminary work, water samples were freeze-dried on a contract basis at the Microbiological Research Establishment, Porton Down, Salisbury. Subsequently, most samples were freeze-dried in a centrifugal freeze-drier (Edwards High Vacuum Ltd, Speedivac Model 30P2) although some samples were prepared in a continuous freeze-drier (Labmark, Borolabs Ltd). The Edwards machine has a cycle time of about 74 h for 6 litres, allowing a throughput of about 12 litres in a working week. We have recently installed a shelf freeze-drier (Edwards High Vacuum Ltd, Minifast 3400) for this work which allows the processing of an 8-litre sample in less than 24 h.

Extraction of Mutagenic Activity

The powdery material produced by the freeze-drying of drinking water samples, broadly equivalent to the total dissolved solids content of the sample, is not immediately amenable to mutagenicity testing, however, and a further step of isolating the organic content from the residue is required.

Choice of Solvent

Petroleum ether, diethyl ether and methanol were compared for efficiency in extracting mutagenicity from freeze-dried concentrates. The mutagenicity assay procedure is given on page 378. Freeze-dried residues of a sample of a lowland surface water (site "X") were extracted in each solvent by shaking for 4 h. In addition, Soxhlet extraction for 8 h was used with each solvent. The extracts were concentrated to small volumes (see *Methanol Extraction Procedure*, following) and tested in a range of amounts equivalent to up to 100 ml original water sample/ml assay mixture (see p. 379).

The results of the mutagenicity test showed the majority of the activity to be in the methanol extract (Table 1); the other extracts were either inactive, slightly mutagenic or toxic. No advantage was obtained by Soxhlet extraction. In the light of these results, methanol extracts were used routinely. The use of a pre-extraction step with diethyl ether did not increase the effectiveness of methanol extraction. It was found that ultrasonication was a quick and efficient method of performing the extraction.

TABLE 1

Mutagenicity of solvent extracts from freeze-dried concentrates of drinking water sample 040

Solvent	S. typhimurium TA98 −S9	S. typhimurium TA98 +S9	S. typhimurium TA100 −S9	S. typhimurium TA100 +S9
Petroleum ether	NS[*,†]	NS	NS	NS
Diethyl ether	+[‡]	NS	NS	NS
Methanol	+++[§]	+++	+	+++[†]
Petroleum ether (S)[‖]	NS[†]	NS	NS	NS
Diethyl ether (S)	+++	NS[†]	—[#]	NS
Methanol (S)	NS	NS	NS	NS

[*] NS, Not statistically significant.
[†] Results of one experiment only.
[‡] (+), Significant at 5% level.
[§] (+++), Significant at 0.1% level.
[‖] (S), Soxhlet extraction.
[#] (—), Markedly inhibitory to test strain.

Methanol Extraction Procedure

The freeze-dried residue was extracted three times with spectrophotometric grade methanol (Rathburn Ltd); the residue and methanol were sonicated while held in ice for about 12 min and then clarified by centrifugation. The volume of methanol was subsequently reduced by rotary evaporation, and then finally under a stream of N_2, with further clarification by centrifugation. For testing for mutagenic activity, the volume of the extract was made up to 100 µl for each litre of original sample, i.e. a concentration factor of 10^4.

Assay for Mutagenic Activity

The presence of mutagenic substances in drinking water samples was assessed by testing for induced mutations in strains of *Salmonella typhimurium* using the fluctuation test.

Test Strains of Bacteria

Salmonella typhimurium TA98 and TA100 (Ames *et al.*, 1975) were used. Cultures were checked regularly for induced and spontaneous mutation rates and the presence of the appropriate genetic markers, i.e. histidine requirement, "deep rough" (*rfa*) mutation, presence of plasmid pKM101, and the deletion of the *uvrB* region. Permanent stock cultures were stored over liquid N_2. The strains were obtained from the Medical Research Council (MRC) Cell Mutation Unit, University of Sussex. Estimates of bacterial numbers were made by serial dilution and replicate plating on nutrient agar medium.

Metabolic Activation

Since a number of chemicals are activated to mutagenic forms during mammalian metabolism (Miller 1970; reviewed by Magee 1974), mutagenicity tests must be conducted both in the presence and absence of a metabolising system. The most widely used system consists of the post-mitochondrial supernatant fraction (S9) of rat liver homogenate (pre-induced to increase enzyme levels) and an NADPH-generating system. The complete system is referred to as "S9 mix". The rat liver S9 fraction was prepared regularly following closely the method of Ames *et al.* (1975), using pooled, young male Sprague–Dawley rats, treated with 500 mg/kg Aroclor 1254 in corn oil. Aliquots of S9 fraction were stored in liquid N_2 and normally used within 3 months of production. The S9 mix was prepared from S9 fraction and cofactors, following exactly the method of Ames *et al.* (1975). For routine testing 100 μl S9 fraction/ml S9 mix were used.

Fluctuation Test

The fluctuation test procedure used was that described by Green *et al.* (1976, 1977). The principles of the test procedure will be briefly described.

The fluctuation test offers a way of measuring the rate of reversion to histidine prototrophy in histidine-requiring test strains. A set of 50 replicate cultures of the test strain is incubated in a liquid medium containing test extract and a carefully measured amount of histidine sufficient for growth of a small number of generations. At the end of the period of growth, a selective liquid medium is added, in which only non-histidine-requiring revertants can grow. The occurrence of revertant mutation is detected by the presence of growth after a few days of incubation as shown by a colour change of the pH indicator dye incorporated in the selective medium. The number of replicate cultures which have produced mutants is recorded and the values obtained in the treated series are compared with those obtained in the "control" series. A statistically significant increase in the number of replicates containing mutants is taken to indicate that the test extract is mutagenic. The rate of mutation may be calculated using figures for the number of bacteria present at the beginning and end of the period of growth.

Safety Precautions

Since this work involves the use of mutagenic/carcinogenic compounds, it is essential that any laboratories intending to undertake such tests should familiarise themselves with the safety hazards involved. Safe operating procedures should be developed for the storage, handling, pipetting and disposal of mutagenic compounds and for dealing with leaks, spills and other excursions (Anonymous 1981).

Amount of Sample Tested

A total amount of extract equivalent to 500 ml original sample water added to one set of 50 test tubes for the fluctuation test was sufficient to give a strong response in the assay. As each set of test tubes contained 5 ml incubation mixture, the amount of sample in the incubation mixture is equivalent to 100 ml original water sample/ml assay mixture. In other words, a final concentration factor of only 100-fold was required to obtain a strong response. Subsequent work confirmed that the amounts of extract equivalent to 20–200 ml water sample/ml mixture are a useful range in which to work, i.e. total amounts of extract tested equivalent to 100–1000 ml water sample.

Effects of Extract on Test Strains

For the valid functioning of the fluctuation test, the numbers of bacteria should be the same in the test and the control series. This assumption may not be met if the sample either inhibits or enhances the growth of the test bacteria. If the extract is inhibitory, it may not be possible to test for mutagenic activity; such inhibition will be detected by the test as a lower response than is seen in the controls. Less marked inhibition may result in only a slight underestimate of the mutagenic activity of the sample.

Enhanced yields of bacteria in the test series compared with the controls can lead to spurious "positive" results. Since the observed number of mutants is the product of the number of bacteria at risk and the mutation rate, an increased yield of bacteria will result in a greater observed number of mutants in the fluctuation test, purely as a result of the spontaneous mutation rate. The simplest explanation of an enhanced yield is the presence of histidine in the material under test, but we have observed an enhanced yield with water samples in which we could not demonstrate the presence of significant quantities of histidine (Forster *et al.* 1981).

For these reasons we have examined the bacterial yields in the fluctuation test after the period of histidine-limited growth. The extracts were found to be mildly inhibitory but rarely unsuitable for assay (Table 2); part of the inhibition may be explained by the precipitation caused by the extracts in the presence of inorganic phosphates.

Test Procedure

Each test point consisted of 50 replicates incubated in test tubes; multi-well systems, e.g. plastic Dispostrays or Microtitre trays, are also suitable.

TABLE 2

Inhibitory effect of drinking water extracts* on Salmonella typhimurium *strain TA100 in the fluctuation test*

Amount of extract (μl/ml test mixture)	Yield† of strain TA100 as percentage of control value	
	Sample no. 042	Sample no. 119/1
20	95.9	82.8
50	64.1	65.0
100	65.2	50.6
200	59.7	33.7

* Methanol extract of freeze-dried concentrate; tested without S9 mix.
† Results of eight replicate cultures from each of two experiments.

Each experiment included replicate negative controls with no test extract and a positive control as well as a range of amounts of the extracts under test. The positive control compounds for each test strain of *S. typhimurium* were 4-nitro-*o*-phenylenediamine (for strain TA98 without S9 mix activation); sodium azide (for strain TA100 without S9 mix activation); ethidium bromide (for strain TA98 with S9 mix activation); 2-aminoanthracene (for strain TA100 with S9 mix activation).

Each test point requires 5 ml 2% (v/v) Vogel–Bonner medium (Vogel & Bonner 1956) supplemented with glucose (4 g/litre), biotin (10 μg/ml) and histidine (1.5 μg/ml for test strain TA98, or 0.5 μg/ml for test strain TA100). To this is added 30 μl overnight culture of test strain TA98, or 30 μl overnight culture of test strain TA100 diluted 10- or 100-fold to reduce the numbers of pre-existing mutants present. After addition of test extract the mixture is dispensed in 100-μl aliquots into 50 test tubes and incubated overnight at 37°C. To each tube is then added 2 ml selective medium (2% v/v, Vogel–Bonner medium supplemented with glucose, 4 g/litre, and bromocresol purple, 5 μg/ml) and the results recorded after incubation for a further 72 h. In experiments using S9 mix activation, 1 ml S9 mix is substituted for 1 ml Vogel–Bonner medium in the initial incubation mixture.

Interpretation of Results

The results of the fluctuation tests were tested for statistical significance using an analysis based on generalised linear interactive modelling (GLIM) (Baker & Nelder 1978). Taking the data from a full range of different amounts of samples for one extract, including the negative controls, GLIM tests whether the variability exhibited by the observed numbers of revertant-bearing wells is significantly greater than would be expected on the assumption of binomial variation. The p value is calculated to test if the observed results could have arisen by chance; p values of 0.05 or less (after adjustments for multiple comparisons) are taken to indicate significant mutagenicity.

The adjustment for multiple comparisons takes account of the fact that four sets of experimental conditions were used with each extract, i.e. test strains TA98 and TA100 with and without S9 mix activation, each of which was assumed to be an independent test. Accordingly, to accept a result as significant at the 5% level a p value of 0.0127 or less is required.

For comparative purposes it is desirable to have a measure of the mutagenicity of the extracts. Therefore, data were analysed using a modified GLIM which attempted to fit a linear component to the results. From this analysis an estimated slope value was obtained for the sample dose–response in terms of Mi (mean induced mutants per tube) per unit concentration of sample tested (millilitres of water sample per millilitres of test mixture). In

TABLE 3

Mutagenicity of extracts* of 16 UK drinking water supplies

Sample no.	S. typhimurium TA98 −S9	S. typhimurium TA98 +S9	S. typhimurium TA100 −S9	S. typhimurium TA100 +S9
Ground water				
100	+++[†]	NS[‡]	+[§]	NS
101	NS	NS	NS	NS
102	NS	NS	++[‖]	NS
113	NS	NS	NS	NS
Upland surface water				
110	NS	NS	++	NS
112	NS	NS	+	NS
115	++	NS	+++	NS
Lowland surface water[#]				
116	+++	NS	+++	NS
114	+++	NS	+++	NS
108	+	NS	+++	NS
109	−[**]	NS	+++	NS
111	+++	NS	+++	NS
091	+++	NT[††]	+++	+++
103	+++	NS	++	NS
104	+++	++	+++	NS
105	+++	++	+++	NS

* Methanol extract of freeze-dried concentrate.
[†] (+++), Significant at 0.1% level.
[‡] NS, Not statistically significant.
[§] (+), Significant at 5% level.
[‖] (++), Significant at 1% level.
[#] Lowland water samples arranged in order of increasing sewage pollution.
[**] (—), Markedly inhibitory to test strain.
[††] NT, Not tested.

the data presented in Tables 4 and 5 and Fig. 1, slope values are given as (Mi per unit concentration of sample tested) × 200. Not all of the data were amenable to this treatment; in a number of cases the GLIM analysis showed that the linear assumption was not appropriate to the data, giving in these cases a poor estimate of the slope value.

Examination for a Contribution of the Extraction Procedure to Mutagenic Activity

Procedural blanks (p. 376) were used to demonstrate that the procedure for extract preparation did not contribute to the observed mutagenicity. In initial studies, procedural blanks prepared using deionised water as the test

TABLE 4

Effect of metabolic activation on the mutagenic response of Salmonella typhimurium *strain TA98 to an extract of drinking water sample 042 in the fluctuation test*

Volume water samples tested[*] per experiment (ml/ml test mixture)	Mean revertant-bearing tubes	Mi	Significance (p) Mutagenicity	Non-linear component in data	Slope of sample response curve Value[‡]	95% confidence limits
Without S9 mix activation						
0	8.0	0.174				
20	20.0	0.511	<0.0001	0.0417[§]	2.544	1.670–3.418
50	31.5	0.994				
100	33.5	1.109				
With S9 mix activation						
0	12.8	0.294				
20	16.5	0.400	<0.0001	<0.0001	8.662	6.352–1.972
50	29.5	0.892				
100	49.5	4.605				

[*] Methanol extract of freeze-dried concentrate.
[†] Two replicate experiments.
[‡] {[Mean induced mutants/tube (Mi)]/(ml water tested/ml test mixture)} × 200.
[§] Not significant.

material, Soxhlet extractions with petroleum ether, diethyl ether and methanol demonstrated some mutagenicity to both test strains. Subsequently, however, satisfactory non-mutagenic blanks have repeatedly been produced for extracts prepared by direct extraction with methanol. For the survey of drinking waters (see below) it was hoped that a groundwater supply would serve as the procedural blank. On extensive testing, however, samples from the groundwater sites we had hoped to use (100, 101) showed weak but reproducible mutagenic activity.

Assay for Histidine in Test Extracts

It is necessary to test whether the water sample extracts to be examined for mutagenic activity contain sufficient histidine to interfere with the *Salmonella* reversion assays.

Assay Methods
Biological Assay. Advantage was taken of the fact that the final yield of *S. typhimurium* TA100 in a culture is linearly related to the initial concentration of histidine available to the bacteria. A series of six flasks was prepared containing minimal (Vogel–Bonner) medium supplemented with 0–0.2

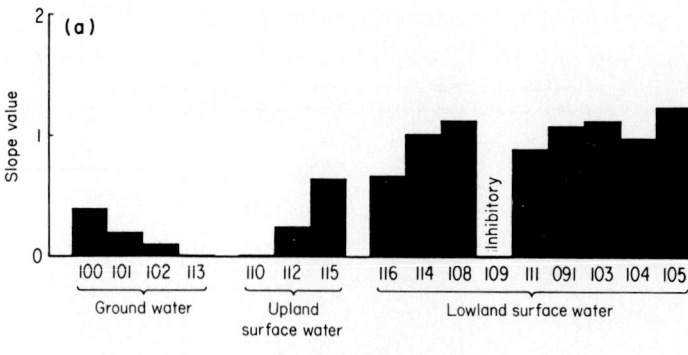

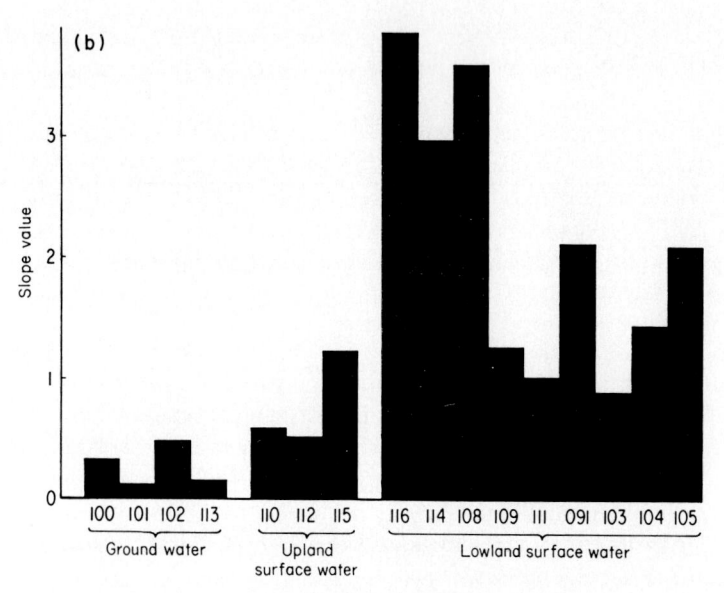

FIG. 1. The mutagenicity slope values of extracts of 16 UK drinking water supplies using *Salmonella typhimurium* TA98 (a) and *S. typhimurium* TA100 (b). Extracts are methanol extracts of freeze-dried concentrate; tested without S9 mix metabolic activation. Slope value = {[mean induced mutants/tube (Mi)]/(ml water tested/ml test mixture)} × 200. Lowland water samples are arranged in order of increasing sewage pollution, left to right.

μg/ml histidine as standards for the assay and excess biotin. Test strain TA100 was added to give a final concentration of 10^5 organisms per millilitre. A range of quantities of the test material was added to a similar series but containing no added histidine. Where possible, the amounts of test material added were estimated to give histidine concentrations midway in the range of the standard series. After overnight growth in a shaking water bath at 37°C, the number of viable organisms was estimated by plating on nutrient agar medium. The concentration of histidine in the test material was then estimated using the regression line of bacterial numbers on histidine concentration obtained from the standards.

Chemical Assay. A modification of the *ortho*-phthalaldehyde method for the assay of histidine (Ambrose *et al.* 1969) was also used for some samples. The fluorescence of the derivatised histidine was measured only at the appropriate retention time during HPLC separation in order to avoid interference by other compounds derivatised by the procedure.

Levels of Histidine Detected

Although histidine is slightly soluble in methanol, the drinking water extracts did not appear to contain histidine in quantities sufficient to interfere with the *Salmonella* reversion assays. The chemical method detected only very small quantities of histidine (0.1–0.8 μg/ml extract) in four methanol extracts. It remained possible, however, that histidine was available, e.g. through the action of bacterial proteases and peptidases on oligopeptides, but not detected by the chemical method. However, using the biological method the histidine content of four further samples was less than the detection limit of 0.005 μg/ml, a negligible concentration. These results suggest that histidine will not interfere with the use of methanol extracts for mutagenicity studies.

Mutagenic Activity in Drinking Water

Survey of Drinking Water Supplies

Using these methods we surveyed the mutagenicity of 16 UK drinking water supplies, chosen to represent a range of water type, i.e. ground water, and upland and lowland surface-derived waters. The lowland supplies included sites subject to various degrees of sewage pollution. Since sample sizes were limited by practical consideration we used in this exercise amounts of test extract equivalent to 20, 50 and 100 ml original water sample/ml fluctuation test mixture. Mutagenic activity was readily detected in this set of extracts (Table 3; Fig. 1). Only two samples did not show any mutagenicity at the levels of extract used; both were ground waters, and one (101) was unchlorinated. However, sample 101 exhibited some mutagenicity on more extensive testing.

The predominant activity of these extracts was to test strain TA100 in the absence of S9 mix activation, although mutagenic activity to strain TA98 in

TABLE 5

*Repeated sampling of drinking water at two sites: test for mutagenicity of extracts** using Salmonella typhimurium *strain TA100*

Sample no.	Date of sampling	Significance of non-linear component in data	Slope Value†	5% confidence limits
Site X				
040	Aug 1978	NS‡	0.676	0.166–1.186
074	Jan 1980	+++§	6.626	5.123–8.129
076	Feb 1980	+‖	1.156	0.785–1.527
079#	Apr–May 1980	+	2.135	1.722–2.548
082#	June–July 1980	NS	1.795	1.118–2.472
093#	Aug–Nov 1980	NS	1.954	1.530–2.378
097#	Feb–Nov 1981	NS	1.421	1.251–1.591
119/I	Nov 1981	+	2.227	1.649–2.760
119/a#	Nov 1981–Jan 1982	NS	1.471	1.167–1.775
Site M				
100a	Nov 1980	NS	0.322	0.050–0.593
100b#	Jan–Feb 1981	NS	0.472	0.286–0.657
100c	Apr 1981	NS	0.281	0.131–0.432

* Methanol extract of freeze-dried concentrate; tested without S9 mix mutagenicity activation.
† {[Mean induced mutants/tube (Mi)]/(ml water tested/ml test mixture)} × 200.
‡ NS, Not statistically significant.
§ (+++), Significant at the 0.1% level.
‖ (+), Significant at the 5% level.
Composite sample collected over period stated.

the absence of S9 mix activation was also common. In tests with S9 mix-mediated metabolic activation, mutagenicity was invariably either abolished or diminished. There were, however, two lowland surface supplies in which mutagenicity to strain TA98 persisted in the presence of S9 mix activation.

Although S9 mix activation attenuated mutagenic activity for this particular set of samples, we have encountered samples in which activation increased mutagenicity. This is illustrated in Table 4 for a surface water sample tested with strain TA98.

An examination of Fig. 1 shows that in general the mutagenic activity of the lowland surface waters was greater than that of the upland surface waters, which in turn was greater than that of the groundwater supplies. This apparent difference may in part be influenced by the choice of sites for sampling; the lowland supplies were predominantly from sites with substantial levels of sewage pollution, while groundwater samples were intentionally selected sites known to be free of contamination problems since it was hoped that groundwater samples would serve as procedural blanks, a hope that was not realised.

Repeated Sampling at One Site

During the course of these studies, methanol extracts of samples from one site (X) were prepared on nine occasions. The results obtained on testing these extracts with strain TA100 without S9 mix activation are presented in Table 5. Individual samples and composite samples (bulked from samples taken over periods of up to 9 months) are included in these results. In qualitative terms the results are consistent in that all the samples were mutagenic to strain TA100. In quantitative terms, there is some variation in the mutagenicity as expressed in terms of the slope of the dose–response curve. The slope values appear to vary substantially but, apart from one sample, the results vary by only a factor of two, implying a difference between, say, 250 and 500 ml to produce the same mutagenic effect. Chemical mutagens, however, may differ in potency by factors of more than six orders of magnitude (McCann & Ames 1976), i.e. from nanogram to milligram quantities. Seen in this perspective, these repeated samples show satisfactory consistency. At another site (M), sampled on three occasions over a period of 5 months, slope values differed by a factor of less than two-fold (Table 5).

There are two sources for the variation in the results obtained with repeated sampling. One is the variation resulting from genuine differences between the samples; for example, it seems likely that there will be seasonal effects on the mutagenic activity of the samples (Grimm-Kibalo *et al.* 1981). In addition, occasional alterations in water treatment practice are likely to influence the mutagenicity of the extracts. The other source of variation is the accuracy of the experimental determination of the mutagenicity. The response with some samples was clearly non-linear and not amenable to the linear analysis; in these cases the quoted slope value has very wide confidence limits. Even with responses to which linear analysis was applicable, further replication of testing would have increased the resolution of the slope value. The contribution of this factor to the variation is probably relatively small.

Purification of Mutagenic Compounds of Freeze-Dried Methanol Extracts

A subsidiary effort was given to the purification of mutagens in freeze-dried methanol extracts of chlorinated water from one site (M) using physicochemical methods of partitioning, separation and fractionation, coupled with mutagenicity testing using strain TA100 without S9 mix activation. The objective of these studies was to produce purified fractions for analysis by gas chromatography–mass spectrometry or liquid chromatography–mass spectrometry.

Initially, partitioning was attempted using an acid–base–neutral separation scheme with an ether extract of the methanol extract. No strong mutagenicity was seen either in any separate fractions or when the fractions were recombined. However, a water-soluble extract of the ether extract was strongly mutagenic. As activity was stable at pH 1 but substantially reduced at pH 13, attempts to separate the water-soluble extract components by an acid–base scheme were confined to separation into acid and residual fractions. Most of the mutagenic activity was found in the residual fraction. From this observation, together with the extraction of activity from the freeze-dried residue by methanol but not by petroleum ether or diethyl ether, it appears that the major mutagenic activity to strain TA100 is due to polar compounds which are labile at alkaline pH values. The use of HPLC is showing some promise in giving good separation of the mutagenicity components of some fractions of the extracts.

Discussion

The results presented here demonstrate that concentration by freeze-drying and extraction with methanol provides an effective method for preparing drinking water samples for examination for the presence of mutagens. The mutagenic activity of extracts is the only measure of the efficiency of extraction, since there is no absolute criterion against which to compare recovery of mutagens. In our experience, smaller volumes of sample are required to demonstrate mutagenicity using this method than are required using XAD resins. Although Crathorne *et al.* (1979) used freeze-drying for extracting non-volatile organic compounds from drinking water, little information is available from the literature about the performance of freeze-drying techniques in recovering mutagenic activity, and assessment of the information is made difficult by the different methods employed. Maruoka & Yamanaka (1982) reported that freeze-dried concentrates were more mutagenic than XAD-2 + diethyl ether extracts of unchlorinated river water. On the other hand, Kool *et al.* (1981) found that XAD-4 + 8-dimethylsulphoxide (DMSO) extracts of untreated Rhine water were at least as effective as freeze-dried extracts; their freeze-dried extracts were prepared by packing the solid residue of a Rhine water sample in a column and eluting organic material with DMSO.

The profusion of methods employed by different investigators of water-borne mutagens may appear to be only a source of confusion and a hindrance to the attempts to compare results. However, in view of the large number of unknown factors which still surround these investigations, the use of a variety of methodological approaches will probably lead to more rapid progress in understanding and should be encouraged at this early stage in the development of the subject.

Measurements of organic material in methanol extracts suggest that on the basis of TOC in redissolved freeze-dried solids and marker compounds in the extracts, recovery rates of 70–90% are achieved (Forster & Wilson 1981). Losses are accounted for mainly by the volatile compounds in water.

Although the usefulness of an extract for mutagenicity testing depends on the apparent advantages of the concentration and extraction steps, the extract also must be suitable for the bioassay method; in this respect the methanol extracts work well with the fluctuation test. The extraction procedure is effective and does not introduce any significant mutagenic activity, the extracts are relatively non-inhibitory to the test strains of bacteria at mutagenic doses and do not contain significant amounts of histidine and the observed mutagenic activity appears not to be an artefact. However, the use of freeze-drying for concentrating the samples may introduce some artefacts; Miller & Stoltz (1978) found that the mutagenicity of isoniazid-treated rat urine was greatly increased by this procedure. The mechanism of the effect is not understood. Also, it remains a possibility that constituents of the extracts such as amino acids may enhance the expression of mutagenic activity without themselves being mutagenic (Thomas *et al.* 1979), but this mechanism is not readily amenable to experimental investigation. It is unlikely that mutagenic activity observed in the extracts resulted from inorganic materials because of the small concentrations at which most of the inorganic materials known to be mutagenic were present.

The only practical difficulty encountered was that the extracts contained some inorganic material which formed a milky precipitate when the extract was added to Vogel–Bonner growth medium. Although satisfactory results were obtained from the mutagenicity tests, the precipitate may have contributed to the mild inhibitory effect on test strain TA100. It also has an effect, yet to be quantified, on the buffering and growth-supporting properties of the medium. Further studies to establish the human health hazard posed by water-borne mutagens will involve tests with mammalian cells in tissue culture, which are much more sensitive to small changes in growth media and other conditions. Therefore, efforts will have to be made to circumvent this problem of precipitation.

The other disadvantages of this method are that the initial capital costs are high and processing times are relatively lengthy. There are a number of possibilities for improving throughput by pre-concentration steps which have yet to be evaluated for this kind of work, e.g. reverse osmosis as a rapid pre-concentration step might offer very substantial improvements, possibly reducing processing times by about 90%.

Our findings with respect to drinking water have demonstrated that UK drinking water samples contain directly acting mutagenic activity, i.e. not requiring S9 mix activation, which, in most cases, is attenuated in the presence of mammalian metabolising activity. These results are consistent with the consensus of findings of other workers (reviewed by Loper 1980). Thus, many questions are raised relating to the sources of mutagens in drinking water, their possible identity and their effects in higher organisms including humans. The implications depend upon investigations of such aspects as the mutagenicity of extracts to mammalian cells *in vitro* and in whole animals, the fate of the organic mutagens in the digestive tract and during mammalian metabolism and the extrapolations that may be made from studies on concentrated extracts of organic material to the long-term effects of the ingestion of trace quantities. The presence of bacterial mutagens is a serious indication of a potential health hazard, but until a great deal more is known it would be premature to draw any conclusions about risks to health.

Acknowledgements

The author acknowledges the large contributions made to this study by members of the toxicology and organics sections of the WRC Medmenham Laboratory. Some of the work was undertaken at the MRC Cell Mutation Unit, University of Sussex, under contract to WRC. Permission from the Department of the Environment and WRC to publish this work is acknowledged. The project was entirely funded by the Department of the Environment.

References

AMBROSE, J. A., GRIMM, A., BURTON, J., PAULLIN, K. & ROSS, C. 1969 Fluorometric determination of histidine. *Clinical Chemistry* **15**, 361–366.

AMES, B. N., MCCANN, J. & YAMASAKI, E. 1975 Methods for detecting carcinogens and mutagens with the salmonella/mammalian microsome mutagenicity test. *Mutation Research* **31**, 347–364.

ANONYMOUS 1981 Guidelines for Work with Carcinogenic Chemicals in Medical Research Council Establishments. London: Medical Research Council.

BAKER, R. J. & NELDER, J. A. 1978 *The GLIM System: Release 3- Generalised Linear Interactive Modelling*. London: Royal Statistical Society.

CRATHORNE, B., WATTS, C. D. & FIELDING, M. 1979 Analysis of non-volatile organic compounds in water by high-performance liquid chromatography. *Journal of Chromatography* **185**, 671–690.

FIELDING, M. & PACKHAM, R. F. 1977 Organic compounds in drinking water and public health. *Journal of the Institution of Water Engineers and Scientists* **31**, 353–375.

FORSTER, R. & WILSON, I. 1981 The application of mutagenicity testing to drinking water. *Journal of the Institution of Water Engineers and Scientists* **35**, 259–274.

FORSTER, R., GREEN, M. H. L., GWILLIAM, R. D., PRIESTLEY, A. & BRIDGES, B. A. 1983 The use of the fluctuation test to detect mutagenic activity in unconcentrated samples of UK drinking waters. *Proceedings of the Fourth Chlorination Conference: Environmental Impact and Health Effects*. Ann Arbor: Ann Arbor Science Publ. (In press.)

GREEN, M. H. L. & MURIEL, W. J. 1976 Mutagen testing using TRP+ reversion in *Escherichia coli*. *Mutation Research* **38**, 3–32.

GREEN, M. H. L., MURIEL, W. J. & BRIDGES, B. A. 1976 Use of a simplified fluctuation test to detect low levels of mutagens. *Mutation Research* **38**, 33–42.

GREEN, M. H. L., BRIDGES, B. A., ROGERS, A. M., HORSPOOL, G., MURIEL, W. J., BRIDGES, J. W. & FRY, J. R. 1977 Mutagen screening by a simplified bacterial fluctuation test: use of microsomal preparations and whole liver cells for metabolic activation. *Mutation Research* **48**, 287–294.

GRIMM-KIBALO, S. M., GLATZ, B. A. & FRITZ, J. S. 1981 Seasonal variation of mutagenic activity in drinking water. *Bulletin of Environmental Contamination and Toxicology* **26**, 188–195.

JOLLEY, R. L. 1981 Concentrating organics in water for biological testing. *Environmental Science and Technology* **15**, 874–880.

KOOL, H. J., VAN KREIJL, C. F., VAN KRANEN, H. J. & DE GREEF, E. 1981 The use of XAD resins for the detection of mutagenic activity in water I. Studies with surface water. *Chemosphere* **10**, 85–98.

LOPER, J. C. 1980 Mutagenic effects of organic compounds in drinking water. *Mutation Research* **76**, 241–268.

MCCANN, J. & AMES, B. N. 1976 Detection of carcinogens and mutagens in the *Salmonella*/microsome test: assay of 300 chemicals: discussion. *Proceedings of the National Academy of Sciences of the United States of America* **73**, 950–954.

MAGEE, P. N. 1974 Activation and inactivation of chemical carcinogens and mutagens in the mammals. *Essays in Biochemistry* **10**, 105–136.

MARUOKA, S. & YAMANAKA, S. 1982 Mutagenicity in *Salmonella typhimurium* tester strains of XAD-2-ether extract, recovered from Katsura River water in Kyoto City, and its fractions. *Mutation Research* 102, 13–26.

MILLER, C. T. & STOLTZ, D. R. 1978 Mutagenicity induced by lyophilisation or storage of urine from isoniazid treated rats. *Mutation Research* 56, 289–293.

MILLER, J. A. 1970 Carcinogenesis by chemicals—an overview. *Cancer Research* **30**, 559–576.

NAZAR, M. A. & RAPSON, W. H. 1982 pH stability of some mutagens produced by aqueous chlorination of organic compounds. *Environmental Mutagenesis* **4**, 435–444.

PAGE, T., HARRIS, R. H. & EPSTEIN, S. S. 1976 Drinking water and cancer mortality in Louisiana. *Science* **193**, 55–57.

THOMAS, H. F., BROWN, D. L., HARTMAN, P. E., WHITE, E. H. & HARTMAN, Z. 1979 Arylmonoalkyl and cyclic triazines: direct acting mutagens. *Mutation Research* **60**, 25–32.

VOGEL, H. J. & BONNER, D. M. 1956 Acetylornithase of *Escherichia coli*: partial purification and some properties. *Journal of Biological Chemistry* **218**, 97–106.

WATTS, C. D., CRATHORNE, B., CRANE, R. I. & FIELDING, M. 1981 Development of techniques for the isolation and identification of non-volatile organics in drinking water. In *Advances in the Identification and Analysis of Organic In Advances in the Identification and Analysis of Organic Pollutants in Water* Keith, L. H., Vol. I. Ann Arbor: Ann Arbor Science Publ.

ZIMMERMANN, F. K. 1977 Genetic effects of nitrous acid. *Mutation Research* **39**, 127–148.

Use of Molluscs to Monitor Bacteria in Water

D. R. TROLLOPE

Department of Botany and Microbiology, University College of Swansea, Swansea, UK

Molluscs, such as the oyster, mussel, whelk and cockle and other shellfish are sometimes a health hazard when eaten because of their content of toxic substances or pathogenic microorganisms derived from the environment (Wood 1972; Anonymous 1974; Ayres *et al.* 1978). The microbial content of filter-feeding molluscs is derived from the large volumes of water filtered during normal feeding, e.g. a 7-cm-long mussel, *Mytilus edulis,* filters 4 litres/h, equivalent to 20 gal/day. A quantitative correspondence between the bacterial content of a mollusc and the water in which it was feeding varies between different types of mollusc, depending upon the different effects of environmental factors, e.g. salinity and temperature.

Although they can be likened to disposable pumps, these animals are not entirely passive towards microorganisms and a considerable concentration effect may be produced in the mollusc tissue. This is advantageous in environmental monitoring since although the bacterial content of water may be small, e.g. 1 bacterium/100 ml, after 2 h a mussel tissue homogenate may contain approximately 1 bacterium/ml (Webber 1982). Retained bacteria may be digested by the mollusc. Feeding live bacteria to California mussels resulted in significant weight gains (ZoBell & Landon 1937), and the animals were maintained exclusively on bacteria for several months (ZoBell & Feltham 1938). However, "as molluscan shellfish have the ability to concentrate bacteria they are useful tools in surveying areas for evidence of faecal pollution and represent an extra dimension in sampling which often cannot be achieved by bacteriological sampling of water alone" (Ayres *et al.* 1978). Since the studies of Dodgson (1928), most microbiological investigations of shellfish and their environment have been bacteriological but increasingly attention is being given to the use of molluscs to detect animal viruses, e.g. Metcalf (1978).

Techniques in bacterial monitoring to be discussed in this chapter will be mostly selected from studies made in the author's laboratory. Swansea and the surrounding area provides a wide range of mollusc types and coastal water quality ranging from excellent to badly polluted (Trollope 1982). The data available are entirely marine apart from those by Cooke (1976) on the freshwater mussel *Hydridella menziesii* in New Zealand and some investigations on the swan mussel *Anodonta cygnea* (Al-Jebouri & Trollope 1984).

Preparation of Mollusc Tissue Suspension

Mollusc Collection

The animals should be collected at random as soon as they are uncovered by the tide. Mussels are readily gathered by hand, whereas the limpet needs to be removed from the rock with a sharp side-ways blow; cockles are best gathered with a short rake. *Scrobicularia plana* and clams make characteristic marks on the surface of the sand and mud and require to be dug up carefully using spade and trowel. The freshwater *Anodonta cygnea* are collected with a long-handled sampling net whilst wading. For the smaller animals at least 10 are needed. The standard technique with mussels uses two batches each of 5 animals that are 5–7 cm long. With large molluscs such as the clam and swan mussel (10–16 cm long) we have used single animals examined in duplicate. The molluscs are transferred to the laboratory in plastic bags without added water or seaweed and if not examined within about 2 h are transferred to a fresh dry bag since the shell water gradually seeps out. Before opening or storage any barnacles should be removed from the shell with a blunt scalpel, the shell scrubbed with a small nail brush, rinsed in cold, running tap water and dried on absorbent paper.

Mollusc Storage

Most workers advocate that shellfish be opened and analysed within 2 h of collection (Thomas & Jones 1971; Trollope 1982), but the molluscs listed in Table 1 can be stored dry at 4°C for several days. Dodgson (1928) showed that numbers of *Escherichia coli* and other coliforms were virtually constant in mussels stored at 4°C over an extended period (see also Table 2). However, marked changes in the bacterial flora have been reported during storage at 20–25°C for the oysters *Crassostrea commercialis* (Son & Fleet 1980) and *Ostrea edulis* (Ayres *et al.* 1978).

TABLE 1

Bacteria isolated from marine molluscs collected from the shores of Swansea Bay, Gower and the Burry Estuary, 1972–1982

Bacteria	Mollusc*
Aeromonas hydrophila	4
Acinetobacter calcoaceticus var. *lwoffi*	4
Bacillus spp.	2, 4, 5
Campylobacter jejuni	4
Clostridium perfringens	1, 3, 4, 6
Corynebacterium sp.	4
Escherichia coli	1, 3, 4, 6
Enterobacter cloacae	4
Erwinia herbicola	4
Faecal streptococci	1, 3, 4, 6
Flavobacterium sp.	4
Klebsiella pneumoniae	4
Kurthia sp.	4
Micrococcus sp	4
Pasteurella sp.	4
Pseudomonas spp.	1, 3, 4, 6
Salmonella hadar	4
Serratia sp.	4
Shigella dysenteriae	4
Staphylococcus sp.	4
Vibrio parahaemolyticus	4
Yersina enterocolitica	4

* 1, *Cerastoderma edule* (cockle); 2, *Littorina littoralis* (periwinkle); 3, *Mya arenaria* (soft clam); 4, *Mytilus edulis* (mussel); 5, *Patella vulgata* (limpet); 6, *Scrobicularia plana* (peppery furrow shell).

Opening the Shells

Considerable force is needed to open the bivalve shells. For *Mytilus edulis*, insert a sterile blunt scalpel (the type with integral blade) between the two component shells at the anterior end where the off-white byssus threads emerge and force the shells apart with a slight twisting movement of the scalpel. The small volume of shell water is allowed to escape, the adductor muscles are cut with the blades of sterile, fine-pointed scissors and the shells are opened fully (Trollope 1982). Scissors are used to cut the mussel flesh free from the shells and the flesh is combined in the lower shell, chopped up and placed into a sterile measuring cylinder. The flesh from five mussels constitutes one pool of tissue; 2 volumes of sterile diluent (0.1% w/v peptone in 3% w/v NaCl) are added to the measuring cylinder. For the periwinkle and limpet, the flesh can be removed with scissors after the shell is cut with scissors or broken using coarse forceps.

Tissue Suspension

For the aseptic preparation of a homogenous tissue suspension, maceration by hand has been superseded by mechanical blending, e.g. with an Atomix Blender (MSE Scientific Instruments, Manor Royal, Crawley, West Sussex) which gives an improved recovery of *E. coli* (Ayres 1975a). Jars should be sterilised by steaming. The Colworth Stomacher (A. J. Seward, Bury St. Edmunds, Suffolk) is ideal for multiple samples; several sample bags can be stomached simultaneously. Two sterile bags should be used, one inside the other, because of the possibility of rupture by grit or shell fragments. The suspension of chopped flesh in diluent is stomached for 30–60 s, returned to the measuring cylinder for 2–5 min for the large particles to settle out and the homogenised liquor is plated. Some comparative counts for a mechanical blender (Atomix) and the stomacher for mussels stored dry at 4°C for 8 days are shown in Table 2. Counts were made using four replicate plates and single plates. The overall stomacher:blender ratio of average daily counts for both plating methods was 1.88 ± 0.90:1; the ratio with only the preferred "four plate method" was 1.97 ± 0.92:1.

TABLE 2

Comparative counts* of lactose-fermenting colonies from mussel tissue on MacConkey Agar Medium following maceration with the Colworth Stomacher and the MSE Atomix Blender during dry storage of mussels at 4°C

Storage time (days)	No. replicate plates:	44°C 4	44°C 1	37°C 4	37°C 1	25°C 4	25°C 1
0		2.57	2.53	5.11	3.12	nt[†]	nt
1		1.92	1.71	1.87	1.59	nt	nt
2		1.0	0.95	1.77	1.41	1.97	nt
6		1.19	0.88	1.05	1.66	1.21	1.05
7		nt	nt	1.69	1.08	nt	nt
8		2.22	4.33	1.66	1.71	2.46	1.42
Mean values		1.78	2.08	2.19	1.76	1.88	1.24

* Expressed as count with the stomacher relative to that with the blender which is given the value 1.0.
[†] nt, Not tested.

In a test with oyster homogenates, Son & Fleet (1980) reported a recovery rate of >85% for two *Salmonella* spp. and *Bacillus cereus*, >70% for *Vibrio parahaemolyticus* and approximately 50% for *Clostridium perfringens*.

Isolation of Bacteria

Range and Types

Most marine shellfish occur in estuaries near large urban populations discharging untreated sewage effluent. Therefore, many bacteria found in molluscs (Table 1) are derived from sewage. *Escherichia coli* is used as a faecal indicator during routine monitoring of oysters and mussels (Thomas & Jones 1971; Wood 1972) whilst *Clostridium perfringens* and faecal streptococci have a complementary value (Ayres 1975b). Other bacteria likely to be of sewage origin are *Campylobacter jejuni*, *Enterobacter cloacae*, *Salmonella hadar*, *Shigella dysenteriae* and *Yersinia enterocolitica* (Al-Jebouri 1980). Some bacteria are typically marine, e.g. *Vibrio parahaemolyticus*, whilst others are more often isolated from fresh water (*Aeromonas hydrophila*), associated with vegetation (*Erwinia herbicola*), or widely distributed (*Acinetobacter calcoaceticus*, *Bacillus* spp., *Klebsiella pneumoniae*, *Pseudomonas* spp.). Most of the molluscs listed in Table 1 were sampled from shores receiving various amounts and types of urban or industrial effluents (Welsh Office 1974a,b; Chubb *et al.* 1980) with considerable freshwater run-off that is not free from urban effluent.

Media

The standard media and conditions for sewage bacteria are used, e.g. MacConkey Agar No. 3 (Oxoid) incubated at 44°C for *E. coli* (Reynolds & Wood 1956; Thomas & Jones 1971; Trollope 1982), Selenite Broth (Oxoid) as enrichment medium for salmonellae, Slanetz & Bartley Agar (Oxoid) for faecal streptococci and neomycin blood agar medium for *Cl. perfringens* (Finegold *et al.* 1971). Lactose-fermenting bacteria are enumerated on MacConkey Agar No. 3 after 24 h at 37 and 25°C as "37° coliforms (total coliforms)" and "25° coliforms (coli-aerogenes bacteria)", respectively. In North America "faecal coliforms" are enumerated. This group is similar to, but not identical with, *E. coli* (Wood 1972). The cystine–lactose electrolyte-deficient agar medium of Bevis (1968) gives excellent differentiation of gram-negative and gram-positive bacteria from *Mytilus edulis* (Al-Jebouri & Trollope 1978; Al-Jebouri 1980). Al-Jebouri (1980) found that the modified bromothymol blue–salt Teepol medium (BTBST) of Sakazaki (1973) yielded greater numbers of *V. parahaemolyticus* from *Mytilus edulis* than the thiosulphate–citrate–bile salt–sucrose medium (TCBS) of Sakazaki (1973). For the enumeration of marine heterotrophs ZoBell Marine Agar 2216E (Difco) is usually used (Ayres *et al.* 1978) but reduced salts concentrations to simulate estuarine conditions have also been considered (Sayler *et al.* 1975; Trollope & Webber 1977). *Mytilus edulis* gave 5- to 20-fold greater counts on ZoBell Marine Agar medium than on modified Upper Bay Yeast Extract Agar Medium (Sayler *et al.* 1975). Weiner *et al.* (1980) compared two estuarine media for the isolation of heterotrophic bacteria from the oyster, *Crassostrea virginica*, and reported that one medium revealed five genera, representing 74% of the total colonies, not isolated on the other medium. See Appendix for descriptions of media.

Enumeration Methods

For counting bacteria in the shellfish tissue, the most probable number technique (Dodgson 1928) or a simplified form, the percentage clean method (Knott 1951; Thomas & Jones 1971; Al-Jebouri & Trollope 1981), roll tubes (Clegg & Sherwood 1947; Reynolds & Wood 1956) and surface and pour plates (Thomas & Jones 1971) have been used. Similar counts of *E. coli* on MacConkey Agar No. 3 from mussels receiving considerable faecal pollution were obtained with pour plates (four replicates) and duplicate spread plates (Al-Jebouri & Trollope 1981). However, with the mussels from a relatively clean shore, the spread plate technique produced nine false negative results from 20 samples of mussel tissue. In a review of methods, including rapid ones, for counting *E. coli* in the oyster *Crassostrea commercialis*, Yoovidhya & Fleet (1981) reported favourably on the membrane-overlay agar plate method of Anderson & Baird-Parker (1975). The droplette technique (A. J. Seward, Bury St. Edmunds, Suffolk) offers a considerable saving in time and medium but yields smaller and more variable counts of heterotrophic bacteria from *Mytilus edulis* on nutrient agar medium and ZoBell Marine Agar Medium than a spread plate method (Webber 1982).

Calculation of Bacterial Numbers in Mollusc Tissue

The original method of Clegg & Sherwood (1947) requires an addition of 2 volumes of diluent to 1 volume of tissue and stipulates that this be interpreted as a dilution factor of 2, on the assumption that, on standing, the hand-macerated tissue settles out leaving the bacteria suspended in 2 volumes of diluent (Thomas & Jones 1971). However, from theoretical considerations and macroscopic observations, this interpretation does not hold with mechanical preparation because much of the tissue suspension is a fine suspension that does not settle out (Ayres 1975a). Therefore, a dilution factor of 3 is more appropriate (Thomas & Jones 1971; Al-Jebouri & Trollope 1981).

Increase of Detection Sensitivity Using Mollusc Dissection

The use of dissected mussel digestive tract instead of the whole tissue significantly increases the sensitivity of detection of *E. coli* (Al-Jebouri & Trollope 1981). With animals from a heavily polluted site, the ratio of digestive tract to total tissue count was 3:1, with mussels from a lightly polluted shore, the same ratio was 4:1–6:1 depending upon the counting method used. These results were due to the digestive tract being small in volume but bacteriologically rich in comparison with the bulk of the mussel tissue; >75% of the total bacteria was located in the stomach and alimentary canal (Al-Jebouri & Trollope 1979; Al-Jebouri 1980).

Monitoring of Environmental Changes

Seasonal Effects

The counts of coliforms and faecal coliforms in molluscs vary seasonally, with large numbers in January and November and small numbers in July and August (Vasconcelos *et al.* 1969). However, the greatest concentrating effect in the molluscs studied by these authors occurred when the numbers in water were small, but the smallest numbers in animals were found when the sea water numbers were maximum. The authors concluded that water temperature was affecting animal activity; many studies have shown that feeding, and hence uptake of bacteria, by certain molluscs is affected adversely by low water temperature (Wood 1965). Ayres *et al.* (1978) established a significant correlation ($p = 0.001$) between temperature and *E. coli* numbers in the tissue of *Mercenaria mercenaria* (the hard clam) and *Ostrea edulis* (the European flat oyster), as did Wood (1965) for flat oysters and Portugese oysters ($p = 0.001$). Al-Jebouri (1980) demonstrated significant correlations ($p < 0.001$) between season and the numbers of *V. parahaemolyticus* and of some sewage-derived bacteria in the tissue of *Mytilus edulis*. It should be noted that various species of shellfish have different temperature requirements, e.g. clams have a higher optimum temperature than oysters and *Mytilus edulis* is quite active at 0°C (Ayres *et al.* 1978).

Comparisons between Different Molluscs

Both the commercially important molluscs and those of no known economic value represent actual and potential means of monitoring the aquatic environment, particularly where they occur as indigenous members of the normal fauna. Shellfish examined locally (Table 3) accumulated coliforms and *E. coli* indicating, in the majority of locations, recent sewage pollution.

TABLE 3

Typical counts of faecal indicator bacteria in mollusc tissue from different molluscs from various locations

| | | | No. bacteria/ml mollusc tissue ||||
Mollusc	Location	Date sampled	37° coliforms	E. coli	Cl. perfringens	Faecal streptococci
Anodonta cygnea (swan mussel)	Brynmill	24.5.80	26250	173	60	45
		3.6.80	6360	105	130	43
	Fairwood	5.6.80	10080	4	0	4
		10.6.80	1800	0	0	0
Cerastoderma edule (cockle)	Salthouse Point	25.7.80	585	68	4	488
		29.7.80	420	105	16	240
	Swansea, Blackpill	5.6.73	251	25	nt*	nt
		6.6.73	128	10	nt	nt
Mya arenaria (clam)	Salthouse Point	25.7.80	540	190	38	465
		29.7.80	780	218	23	270
Mytilus edulis (mussel)	Loughor Bridge	10.7.80	6015	147	18	34
		15.7.80	3345	259	79	106
	Oxwich	16.2.72	43	11	nt	nt
		14.1.76	0	0	nt	nt
	Swansea, Mumbles	16.6.80	1500	200	62	120
		25.6.80	60	16	116	49
	Worms Head†	10.10.73	2	0	nt	nt
		23.7.81	nt	0	nt	0
Scrobicularia plana (peppery furrow shell)	Loughor Bridge	10.7.80	5535	352	173	165
		15.7.80	14215	990	285	608
	Salthouse Point	13.7.76	284	93	nt	nt

* nt, Not tested.
† Remote from known sewage effluent outfall.

Variation in bacterial numbers reflected changes in effluent concentration and distance from known sewage outfalls. However, the ability of different shellfish at the same or nearby sites to concentrate the same bacteria to different degrees has been noted by Trollope & Webber (1977), who observed that coliforms were accumulated in greater numbers by *Scrobicularia plana* than by mussels, and by Vasconcelos *et al.* (1969), who reported that coliforms and faecal coliforms were present in greater number in clams than in oysters.

The examination of the freshwater mussel *Anodonta cygnea* showed day-to-day variation between animals to be no greater than variation found between separate animals collected on the same day (Al-Jebouri & Trollope 1984).

TABLE 4

Correlation coefficients (C) and significance levels (p) between numbers of bacteria in sea water and in homogenised tissue of Mytilus edulis*

Sea water	Mussel tissue					
	Escherichia coli		37° coliforms		25° coliforms	
	C	p	C	p	C	p
Escherichia coli	0.8180	<0.001	0.2854	<0.05	0.4725	<0.001
37° coliforms	0.4015	<0.001	0.3891	<0.01	0.4976	<0.001
25° coliforms	0.5598	<0.001	0.3585	<0.01	0.4654	<0.001

* Webber (1982).

Distinct differences were observed in animals from urban and rural sites, but the counts of *E. coli* and 37° coliforms from the rural site were considerably greater than comparable values reported for *Hydridella menziesii* sampled in three New Zealand freshwater lakes (Cooke 1976).

A bacteriological similarity between *Mytilus edulis* and the overlying sea water is well documented (Dodgson 1928; Wood 1957). However, Webber (1982) demonstrated significant correlations for counts of faecal indicator bacteria (Table 4) but not for heterotrophic bacteria. Furthermore, *E. coli* appears to accumulate to a smaller extent than other coliforms in Pacific oysters (*Crassostrea gigas*) and Manila clams (*Tapes japonica*) (Vasconcelos *et al.* 1969) and in *Mytilus edulis* (Webber 1982). Vasconcelos *et al.* (1969) considered that field conditions, with variations in tide, climate, wave action and numbers of bacteria, contribute to the greater accumulation of bacteria than occurs in laboratory studies where there are constant conditions and much greater concentrations of bacteria. The concentration of organic matter in the surrounding water is particularly relevant where sewage effects are being studied. Ayres *et al.* (1978) found that oyster activity was stimulated by organic matter from sewage, resulting in removal of bacteria from the oyster gut at a rate which exceeded uptake. More research is needed in this area.

Transfer of Molluscs between Sites

Marine Shellfish

Wood (1957) relaid "bacteriologically clean" mussels and oysters in weighted and buoyed cages on the shore just below low water "in order that they might assume the bacterial characteristics of the ground." During a minimum total immersion period of 24 h, examination of shellfish and water samples at hourly intervals showed a close similarity in levels of *E. coli*.

There was a large amount of sewage pollution in the estuary for about 4 h on each tide. The "clean" mussels reflected the increase in numbers of bacteria with a 1- to 2-h delay and became cleansed or showed a marked reduction in counts in the same period. Webber (1982) transferred mussels between two relatively clean sites and observed that after 48 h, i.e. four 6-h tide immersions, the introduced mussels had achieved a bacteriological status similar to that of indigenous mussels irrespective of their initial degree of faecal contamination. Relaying for one tide, i.e. 6 h, was equally satisfactory and, particularly when it was overnight, had the added advantage that there was little likelihood of vandalism to the container needed to keep the mussels captive.

Freshwater Shellfish

The freshwater mollusc *Anodonta cygnea* has a lower pumping rate than the marine mussel and oyster (De Bruin & Davids 1970). Consequently, when samples of *Anodonta* were transferred from a polluted freshwater lake in the Swansea area to one remote from the urban population, it was 48 h before the bacteriological changes appeared to be stabilised although faecal indicator bacteria were reduced at 6 h and not detected after 24 h (Al-Jebouri & Trollope 1984).

The Use of Bacteriologically Cleansed Molluscs

The "cleansing" (or depuration) of shellfish is practised commercially on a large scale in many countries and the technology has been reviewed by Furfari (1979). Some of the commercial procedures (Wood 1961) used to render molluscs safe for human consumption (Dodgson 1928; Ayres *et al.* 1978) have been used to produce molluscs for environmental monitoring (Trollope & Webber 1977).

The continuous immersion of *Mytilus edulis* for 48 h in clean sea water or an aquarium resulted in the removal of *E. coli* but increases in counts of heterotrophic marine bacteria (Trollope & Webber 1977). Examination of cleansed mussels after transfer to sewage-polluted shores suggested that these immersion treatments were not enabling typical counts of sewage-derived bacteria to be established. When mussels were cleansed in UV-irradiated sea water (Fig. 1) according to a simulated tidal regime with alternate 6-h periods of air and sea water (Webber 1982), numbers of faecal indicator bacteria were greatly reduced, thereby providing mussels of acceptable commercial quality on bacteriological criteria (Ayres *et al.* 1978). Furthermore, almost total elimination of 37° coliforms and 25° coliforms occurred without any significant changes in plate counts of total bacteria and heterotrophic marine bacteria

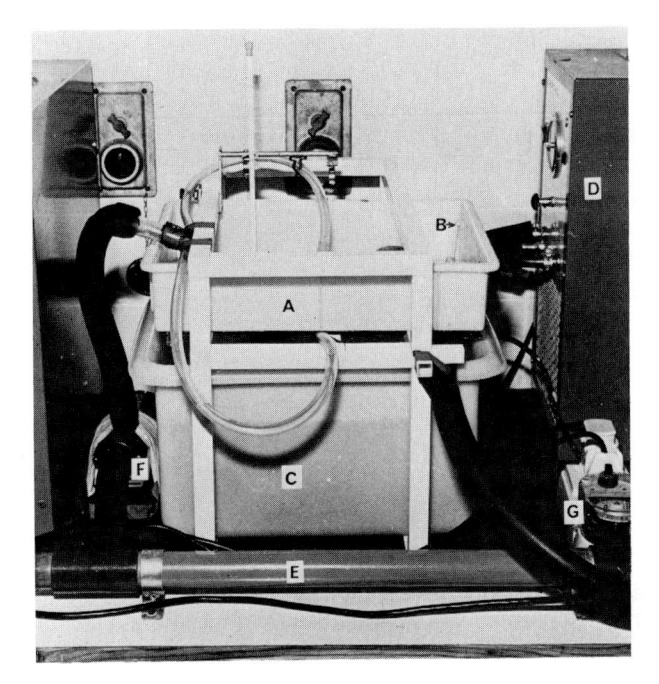

FIG. 1. Apparatus for the bacteriological cleansing (depuration) of molluscs. (A) Plastic tray containing molluscs; (B) overflow tube to produce constant depth of sea water in (A); (C) plastic container of sea water maintained at a constant temperature by the refrigerant from a chiller-circulator (D) flowing through a stainless steel coil; (E) enclosed UV steriliser; (F) pump; (G) time switch to control pump and UV steriliser.

(Webber 1982). Mussels cleansed by this simulated tidal regime and relaid showed close similarities in numbers of faecal indicator bacteria to indigenous mussels after immersion for 6 h (Trollope & Webber 1977).

When *Scrobicularia plana* was placed in the running water of a marine aquarium, elimination of faecal indicator bacteria occurred slowly, requiring 4–7 days continuous treatment for *E. coli* to be eliminated (Trollope & Webber 1977). The bacterial flora of partially cleansed *Scrobicularia* that had received 3–4 days treatment in the aquarium became similar to that of indigenous animals 1–2 days after being relaid on the original polluted shore.

In the cleansing experiments it was common to find that the greatest reduction in bacterial numbers occurred during initial depuration. Son & Fleet (1980) experienced this effect with the oyster *Crassostrea commercialis*. *Salmonella typhimurium*, *S. senftenberg*, *B. cereus*, *Cl. perfringens* and *V. parahaemolyticus* were readily removed under conditions which promoted removal of *E. coli*.

Other Specific Environmental Applications

Commercial Shellfish Production

Bacteriological analysis of mollusc species can be used to aid safe commercial production of edible shellfish. Ayres *et al.* (1978) stated that such examination could be used "to determine the sanitary assessment of an area of existing of potential shellfish culture to determine whether an area warrants control under the Public Health (Shellfish) Regulations 1934". Depurated clams and oysters were used by Vasconcelos *et al.* (1969) to check for contamination of other saleable oysters and clams by faecal bacteria during temporary relaying ("wet storage"). During periods up to 24 h the clams yielded the greater shellfish:water faecal bacteria ratio. Changes in the shellfish rapidly reflected changes in the surrounding water, but the magnitude and frequency of changes in the shellfish were less than those of the water.

Bacterial Tracers of Water Movement

The use of the coloured bacterium *Serratia marcescens* (syn. *indica*) pioneered by Robson (1956) to follow the movement of water masses has been extended by Pike *et al.* (1969) to the addition of *Bacillus subtilis* var. *niger* (syn. *Bacillus globigii*) for tracing effluent and sewage movement and survival of terrestrial organisms in the sea. Al-Salihi & Trollope (1978) and Al-Salihi (1980) showed that *Serratia marcescens* and *B. subtilis* var. *niger*, after release on the marine shore, were recovered from indigenous mussels in far greater numbers and for much longer periods than from sea water samples. The pattern of the spatial distribution of tracer organisms within the mussel bed was similar to that observed for coliforms. Water movements were followed for up to 250 m; previous studies using bacterial tracers have followed water movements over 16 km (Rippon 1963) and 28 km (Gameson *et al.* 1968). This technique has considerable potential for the labelling of point sources and the tracking of zones of influence of an outfall using indigenous or introduced molluscs.

Sewage Outfall Assessment

The bacteriological analysis of oysters assisted Ayres *et al.* (1978) (Table 5) in detecting small levels of water-borne organisms to effect a comparison of faecal bacterial numbers in an estuary before and after the introduction of a sewage works improvement scheme. Samples were taken at intervals over a 6-month period to minimise seasonal effects and at or near low water when pollution levels were judged to be maximal. It was possible to show signifi-

TABLE 5

Counts of Escherichia coli *in water and oyster tissue from an estuary before and after the introduction of a sewage works improvement scheme*[*]

Distance from discharge point of sewage works (miles)	No. bacteria/ml	
	Water	Oyster
Before scheme		
0–1	5.6	>20
1–2	2.3	5.0
2–4	<1.0	2.0
After scheme		
0–1	2.6	5.0
1–2	<1.0	0.8
2–4	Undetected	<1.0

[*] After Ayres *et al.* (1978).

cant improvements in water quality resulting from sewage plant modernisation and resiting of the discharge point.

Acknowledgements

The author gratefully acknowledges the help of Dr. W. Kwantes and Mr. R. Brooks, Public Health Laboratory Service, Swansea in identifying *Salmonella hadar* and Dr. D. L. Webber for discussions and data.

Appendix

Cystine–Lactose Electrolyte-Deficient Medium (CLED)

This medium contains (g/litre) peptone, 4.0; beef extract, 3.0; tryptone (Oxoid), 4.0; lactose, 10.0; cystine, 0.128; bromothymol blue (0.2% w/v aq.), 10.0 ml; Andrade indicator, 10.0 ml; agar, 15.0; water, 1 litre. The cystine is ground to a fine powder and dissolved in boiling water and added to the other dissolved ingredients. The pH is adjusted to 7.4.

Media for Vibrio parahaemolyticus

Modified BTBST medium contains (g/litre) peptone, 10.0; beef extract, 3.0; sucrose, 10.0; NaCl, 40.0; bromothymol blue, 0.04; thymol blue, 0.04;

tergitol 7 (B.D.H.), 2.0 ml; agar, 15.0; water, 1 litre; final pH 7.8. TCBS medium contains (g/litre) peptone, 10.0; yeast extract, 5.0; sodium citrate, 10.0; $Na_2S_2O_3 \cdot 5H_2O$, 10.0; NaCl, 10.0; sucrose, 20.0; sodium taurocholate (B.D.H.) 5.0; ferric citrate, 1.0 g; bromothymol blue, 0.04; thymol blue, 0.04; agar, 15.0; water 1 litre; final pH 8.6.

Medium for Campylobacter jejuni

The medium contains blood agar base (Oxoid); horse blood, 10% by volume; trimethoprim, 5 µg/ml; polymyxin B, 3 µg/ml; vancomycin, 10 µg/ml. Incubation is at 43°C for 24–72 h in an atmosphere of 5% O_2, 10% CO_2 and 85% N_2.

Upper Bay Yeast Extract Agar Medium

This medium contains (g/litre) peptone (Oxoid L46), 1.0; yeast extract 1.0; NaCl, 5.0; KCl, 0.16; $MgSO_4 \cdot 7H_2O$, 1.5; agar 20.0.

References

AL-JEBOURI, M. M. H. M. 1980 Bacteria isolated from separated organs of *Mytilus edulis* collected from sewage polluted shores. Ph.D. Thesis, University of Wales.

AL-JEBOURI, M. M. & TROLLOPE, D. R. 1978 The enumeration of enterobacteria from *Mytilis edulis* using CLED medium. *Society for General Microbiology Quarterly* 6, 29.

AL-JEBOURI, M. M. & TROLLOPE, D. R. 1979 The effects of season and site on the numbers of enteric bacteria from mussel organs: an analysis including multivariate analysis of variance. *Journal of Applied Bacteriology* 47, 11.

AL-JEBOURI, M. M. & TROLLOPE, D. R. 1981 The *Escherichia coli* content of *Mytilus edulis* from analysis of whole tissue or digestive tract. *Journal of Applied Bacteriology* 51, 135–142.

AL-JEBOURI, M. M. & TROLLOPE, D. R. 1984 Indicator bacteria in freshwater and marine molluscs. *Hydrobiologia* 111, 93–102.

AL-SALIHI, S. B. S. 1980 The uptake by *Mytilus edulis* of bacteria used in tracing water movement. Ph.D. Thesis, University of Wales.

AL-SALIHI, S. & TROLLOPE, D. R. 1978 The uptake of *Serratia marcescens* and *Bacillus subtilis* var. *niger* by *Mytilus edulis* on a sewage-polluted shore. *Society for General Microbiology Quarterly* 6, 29.

ANDERSON, J. M. & BAIRD-PARKER, A. C. 1975 A rapid and direct plate method for enumerating *Escherichia coli* biotype 1 in food. *Journal of Applied Bacteriology* 39, 111–117.

ANONYMOUS 1974 *Fish and Shellfish Hygiene*. Rome: FAO, UN.

AYRES, P. A. 1975a Recovery of *Escherichia coli* and coliforms from macerated shellfish. *Journal of Applied Bacteriology* 39, 353–356.

AYRES, P. A. 1975b The quantitative bacteriology of some commercial bivalve shellfish entering British markets. *Journal of Hygiene* 74, 431–440.

AYRES, P. A., BURTON, H. W. & CULLUM, M. L. 1978 Sewage pollution and shellfish. In *Techniques for the Study of Mixed Populations* Society for Applied Bacteriology Techical Series No. 11 ed. Lovelock, D. W. & Davies, R., pp. 51–62. London: Academic Press.

BEVIS, T. D. 1968 A modified electrolyte-deficient culture medium. *Journal of Medical Laboratory and Technology* **25**, 38–41.

CHUBB, C. J., DALE, R. P. & STONER, J. H. 1980 Inputs to Swansea Bay. In *Industrialised Embayments and their Environmental Problems* eds. Collins, M. B., Banner, F. T., Tyler, P. A., Wakefield, S. J. & James, A. E., pp. 307–327. Oxford: Pergamon Press.

CLEGG, L. F. L. & SHERWOOD, H. P. 1947 The bacteriological examination of molluscan shellfish. *Journal of Hygiene* **45**, 504–521.

COOKE, M. D. 1976 Antibiotic resistance among coliform and faecal coliform bacteria isolated from the freshwater mussel *Hydridella menziesii*. *Antimicrobial Agents and Chemotherapy* **9**, 885–888.

DE BRUIN, J. P. C. & DAVIDS, C. 1970 Observations on the rate of water pumping of the freshwater mussel *Anodonta cygnea zellensis* (Gmelin). *Netherlands Journal of Zoology* **20**, 380–391.

DODGSON, R. W. 1928 Report on mussel purification. *Ministry of Agriculture and Fisheries, Fishery Investigations Series 2* **10**, No. 1.

FINEGOLD, S. M., SUGIHARA, P. T. & SUTTER, V. L. 1971 Use of selective media for isolation of anaerobes from humans. In *Isolation of Anaerobes*. Society for Applied Bacteriology Technical Series No. 5 eds. Shapton, D. A. & Board, R. C., pp. 99–108. London: Academic Press.

FURFARI, S. A. 1979 Shellfish purification: a review of current technology. In *Advances in Aquaculture* eds. Pillay, T. V. R. & Dill, W. A. pp. 385–394. Farnham, Surrey: FAO & Fishing News Books.

GAMESON, A. L. H., PIKE, E. B. & BARRETT, M. J. 1968 Some factors influencing the concentration of coliform bacteria on beaches. *Revue Internationale d'Oceanographie Medicale* **9**, 255–280.

KNOTT, F. A. 1951 *Memorandum on the Principles and Standards Employed by the Worshipful Company of Fishmongers in the Bacteriological Control of Shellfish in the London Markets*. London: Fishmongers' Company.

METCALF, T. G. 1978 Indicators of viruses in shellfish. In *Indicators of Viruses in Water and Food* ed. Berg, G., pp. 383–415. Michigan: Ann Arbor Science Publ.

PIKE, E. B., BUFTON, A. W. J. & GOULD, D. J. 1969 The use of *Serratia indica* and *Bacillus subtilis* var. *niger* spores for tracing sewage dispersion in the sea. *Journal of Applied Bacteriology* **32**, 206–216.

REYNOLDS, N. & WOOD, P. C. 1956 Improved techniques for the bacteriological examination of molluscan shellfish. *Journal of Applied Bacteriology* **19**, 20–25.

RIPPON, J. E. 1963 The use of a coloured bacterium as an indicator of local water movement. *Chemistry and Industry, London* p. **445**.

ROBSON, J. G. 1956 Bacterial method for tracing sewage pollution. *Journal of Applied Bacteriology* **19**, 243–246.

SAKAZAKI, R. 1973 Control of contamination with *Vibrio parahaemolyticus* in seafoods and isolation and identification of the vibrio. In *The Microbiological Safety of Food* eds. Hobbs, B. C. & Christian, J. H. B., pp. 375–384. London: Academic Press.

SAYLER, G. S., NELSON, J. D., JUSTICE, A. & COLWELL, R. R. 1975 Distribution and significance of faecal indicator organisms in the upper Chesapeake Bay. *Applied Microbiology* **30**, 625–638.

SON, N. T. & FLEET, G. H. 1980 Behaviour of pathogenic bacteria in the oyster, *Crassostrea commercialis*, during depuration, re-laying, and storage. *Applied and Environmental Microbiology* **40**, 994–1002.

THOMAS, K. L. & JONES, A. M. 1971 Comparison of methods of estimating the number of *Escherichia coli* in edible mussels and the relationship between the presence of salmonellae and *E. coli*. *Journal of Applied Bacteriology* **34**, 717–725.

TROLLOPE, D. R. 1982 Faecal contamination of the common mussel. In *Sourcebook of Experiments for the Teaching of Microbiology* eds. Primrose, S. B. & Wardlaw, A. C., pp. 531–539. London: Academic Press.

TROLLOPE, D. R. & WEBBER, D. L. 1977 Shellfish bacteriology: coliform and marine bacteria in cockles (*Cardium edule*), mussels (*Mytilus edulis*) and *Scrobicularia plana*. In *Problems of a Small Estuary* eds. Nelson-Smith, A. & Bridges, E. M., pp. 6:3/1–6:3/18. Swansea: University College.

WEBBER, D. L. 1982 The accumulation of faecal indicator bacteria by the mussel, *Mytilus edulis*. Ph.D. Thesis, University of Wales.

WEINER, R. M., HUSSONG, D. & COLWELL, R. R. 1980 An estuarine agar medium for enumeration of aerobic heterotrophic bacteria associated with water, sediment, and shellfish. *Canadian Journal of Microbiology* **26**, 1366–1368.

WELSH OFFICE 1974a Report of the Working Party on Possible Pollution in Swansea Bay, Vol I. Welsh Office.

WELSH OFFICE 1974b Report of the Working Party on Possible Pollution in Swansea Bay, Vol II, Technical Reports. Welsh Office.

VASCONCELOS, G. J., JAKUBOWSKI, W. & ERICKSEN, T. H. 1969 Bacteriological changes in shellfish maintained in an estuarine environment. *Proceedings of the National Shellfisheries Association* **59**, 67–83.

WOOD, P. C. 1957 Factors affecting the pollution and self-purification of molluscan shellfish. *Journal du Conseil, Conseil International pour l'Exploration de la Mer* **22**, 200–208.

WOOD, P. C. 1961 The principles of water sterilisation by ultra-violet light, and their application in the purification of oysters. *Ministry of Agriculture Fisheries and Food Fishery Investigations Series 2* **23**, No. 6.

WOOD, P. C. 1965 The effect of water temperature on the sanitary quality of *Ostrea edulis* and *Crassostrea angulata* held in polluted waters. Pollutions marines par les micro-organismes et les produits petroliers, Symposium de Monaco (Avril 1964). *Monaco: Commision Internationale pour l'Exploration Scientifique de le Mer Mediterranee* pp. 307–317.

WOOD, P. C. 1972 The principles and methods employed for the sanitary control of molluscan shellfish. In *Marine Pollution and Sea Life* ed. Ruivo, M., pp. 560–565. West Byfleet & London: Fishing News (Books).

YOOVIDHYA, T. & FLEET, G. H. 1981 An evaluation of the A-1 most probable number and the Anderson and Baird-Parker plate count methods for enumerating *Echerichia coli* in the Sydney Rock oyster, *Crassotrea commercialis*. *Journal of Applied Bacteriology* **50**, 519–528.

ZOBELL, C. E. & LANDON, W. A. 1937 Bacteriological nutrition of the Californian mussel. *Proceedings of the Society for Experimental Biology and Medicine* **36**, 607–609.

ZOBELL, C. E. & FELTHAM, C. B. 1938 Bacteria as food for certain marine invertebrates. *Journal of Marine Research* **1**, 312–327.

Index

A

Absorbed cells, 226
Acaulospora spp., 97
Acetate, utilization of, 140, 148, 149, 152, 154, 155, 238, 241, 242, 260, 276
Acetoclastic methane bacteria, 162
Acetogenic bacteria, 162
Acetylene reduction test, 70, 71
Acid-forming bacteria, of the sludge digestion process, 161, 162
Acid hydrolysis, of lignin, 46
Acidification, in digester systems, 124
Acinetobacter calcoaceticus, 397
Actinomycetes, lignin degradation by, 50–52, 54, 55, 62
"Activated oxygen species" hypothesis, 49
Activated sludge, 169–180, 200, 201, 214, 215, 220
   ATP control of, 183–195
   laboratory scale simulation of, 198, 199, 211
   rate of recycle, 184
   rate of wastage, 183, 184
   returned, 189
ADAS, 359
Adenosine triphosphate, 186, 187
   ssays for, 6, 7
   calibration curve for, 187, 188
   measurement of in the activated sludge process, 185, 195
   standards for, 187
Aeration tanks, 175, 184
Aeromonas hydrophila, 397
Affinity coefficient, substrate, 304
Agaricus bisporus, 10, 13, 14
Agricultural Development and Advisory Service, 359, *see also* ADAS
Agrobacterium, 85, 86
Agrobacterium tunefaciens, 90
Algae, in a photobioreactor, 313–321

Alginate, immobilisation of microbial cells in, 221–230
Alginic acid, 221, 222
Aliphatic acids, 276
Alka-Seltzer tablets, use of in the isolation of sulphate-reducing bacteria, 244
Alkaline phosphatase, 350, 351, 354, 358
All-glass chemostat, 283
American Petroleum Institute, 235, 236, 243–245, 253
Ammonia in farm wastes, 128, 134
α-Amylase, 227
Amylase assays, 261, 267, 268
Amylolytic bacteria, 136
Anastomosis in VA mycorrhiza, 109
Aniline, 284, 286
Anodonta cygnea, 394, 400, 402
Anoxic ecosystems, 275
Anthranilic acid, pathway for, 284–286
Antibiotic resistance, in rhizobia, 83, 85
Autotrophic growth, 237
Axenic conditions, growth of fungi in, 6–8
Azotobacter, 221

B

Bacillus cereus, 397, 403
Bacillus spp., 397
Baculoviridae, 323
Baculoviruses, 323–345
   advantages and disadvantages of, 342, 343
   as pesticides, 324
   counting of, 337
   ecological aspects of, 328, 329
   economic competitiveness, 342, 343
   environmental acceptability, 343
   for pest control, 341, 342
   identification of, 334, 335
   in mosquitoes, 324

Baculoviruses *(cont.)*
  nomenclature, 326
  replication of in the insect, 326–328
  residues on plants, 329
  safety tests, 340, 341
  serological methods for detection of, 336, 337
  strain differences, 329
  survival of, in soil, 329
Bacteria
  acetoclastic methane, 162
  acetogenic, 162
  acid-forming, 161, 162
  anaerobic, 119–121, 154, 159–167
  biodegradation of lignin by, 50–52
  coliform, 397
  enumeration of in shellfish, 398
  faecal, 404
  faecal indicator, 259, 401–403
  fermentative, 136, 140
  heterotrophic, 398, 401
  hydrogen-utilising, 162
  lactose-fermenting, 397
  methanogenic, 120, 122, 123, 136, 140, 141, 149–154, 277, 288
  photosynthetic, 296
  sulphate-reducing, 235–255, 288
Bacterial tracers, 404
Barley, effects of microorganisms on growth of seedlings, 76, 77
BASIC, use of in bioreactor control programme, 317
Batch culture, respiration rates in, 177
Beads, alginate, 223, 224
Biochemical oxygen demand, see BOD
Biodegradation of lignin, 47–52
Biogas, 120–122, 125, 139, 159
Biological control of insect pests, 323–345
Biomass, fungal estimates of, 7–15
  activated sludge, 198, 211
  constant, in a photobioreactor, 313–321
  density of, 314, 316, 321
  metal removal by, 210–212
  microbial estimates of, 185
Biopesticides, 323
Biotechnology, definition of, 1, 2
  soil, 2
  environmental, 2, 3
Björkman's lignin, 37, 38
Black pepper, VAM infection on, 107

BOD, 227, 229
  in farm wastes, 135
  in sewage sludge, 184, 186, 188, 189, 192, 194
  of refuse, 287
Bouteloua gracilis, 108
"Bulking" sludge, 179
Butyrate as a carbon source, 238

C

Caddington Sewage Treatment Works, 366
Cadmium–*Klebsiella aerogenes* complex, 209–213, 215
Calvatia gigantea, 338
CAMAC crate, 172–174
Campylobacter jejuni, 397
Carcinogenic aromatic amines, 227, 228
κ-Carrageenan, 225, 226
"Cascade hybridisation," 56
Cattle wastes, 122, 123, 125, 127, 128, 133
Cefoxitin, 151
Cellobiose quinone oxidoreductase, 53
Cellophane, use in cellulase assays, 262, 263, 271, 272
Cell suspension assay, 365–374
Cells, immobilised, use of in waste water treatment, 219–231
Cellulase assays, 261–263, 265, 267–269, 271
"Cellulase" complex secreted by *T. reesei*, 22
Cellulolytic bacteria, 136, 137
Cellulose acetate, cell entrapment in, 226
Cellulose, 69, 70, 121, 135, 229, 266, 269, 276, see also Hemicellulose, Lignin, Lignocellulose
  dynamics of decomposition of, 19–30, 33, 34
Cellulolysis, 70, 262
Centrifugation, as a method of polymer extraction, 199–201
CH_4 (methane)
  conversion of organic matter to, 139–155, 279
  fermentation of, 159, 229
  generation rates, 147–149
  oxidation, 291
  production, 288, 290
Characterisation of methanogenic bacteria, 153, 154

INDEX

Cheese-making industry, BOD waste from, 229
"Chemolithotrophic growth," 237
Chemostat, 71, 72, 172, 295, 296, 301, *see also* Gradostat
Chorella, 313, 321
Citrus, VAM infection on, 96, 107
Clams, 399, 401, 404
Cleansing of shellfish, 402, 403
Clingfilm, 339, 353, 354
Cloned genes, 56, 58
Closed-culture fermenters, 281–283
Closed-enrichment apparatus, 281–283
Clostridium acetobutylicum, 306
Clostridium butyricum, 77
Clostridium perfringens, 397, 403
Clostridium thermocellum, 262
Clover
  rhizobia on, 80
  VAM infection on, 96, 106–108
Clq assay, 357, 359
^{13}C NMR, 44
CO_2, 276, 279
CO_2–air gas mixture, for the cell suspension assay, 369, 373, 374
CO_2 evolution, 244
  relationshp to mycelial growth, 7
"Coarse endophyte mycorrhiza," 101
Cobalt–*Klebsiella aerogenes* complex, 209–213
Cocksackie viruses, 370–373
Coconut rhinoceros beetle, 327
COD in farm wastes, 135, 137, 227
Coliforms in molluscs, 397, 399, 402
Collagen fibres, entrapment between, 226
Colorimetric techniques, of measuring VAM root infection, 104, 105
Colworth stomacher, 396
Competition in a gradostat, 306, 309
Complexipes, 97
Composts, novel, 69–78
Computer Applications in Fermentation Systems, Conference on, 321
Computer control of a photobioreactor, 313–321
Computer-controlled fermenters, 169–180
Computer simulation, *see also* Models
  of gradostats, 300, 301, 303
  of methane fermentations of glucose, 163, 166
  of the activated sludge process, 169–180

Coniferyl alcohols, 41
Conjugation in rhizobia, 87, 88
Contact digester, 122
Continuous flow column, to study cellulose decomposition, 20–22
Continuous illumination, effect on bioreactor control, 319
Copper–*Klebsiella aerogenes* complex, 209, 213
Coppermills Water Treatment Works, 366
Coprosma robusta, 107
Coriolus versicolor, 227
Counts
  of anaerobic bacteria in farm wastes, 136, 137
  haemocytometer, 337
  proportional, 338
  of VAM spores, 101–103
Crassostrea commercialis, 394, 398, 403
Crassostrea gigas, 401
"Creep" phenomenon, 223
p-Cresol, 284, 285
Culture media, *see* Media
Cultures, pot, of VA mycorrhiza, 95, 106–108
Cyclohexane, 276
Cydia pomonella, diet for, 345

D

Dalapon, degradation of, 227
DAS ELISA, *see* Double antibody sandwich ELISA
Data collection in computer-controlled fermenters, 174
Day/night cycle, simulation of, 317, 318
Decarboxylation, 276, 285
Decomposition, microbial, of straw, 69–78
Degradation, microbial, of pollutants, 219, 220
Degradation products, microbial, effects of, 76, 77
Dehydrogenation polymers, 39–61
Depuration of shellfish, 402–404
Desulfobulbus propionicus, 242
Desulfonema limicola, 242
Desulfotomaculum spp., 241
Desulfovibrio spp., 240–242
Desulfuromonas spp., 241
Detoxification, of pollutants, 219, 220, 285

Deutsche Sammlung von
    Mikroorganismen, 152
DHP, see Dehydrogenation polymers
Diet, artificial, for insects, 331, 332, 339,
    344, 345
Diethyl ether, for mutagenicity tests, 377
Digester gases, field measurements of H_2
    in, 166
Digesters for farm wastes, 119–140
Digestion, anaerobic
    of farm wastes, 119–121, 154
    of sewage sludge, 159–167, 179
Dissection of mussel tissue, 399
Diurnal illumination, effect on control of a
    photobioreactor, 319, 320
Diurnal variation in light intensity,
    317–321
DMSO (dimethyl sulphoxide) extracts of
    Rhine water, 388
DNA
    cloning of, 56, 58, 85
    preparation of, 60–62
    recombinant, 54, 55, 58
    restriction enzyme analysis of, 334, 335
    Rhizobium, 85–92
Double antibody sandwich ELISA,
    350–359, 361, 362
Drill dilutor extractor, for ELISA, 355, 360
Drinking water
    collection of samples, 375–390
    denitrification of, 228
    mutagenicity testing of, 375–390
Dry films for counting baculoviruses, 337
Dung, sulphate-reducing bacteria from,
    242

E

Echo virus type 4, 367
EDTA, as a polymer extractant, 200–202
Eggs, insect, sterilization of, 331
Electron microscopy for GV granules, 338,
    339
Electrophoresis
    two dimensional, 58
    of virus particles, 334–336
ELISA, see Enzyme-linked immunosorbent
    assay
Elson–Morgan reaction, 9
Embden–Meyerhof pathway, 161, 163

Endogonaceae, 96, 97
Endogone, 96, 109
Endophytes
    mycorrhizal, 100, 101
    screening of, 113, 114
Enrichment cultures, 142, 143, 154, 155,
    281–283, 306, 307
Enteric viruses, detection of, 365–374
Enterobacter cloacae, 397
Enteroviruses, detection of, 365–374
Entrophospora, 97
Enumeration of bacteria in shellfish tissue,
    398, 402, 403
Enzymes
    activity of, in landfills, 268–271
    assay procedures, 260–265
    extracellular, 13, 14, 260, 270
    extraction of, 260
    implicated in lignin degradation, 52–54,
        58, 62
    microbial, degradation of pollutants by,
        219–221, 227–231
Enzyme-linked immunosorbent assay,
    12–15, 336, 349–361, see also
    Radioimmunoassay
    advantages of, 352
    buffers for, 362
    double antibody sandwich, 350–359,
        361, 362
    equipment for, 355, 356
    interpretation of results of, 355
    modifications of, 357–359
    practical applications of, 359–361
    sensitivity and specificity of, 356, 357
    for specific host–virus combinations, 355
Ergosterol, as a measure of fungal biomass,
    10–12, 15
Erwinia herbicola, 397
Erwinia rhapontici, 230
Escherichia coli, 55, 56, 85, 86, 91, 92,
    305, 306, 309, 394–403
Ethanol
    as a carbon source, 238
    generation of, 229, 230
    precipitation of polymers in, 200, 201
Eubacteria, lignin degradation by, 50, 52,
    139, 151
Eukaryotes, gene expression in, 55
Exhaled hydrogen monitor, 166
Extraction of extracellular polymers,
    199–202

INDEX

F

Faecal coliforms in molluscs, 397, 399
Faecal indicator bacteria, 259, 401–403
Faecal streptococci, 397
Fatty acids, volatile, 279
Fed-batch fermentations, methanogenesis in, 145–149, 154
Fermentation
  of farm wastes, 121, 123–126, 134
  of sewage sludge, 159, 161
  systems, computer applications in, 320
Fermentative bacteria, 136, 140
Fermenters
  closed-culture, 281–283
  computer-controlled, 169–180
  for food processing wastes, 141–144
  in a gradostat, 297
  multi-stage, 287
  open-culture, 283–291
  single-stage, 287
  to study anoxic metabolism, 275–291
  for treatment of agricultural wastes, 119–137
Ferrous ions, in media for sulphate-reducing bacteria, 214
Fibre degradation rate, in digesters, 122–124, 135
Fick's law, 73
Field beans, rhizobia on, 79, 80
Field inoculation with VAM, 111, 112
Fixed-film digester, 122, 124
Floc structure, effect of EDTA on, 202
Flocculation tests, 349, 352
Flow rates, effect of on solute concentrations, 300, 301, 309
Fluctuation test, 379, 380, 389
Fluid drilling, with VAM, 112
Fluidised-bed digester, 122
Fluorescence in methanogens, 152, 153
Fluorescent antibodies, to measure fungal biomass, 12
Fluorogenic substrate, for the detection of plant viruses, 358, 359
Fomes annosus, 47
Food-processing wastes, 139–157
Free radical species, production of by fungal enzymes, 54
Freeze-dried extracts, mutagenicity testing of drinking water with, 375–390
Freeze-driers, 376

Freeze-drying
  of inclusion bodies, 333
  of methanogenic bacteria, 152, 153
Freshwater shellfish, 402
Freshwater strains of sulphate-reducing bacteria, 241
Fuels from farm wastes, 119, 121, 139
Fungal growth on solid substrates, 5–15
Fungi
  aerobic cellulolytic, 72
  biodegradation of lignin by, 47–63
  growth of on solid media, 5–15
  white-rot, 33, 59
Fungi Imperfecti, 109
Furrows, VAM inoculation in, 112
Fusarium moniliforme, 14
Fusarium solani, 47, 52, 53

G

Gel diffusion tests, 349, 352
Gel-filtration method, 208–210
Gel-permeation chromatography, 47
Gene libraries, of ligninolytic organisms, 58
Genes
  cloned, 56, 58
  symbiotic, 89
Gene transfer between *Rhizobium* strains, 86–88
Generalised linear interactive modelling (GLIM), 381, 382
Generalist organisms, 309
Gigaspora spp., 97, 101
Glass tower fermenter, 288
Glaziella, 97
GLIM, *see* Generalised linear interactive modelling
Glomus caledonius, 108
Glomus epigaeus, 111
Glomus fasciculatus, 105, 108
Glomus mossae, 108
Glomus spp., 97, 98, 100, 101
Glomus tenuis, 101
Glucosamine determination, 9
Glucose, methane fermentation of, 163, 166, 171, 175, 178, 229
Glucose oxidase, 53
Glucose oxidase–peroxidase method of glucose determination, 176

414 INDEX

Glycolignin, 34
Glycollic acid liberation by *Chorella*, 321
Gradient plate, two-dimensional, 296
Gradostat, 295–311
  antagonism and growth inhibition in, 306
  applications, 308, 309
  competition in, 306
  description of, 296–299
  enrichment culture technique in, 306, 307
  growth in, 303–305
  residence times in, 301, 302
  separation of cells from substrates in, 311
  transfer of materials in, 299–301
  transient states in, 307
  two-dimensional, 310, 311
Granulosis virus, see GV
Grass lignins, 35, 42
Ground water, mutagenic activity in, 385–387
Growth rate in a gradostat, 304, 305
Growth yield coefficients in a gradostat, 303
L-Guluronic acid, 221, 222, 225
GV, 326, 327, 338, 339

H

H_2
  field measurements of in digester gases, 166
  traces of in sewage sludge fermentation, 159–167
Hardwood lignins, 35, 36
Heat treatment for polymer extraction, 200
Heliothis, larvae of, 331
Hemicellulolytic bacteria, 136, 137
Hemicellulose, 34, 69, 121, 135, 266
Herbicides, degradation of, 227
Heterogeneity, spatial and temporal, 296
Heterogeneous ecosystems, 295–311
Heterokaryosis, 109
High Heavens Landfill Site, 267
Higuchi dialysis sac method, 40
Histidine in mutagenicity tests, 380, 383–385
Hoffman's Tobacco Hornworm diet, 331

Host plants for VAM, 96, 106, 107
Hydridella menziesii, 394, 401
Hydrogen transfer reactions, 161, 167
Hydrogen-utilising bacteria, 162
Hyphae, measurement of length of, 9

I

Immobilisation
  collagen, 226
  of microbial cells in calcium alginate, 221–225
Immobilised cells, use of in waste water treatment, 219–231
Immunofluorescence, 245
Immunological method of polypeptide identification, 58
Immunoradiometric assay, 336
IMRA, see Immunoradiometric assay
Inclusion bodies of baculoviruses, 326, 327
  purification of, 322, 332
Indicator particles, 338
"Indulin AT," 38
Infectivity assays, 339, 340
Inocula, for novel composts, 70, 71
Inoculation, field, with VAM, 111, 112
Inoculum, VAM, production of, 110–113
Insects
  diets for, 331, 332
  laboratory practice for work with, 330, 331
  rearing of, 330–332
  virus resistance in, 329
  viruses for the control of, 323–345
Integrity of capsular polymers, 202
Iron in media for sulphate-reducing bacteria, 244
Isotope-labelled lignocellulose, 41, 42

K

Katharometer for fermenter systems, 143
Kinetics, Monod, 29, 30, 303
  of cellulose digestion, 22–30
Kjeldahl nitrogen determination, 6, 8
Klebsiella aerogenes, 199–201, 206–214
Klebsiella pneumoniae, 397
Kraft lignins, 38, 39, 227

INDEX

L

LD_{50}
  relation to host age, 326
  tests for insect infectivity of baculoviruses, 339, 340
Laboratory-activated sludge, 198, 199, 211
Laboratory models, see Models
Lactate as a carbon source, for
  sulphate-reducing bacteria, 238
  in a gradostat, 307
Lactones, 50
Lactose-fermenting bacteria, 397
Lagoon storage of farm wastes, 131, 132
Laminaria, alginate from, 221, 222
Landfill
  microbial activity in, 259–272, 275, 277, 279, 286, 287
  multi-stage model system, 288, 289
Latex test, 349
Leachate, production of, 266, 270, 286, 287
Lepidoptera
  diet for, 332, 344, 345
  viruses of, 326–328, 331, 344
Lettuces, VAM infection on, 111
Light intensity in a photobioreactor, 317, 321
Lignin
  biosynthesis of, 34–37, 42
  bleaching of, 227, 228
  degradation of, 33–63, 276, 277
  DHP, 39–41
  enzymes implicated in degradation of, 52–54
  extraction of, 37
  in farm wastes, 135
  in landfill, 266, 269
  kraft, 38, 39, 227
  labelling of, 41
  milled-wood, 37, 38, 40
  model compounds, 40, 41
  molecular biological approach to degradation of, 54–63
  structural analysis of lignin polymers, 42, 43
  synthetic, 41
Lignocellulose, degradation of, 19, 33, 41, 42, 50, 51, 62, see also Cellulose decomposition, Lignin
Linked culture vessels, 72, 73

Lipase assays, 261
Lowland surface waters, mutagenic activity of, 385–387
Lucern, VAM infection on, 96

M

Macrocystis, alginate from, 221, 222
Maize, VAM infection on, 96
Malate, as a carbon source, 238
Mannans, 34
D-Mannuronic acid, 221
Marine shellfish, 402
Marine strains, of sulphate-reducing bacteria, 241, 242, 244, 248
Mass balance for a photobioreactor, 314
Mass of sludge wasted, 188, 189
Mathematical models, see Models
Media
  for growth of rhizobia, 80, 81
  for sewage bacteria, 397, 398
  for sulphate-reducing bacteria, 235–255
  overlay, 368, 369
  problems with in oil fields, 243–245
  solid, fungal growth on, 5–15, 252
  sulphate-free, 250, 251
  supplements for, 81, 82
Medical Research Council, 378
Medium
  CLED, 405
  for *Campylobacter jejuni*, 406
  for *Vibrio parahaemolyticus*, 405, 406
  Upper Bay yeast extract agar, 406
Mercenaria mercenaria, 399
Mesophilic digestion, 121, 122, 127, 132, 137
Metabolic activity of activated sludge, 185
Metabolic intermediates of aromatic monomer catabolism, identification of, 280
Metal complexation by extracellular polymers, 207–213
Metals, removal of in biological waste water treatment, 197–215
Methane, see CH_4
Methanobacterium sp., 147, 155
Methanogenic bacteria, 120, 122, 123, 136, 140, 141, 149, 277, 288
  characterisation of, 153, 154
  isolation of, 149–152

Methanogensis, 120, 122–124, 139–157, 159–167, 260, 290, 291, *see also* CH$_4$
Methanol extract for mutagenicity tests, 377, 378, 383, 385, 387, 388
Methanosarcina barkeri, 147–149, 151, 154, 155
Methanosarcina mazei, 149
Methanothrix soehngenii, 147–149, 154
N-Methyl-*N*-nitro-*N*-nitrosoguanidine (NTG), 84, 85
MGW Lauda thermocirculator, 285
Michaelis–Menten expression, 29
Microbial activity, measurement of in the activated sludge process, 184, 185
Microbial associations, interacting, 275–291
Microbial degradation products, effects of, 76, 77
Microbiological Research Establishment, 377
Midgut of insect as a site of virus infection, 326, 327
Milled-wood lignin, 37, 38, 40
Ministry of Agriculture, Fisheries and Food, 330
Mixed liquor, *see* ML
Mixed-liquor suspended-solids concentration, *see* MLSS
Mixotrophic growth, 237
ML (mixed liquor), 191–194
MLSS concentration, 186–189, 190
MNPV, 326
Models, experimental, 20–22, *see also* Computer simulation
 landfill multi-stage, 288, 289
 laboratory, 295
 mathematical, 19, 20, 22–24, 125, 126, 134, 136
Modicella, 97
Molecular biological approach to lignin degradation, 54–63
Molluscs
 bacteriologically cleansed, 402, 403
 dissection of, 399
 use of to monitor bacteria in water, 393–406
Monkey kidney cell lines, use of in the cell suspension assay, 366, 367
Monod kinetics, 29, 30, 303

Monod rate equations, 162, 163
Monolayer plaque method, 365, 366, 368, 370, 374
Monoxenic cultures of VA mycorrhiza, 108
Morphometric cytology, 104
Mosquitoes, baculoviruses in, 324
"Most probable number" technique, 106, 243, 398
Multi-channel tubing pump, 298
Multi-stage digesters, 124, 128–130
Mussels, 393–406
 opening of shells, 395, 396
Mutagenic treatments for rhizobia, 84, 85
Mutagenicity testing of drinking water, 375–390
 assay procedure for, 378–385
Mutagens, water-borne, 387, 389, 390
Mutants, *Rhizobium*, isolation of, 83–86
Mutations
 in rhizobia, 84–86
 in VA mycorrhiza, 109
Mycena galopus, 12
Mycorrhiza, vesicular–arbuscular
 genetics of, 109, 110
 identification of, 96–101
 isolation of, 106, 107
 methods for evaluating and manipulating, 95–114
 quantification of, 101–107
Mytilus edulis, 393, 395, 397–399, 401, 402

N

^{15}N Incorporation, measurement of, 8
Nickel–*Klebsiella aerogenes* complex, 209–213
Nitrate
 in a gradostat, 307
 reduction of, 236
Nitrification, study of, 20, 22, 309
Nitrocellulose filters, blotting of a plasmid DNA to, 91
Nitrogen cycle, 309
Nitrogen fixation, 79, 85, 92
NMR, *see* Nuclear magnetic resonance spectroscopy
Nodulation, 79, 80, 85, 92
Non-axenic conditions, growth of fungi in, 9–14

NPV, 326, 327, 330, 341
NTG, 84, 85
Nuclear magnetic resonance spectroscopy, 43–46
Nuclear polyhedrosis viruses, see NPV
Nucleic acids, assays for, 6, 7
Nucleocapsids, storage of, 333, 334
Nutrient/harvest pump, in a photobioreactor, 316, 317

O

Oil fields, media used in, 243–245, 253
Onions, VAM infection on, 96, 105–107
Oryctes rhinoceros, 327, 342
Orycytes virus, 326
Ostrea edulis, 394, 399
Overlay media, for suspended-cell plaque assays, 368
Oxidative degradation of lignin, 46
Oxygen
  reactions linked with, 275, 276, 287
  role of in lignin degradation, 48
Oxygen consumption, 7
Oxygen diffusion column, 73–76
Oysters, 393, 394, 397, 403, 404

P

Paper mills, "white waters" from, 227
Paracoccus denitrificans, 303
Parasexual cycle in fungi, 109
Parathion, degradation of, 227
Partitioning of mutagenic compounds, 387, 388
Peak light intensity, measurement of, 187
Peas
  nitrogen fixation by, 79, 80
  VAM fungi on, 96
Pectins, 34
Pellets, VAM infected, 112
Penicillium corylophilum, 77
Percolating columns, 296
Periodic changes in activated-sludge systems, 169
Pesticides
  chemical, compared with baculoviruses, 342, 343
  degradation of, 227
  resistance to, 344
  safety testing of, 340, 341
Pests, regulations for working with, 330, 331
pH of media for sulphate-reducing bacteria, 240
Phanerochaete chrysosporium, 47, 48, 52, 54–56, 61, 62, 227
Phenol, 282–284
Phenols, degradation of, 228, 229
Phosphate uptake, rate of, 113
Photobioreactor, computer-controlled, 313–321
Photometer, for ATP measurement, 187
Photosynthetic bacteria, 296
o-Phthalaldehyde, 385
PIB, 332, 337, 338
Pig wastes, 123, 125, 127, 128, 130, 132–136
Plant bioassay for microbial degradation products, 76, 77
Plant viruses, 349–361
Plaque assays, 365, 366, 368–374
Plasmids
  elimination of from *Rhizobium* species, 80, 86, 88
  isolation of, 89–91
  P-group, 92
  Q-group, 92
Plasmons, 109
Plate coating in the ELISA process, 352, 353
Plate counts of bacteria in shellfish, 402, 403
Plate washing in the ELISA process, 353
Plug-flow conditions simulated, 178, 179
Plum pox virus, routine detection of, 359, 360
Poinsettia mosaic virus, 360, 361
Poisson distribution, 337
Polio type 3 viruses, 370, 371
Poliovirus type 1, attenuated, 367, 369–373
Pollutants
  removal of, 231
  types of, 219
Pollution
  control of, 119, 120, 121, 215, 285
  from sewage, 385, 402, 404
Polyacrylamide gel for cell immobilisation, 225, 226

Polymers
  assay of, 201
  extracellular, role of in controlling metal removal, 197–215
  integrity of, 202, 206
  purification and recovery of, 200, 201
  soluble and colloidal, 212, 213
Polyporus versicolor, 53
Polystictus versicolor, 47, 53
Pot cultures of VAM, 95, 106–108
Potatoes
  extraction of sap from tubers, 355, 360
  viruses of, 349
Precipitation tests, 352
Pre-cropping as a method of VAM field inoculation, 113
Productivity ratios of viruses, 332
Prosthecochloris aestuarii, 241
Protease assays, 261, 267
Pseudomonas aeruginosa, 87, 88, 306
Pseudomonas denitrificans, 228
Pseudomonas putida, 51, 227, 285
Pseudomonas spp., 92, 221, 229
  lignin degradation by, 51
Psychrophilic sulphate-reducing bacteria, 242
Pulping, 33, 40
Pyridine nucleotides, effect on respiration rates in computer-controlled fermenters, 177, 178
Pyruvate, as a carbon source, 238

Q

Quantification of VAM, 101–107
Quinone-reducing oxidoreductases, 53
Quinone transformation, products of, 49, 50

R

Radioimmunoassay, 12–15, 336
Radiorespirometry, 245
Rate of wastage of activated sludge, 183, 184, 188, 189, 192–194
Recombinant DNA techniques, 54, 55, 58

Recycling, *see* rate of recycle
Redox dyes, 239
Redox-poising agents, 236, 239, 241, 243
Reflux column, 286, 287
Refuse
  age of, 270
  pulverised, 267
Reovirus, 367
Residence times in the gradostat, 301, 302
Respiration rates
  in computer-controlled fermenters, 176, 177
  use of to measure fungal growth, 7
Restriction endonuclease analysis, 334, 335
Retention time in digester systems, 122, 124, 125, 128, 131, 134
Returned activated sludge, 189–193
Reverse osmosis, 390
Rhizobia
  antibiotic resistance in, 83, 85
  fast-growing species of, 79–92
Rhizobium
  genetic manipulation of, 79–92
  media for, 80, 81
  storage of, 81–83
Rhizobium japonicum, 90
Rhizobium leguminosarum, 80, 81, 86–88, 90
Rhizobium meliloti, 80, 87, 90
Rhizobium trifolii, 80, 90
Rhizophagus tenuis, 101
Rhodococcus erythropolis, 53
Rhodophyceae, 225
RIA, *see* Radioimmunoassay
River Thames, dip samples from, 366
River Wear, dip samples from, 366
RNA
  fungal, 55, 56
  messenger, 56, 58
  of eubacteria, 140
  preparation of, 61
Root infections, mycorrhizal, 99–101
Root length, mycorrhizal, 104, 105
Root slide technique, for estimating mycorrhizal infection, 104
Rumen
  anoxic ecosystems in, 275
  microbial activity in, 259
  sulphate-reducing bacteria from, 242
Ryegrass mosaic virus, 361

S

Saccharomyces cerevisiae, 10, 55, 56
Safety precautions for mutagenicity tests, 379
Safety tests for pesticides, 340, 341, 343
Salmonella hadar, 397
Salmonella spp., 397
Salmonella senftenberg, 403
Saltmarsh sediment, 279
Saprolegnia monoica, 52
Sawflies, viruses of, 326–328, 332, 340, 344
Sclerocystis spp., 97
Scrobicularia plana, 394, 400, 403
SDS–polyacrylamide gels, electrophoresis in, 334, 335
Seeds coated with VAM inoculum, 112
"Sellotape" impression method for assessing numbers of PIB on leaf surfaces, 338
Sephadex G-type gels, 207
Serological methods
 for detection and identification of baculoviruses, 336, 337
 for detection of plant viruses, 349–361
Serratia marcescens, 404
Sewage
 bacteria from, 397–401
 pollution caused by, 385, 402
Sewage outfall assessment, 404, 405
Sewage sludge, 142, 159–167, 229, *see also* Activated sludge
 activated, 169–180, 183–195, 200, 201, 214, 215, 220
 synthetic, 198, 199
Shellfish
 as a health hazard, 393–406
 commercial production of, 404
 freshwater, 402
 marine, 401, 402
Shells of mussels, opening of, 395, 396
Shigella dysenteriae, 397
Simulation, *see* Computer simulation
Single-stage digesters, 121–123, 126–128, 136
Sludge, *see* Sewage sludge
Sludge age, 189–192, 198, 214, 215
Sludge loading, 188
Sludge volume index, 189

Slurries
 cattle, 122, 123, 125, 127, 128, 133
 farm, degradation of, 122–125
 feedstock, 125
 pig, 123–125, 127, 128, 130, 132–136
 poultry, 125, 133
SNPV, 326
Softwood lignins, 35, 36
Soil, survival of baculoviruses in, 329
Soil aggregate stability, effects of degradation products on, 77
Soil ecosystems, application of Monod kinetics to, 30
Soil infectivity to VAM, 105, 106
Soil-stabilising agents, production of, 69
Solid substrates, fungal growth on, 5–15
Solids in farm wastes, degradation of, 123, 124, 126–128, 131, 132, 134–136
Solvents for extracting mutagenicity from freeze-dried concentrates, 377
Sorghum, VAM infection on, 106, 107
Specialist organisms, 309
Specific growth rate of algal biomass, 314–316
Spectroscopy
 infrared and UV, 46
 NMR, 43–46
Sphacelia sorghi, 52
Spores
 counts of, 101–103
 mycorrhizal, 95–114
 resting, 96–99
Sporotrichum pulverulentum, 47, 48, 50–53, 55, 61, 62
Spray application of viruses, 342, 343
SSVI, *see* Stirred specific volume index
Stability constants, determination of, 207–210, 214
Sterol content of fungi, 10
Stirred specific volume index, 189
Stirred-tank digester, 122, 123, 125–132
Stock plants for VAM cultures, 95, 107, 108
Storage of methanogenic bacteria, 152, 153
Strawberry, VAM infection on, 106, 107
Streptococci, faecal, 397
Streptomyces, 55
Streptomyces griseus, 42
Streptomyces viridosporus, 46
Stylosanthes spp., 107

Substrate consumption/weight loss, for estimating fungal biomass, 7
Sulphate, *see also* Sulphate-reducing bacteria
  estimation of, 279
  in a gradostat, 307
  reduction of, 288, 290
Sulphate-free media, 250, 251
Sulphate-reducing bacteria, 235–255, 283, 284, 288, 291
  psychrophilic, 242
Sulphide, estimation of, 280
Sulphur cycle, 307, 309
Suspended-cell cultures, 365–374
SVI, *see* Sludge volume index
Symbiotic efficiency, of VA mycorrhiza, 95, 111, 113
Symbiotic properties in *Rhizobium*, 89
Synthetic sewage, 198, 199

T

Tapes japonica, 401
Technicon Autoanalyser, 298
Thames Water Authority, 366, 367
Thermocirculator, Churchill, 288
Thermophilic digestion, 121, 122, 127
Thin-layer chromatography, 280
Thiobacillus ferroxidans, 12
Thioglycollate, 239
Total suspended solids, 185, 189
Toxic metals in waste water, 197, 215, *see also* Metal complexation
Transduction in rhizobia, 86, 87
Transfer of material in a gradostat, 299–301
Transformation in rhizobia, 86
Transient states in a gradostat, 307
Transposon mutation in *Rhizobium*, 85, 86
Trichoderma reesei, 20, 22–24, 29, 30, 265
Trichoderma spp., 7, 20
Trichoderma viride, 20
Trifolium parviflorum, 108
Trifolium repens, 108
TSS, *see* Total suspended solids
TTL logic controllers, 174
"Tubular" digester, 122, 123, 125, 130
Two-phase digester systems, 123, 124, 132

Two-stage digestion of farm wastes, 123, 132–137
TY medium for rhizobia, 80, 82

U

Ultrasonication, 200, 201, 377
Ultraviolet spectroscopy, 46
Upflow sludge-blanket digester, 122
Upland waters, mutagenic activity in, 385–387

V

VA mycorrhiza, *see* Mycorrhiza
Vector systems for *Rhizobium* cloning, 91, 92
Vesicular–arbuscular mycorrhiza, *see* Mycorrhiza
Vibrio parahaemolyticus, 397, 399, 403
Virus particles, storage of, 333, 334
Viruses, *see also* Baculoviruses
  enteric, 365–374
  insect control by, 323–345
  plant, 349–361
  potato, 349, 360
  spray application of, 342
Volatile suspended solids, 185
Volume of sludge wasted, 189, 190
VSS, *see* Volatile suspended solids
VT100 terminal, 172–174

W

Wastage, *see* Rate of wastage
Waste water treatment processes
  computer simulation of, 169, 179, 180, 184
  removal of toxic metals in, 197–215
  use of immobilised cells in, 219–231
Wastes
  agricultural, treatment of, 119–137
  food-processing, 139–157
Water
  drinking, 228, 375–390
  monitoring of bacteria in, 393–406
Water movement, bacterial tracers of, 404
Water Pollution Research Laboratory, 200
Water Research Centre, 159, 375
Watson Marlow flow inducer, 288

Welsh Water Authority, 188
Wet-sieving technique to extract VAM resting spores from soil, 101–103
"White waters," 227
Wiseana spp., 341, 342
World Health Organization, 275

X, Y, Z

Xanthomonas spp., 51
Xenobiotics, 275–291, 300, 309
Xylans, 34
Yersinia enterocolitica, 397
Zoogloea ramigera, 197
Zulaufverfahren, 40
Zutropferfahren, 39, 40

THE SOCIETY FOR APPLIED BACTERIOLOGY TECHNICAL SERIES

General Editor: F. A. Skinner

1 Identification Methods for Microbiologists Part A 1966
  Edited by B. M. Gibbs and F. A. Skinner
  Out of print

2 Identification Methods for Microbiologists Part B 1968
  Edited by B. M. Gibbs and D. A. Shapton
  Out of print

3 Isolation Methods for Microbiologists 1969
  Edited by D. A. Shapton and G. W. Gould

4 Automation, Mechanization and Data Handling in Microbiology 1970
  Edited by Ann Baillie and R. J. Gilbert

5 Isolation of Anaerobes 1971
  Edited by D. A. Shapton and R. G. Board

6 Safety in Microbiology 1972
  Edited by D. A. Shapton and R. G. Board

7 Sampling—Microbiological Monitoring of Environments 1973
  Edited by R. G. Board and D. W. Lovelock

8 Some Methods for Microbiological Assay 1975
  Edited by R. G. Board and D. W. Lovelock

9 Microbial Aspects of the Deterioration of Materials 1975
  Edited by R. J. Gilbert and D. W. Lovelock

10 Microbial Ultrastructure: The Use of the Electron Microscope 1976
   Edited by R. Fuller and D. W. Lovelock

11 Techniques for the Study of Mixed Populations 1978
   Edited by D. W. Lovelock and R. Davies

12 Plant Pathogens 1979
   Edited by D. W. Lovelock

13 Cold Tolerant Microbes in Spoilage and the Environment 1979
   Edited by A. D. Russell and R. Fuller

14 Identification Methods for Microbiologists (2nd Edn.) 1979
   Edited by F. A. Skinner and D. W. Lovelock

15 Microbial Growth and Survival in Extremes of Environment 1980
   Edited by G. W. Gould and Janet E. L. Corry

16 Disinfectants: Their Use and Evaluation of Effectiveness 1981
   Edited by C. H. Collins, M. C. Allwood, Sally F. Bloomfield and A. Fox

17 Isolation and Identification Methods for Food Poisoning Organisms 1982
   Edited by Janet E. L. Corry, Diane Roberts and F. A. Skinner

18 Antibiotics: Assessment of Antimicrobial Activity and Resistance 1983
   Edited by A. D. Russell and L. B. Quesnel

19 Microbiological Methods for Environmental Biotechnology 1984
   Edited by J. M. Grainger and J. M. Lynch